Multilingual Text–to–Speech Synthesis: The Bell Labs Approach

Multilingual Text–to–Speech Synthesis: The Bell Labs Approach

Richard Sproat, editor
Bell Laboratories, Lucent Technologies

With a foreword by
Louis Pols, *University of Amsterdam*

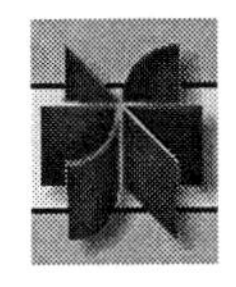

KLUWER ACADEMIC PUBLISHERS
DORDRECHT / BOSTON / LONDON

Distributors for North America:
Kluwer Academic Publishers
101 Philip Drive
Assinippi Park
Norwell, Massachusetts 02061 USA

Distributors for all other countries:
Kluwer Academic Publishers Group
Distribution Centre
Post Office Box 322
3300 AH Dordrecht, THE NETHERLANDS

Library of Congress Cataloging-in-Publication Data

A C.I.P. Catalogue record for this book is available
from the Library of Congress.

Printed on acid-free paper.

Printed in the United States of America

Contents

List of Figures

List of Tables

Contributors

Kazuaki Maeda. Kazuaki Maeda is a doctoral candidate in the Computer and Information Science Department at the University of Pennsylvania. He is also a research assistant in the Phonetics Lab of the Department of Linguistics under the supervision of Dr. Mark Liberman. At Bell Labs, he is working as a consultant on the Japanese TTS system. His research interests include the modeling of Japanese intonation, phonetics and computational linguistics.

Bernd Möbius. Bernd Möbius obtained his Ph.D. in phonetics and linguistics from the University of Bonn, Germany. His research interests include speech production and perception, phonetics, phonology, intonation, and computational linguistics. Since joining Bell Labs in 1993, he has been responsible for the German TTS system. In multilingual TTS he also works on the construction of acoustic inventories and the improvement of voice quality.

Joseph Olive. Joseph Olive has worked on TTS since 1970, when he started working on concatenative synthesis at Bell Laboratories. In 1979 he designed a complete TTS system and implemented it in a real-time hardware device, which was exhibited at Epcot Center for over 10 years. His recent interests include methods for improving the speech quality of concatenative synthesizers.

Chilin Shih. Chilin Shih first came to Bell Labs in 1986 as a postdoc after completing her Ph.D. at UC San Diego, and worked on an early version of our current Mandarin TTS system. After teaching at Cornell and Rutgers Universities for six years, she returned to Bell Labs in 1993. Her recent work has included acoustic inventory construction, duration modeling and intonation modeling for multilingual synthesis.

Richard Sproat. Richard Sproat joined Bell Laboratories in 1985, after finishing his Ph.D. in Linguistics at the Massachusetts Institute of Technology. Since joining the Labs, his research interests have spanned a number of areas, including morphology, articulatory phonetics, prediction of prosody (i.e., accent patterns) from text and the design of text-to-speech systems.

Most of his recent work has addressed text-analysis in multilingual TTS systems.

Evelyne Tzoukermann. After completing her Ph.D. at the University of Paris III, Evelyne Tzoukermann obtained a Fulbright fellowship and pursued a post-doctoral year at Brown University. After spending two years as a visiting scientist at IBM Watson Research Center, she joined Bell Labs in 1989. Her research interests include computational morphology, part-of-speech disambiguation, and text analysis. Her recent work includes the construction of the Bell Labs French TTS system.

Jan van Santen. Jan van Santen came to Bell Laboratories in 1979, after completing his Ph.D. in Mathematical Psychology at the University of Michigan. Since joining the Labs, his research interests have encompassed several areas, including visual perception, user interface design, and, since 1986, text-to-speech synthesis. His recent work has mostly addressed prosody and signal processing in multilingual TTS systems.

Foreword

Louis Pols, *University of Amsterdam*

The editor and the six other contributors of this book are brave enough to compare their product with the influential 1987 book on the MITalk system and to consider this comparison to be a good test of the general progress made in text-to-speech synthesis over the past ten years. Naturally their conclusion is that substantial progress has been made, but that more work is still required.

I tend to agree with them, although I do not consider the so-called Bell Labs Approach to be typical for present-day text-to-speech synthesis, partly because it is simply different, partly because it is actually very advanced.

With some simplification, the MITalk system could perhaps be characterized by the use of morphological decomposition and the Klatt formant synthesizer; likewise the modular Bell Labs TTS system might be characterized by the use of regular relations for text analysis and concatenative synthesis of parameterized (mainly) diphone units. Much of the work is related to carefully selecting and analyzing texts as well as recorded speech material to optimize the system performance.

Personally I do not believe that concatenative synthesis of whatever type, unless highly adaptable, will ultimately be sufficiently flexible to serve the needs for many future applications even if collecting the required concatenative units becomes easier and easier. Still, one would like to be able to switch quickly, not just to new languages, but also to several other speakers/voices, speaking styles and emotions. Perhaps in telecommunication applications *emotion* may not be a major issue (page 245); but it certainly is an issue in other applications, like in reading machines for the visually handicapped, in human-machine dialogs, in speech-to-speech translation, or in normative (second) language pronunciation.

However, the Bell Labs approach is typical of speech technology research in the 90's in as far as many practical and application-oriented problems simply have to be solved, frequently in an elegant way, sometimes in a brute force way. The book gives ample examples of such elegant solutions (e.g. in properly modeling segmental timing, coarticulation and reduction), and the statistically-based solutions are very ingenious. Over the past twenty years, in which more than forty people have worked on this technology at Bell Labs, a wealth of tools and of language-universal and language-specific knowledge has been collected. This book is the first occasion that much of this information is put together in an accessible way.

The most overwhelming extra feature of the Bell Labs Multilingual TTS engine, is its *multilinguality* (using common algorithms for multiple languages). Because of the speech and language database orientation and because of the tools available, it is relatively easy for them to develop a synthesizer for a new language quickly. That this is really true becomes clear from the present list of ten diverse languages for which versions are already operational: American English, German, French, Italian, Castilian and Mexican Spanish, Japanese, Mandarin Chinese, Russian, Navajo, and Romanian. The audio demonstrations on the WWW-page certainly add to this reality. Actually as a European I am a bit shocked that the topic of multilinguality has not been claimed more strongly by researchers on this continent.

I am very pleased with the special chapter on *evaluation*. Proper evaluation will not solve all problems and each test can only serve its own purpose, but then at least developers' prejudices can be avoided and users' satisfaction is more probable. Still the problem remains how to weigh naturalness against intelligibility.

Comparison with ten years ago also raises the question of what further improvements still are required and can be expected ten years from now. I expect speech technology in general, and TTS synthesis more specifically, to become more and more a common part of life. There will be a need for cheap and dirty synthesizers with a limited vocabulary and standard prosody, as well as for highly advanced synthesizers able to generate from a multitude of inputs (concept, text, enriched text, database information, e-mail, WWW, etc.) intelligible and natural-sounding speech in a variety of ways according to the listeners' needs, perhaps supported by visual gestures. The Bell Labs Multilingual TTS engine will then be much cheaper, much more powerful and will have several competitors! TTS research will continue to be a key technology in testing our acquired knowledge about text interpretation and spoken language production.

Preface

About six years ago, Stephen Levinson, who was then the head of the Linguistics Research Department at AT&T Bell Laboratories, suggested that we should write a book describing the Bell Labs Text-to-Speech work; after all, as he noted, there was a book on the MITalk system, and it would be good if we too could produce such a book.

While the suggestion was clearly a good one, there were various reasons why we did not immediately take it up. Probably the most important of these was that we were at the time starting to branch out in a serious way into languages besides American English, which had heretofore been our main staple. (As early as 1987 we had built a prototype Mandarin synthesizer; and starting in 1989 we had worked with Telefónica Investigación y Desarrollo on a Castilian Spanish system; but neither of these projects constituted an attempt to build, totally in-house and with our own tools and techniques, an end-to-end system for a new language.) We felt therefore that it would be better to wait until the multilingual work was more advanced, so that we could report on our efforts in multilingual rather than monolingual synthesis. Now, six years later, while we do not feel that our work in any area of multilingual synthesis is complete, we do believe that it is at a stage where it makes sense to report on it in the form of a book.

The work presented in this book is the result of the efforts of roughly forty people (including the present authors) spread over more than twenty years. The contributors of the thirty-odd researchers who are not contributors to this book are listed and acknowledged in Section 1.4. Forty may seem like a large number, especially if one subscribes to the all-too-prevalent assumption that Text-to-Speech is an "easy" technology (at least compared to such "hard" technologies as Automatic Speech Recognition). But when one considers the range of problems that are faced in constructing Text-to-Speech systems — linguistic analysis of the input text, modeling of segmental durations, intonation, source modeling, concatenative unit selection, waveform synthesis — and when one considers that we have tried to attack at least some of these problems in each of ten languages, forty does not seem enough.

For the preparation of the book itself, we would like to thank Alex Greene, our editor at Kluwer Academic Publishers for his suggestions, as well as his patience, during our work on this book. We would also like to acknowledge the suggestions of three anonymous referees, who commented on our original outline for this book, and from whom we derived some clearer ideas about what the book's focus should be. Finally, I would like to thank Lillian Schwartz for critiquing earlier versions of the cover design.

Richard Sproat
Murray Hill, New Jersey

Chapter 1

Introduction

Jan van Santen, Richard Sproat

Ten years ago, a book in many ways similar to the current book was published: *From Text to Speech: The MITalk System* (Allen, Hunnicutt, and Klatt, 1987). What has been accomplished in the intervening years?

Certainly, text-to-speech synthesis has moved to the main stage of speech technology, a stage shared with speech coding, automatic speech recognition, and speaker identification. Also, many papers have been written, special conferences dedicated to text-to-speech have been organized (First ESCA TTS workshop in Autrans, France, 1990; Second IEEE/AAAI/ESCA TTS workshop in Mohonk, US, 1994; and soon the Third ESCA TTS workshop in Sydney, Australia, 1998), and an increasing convergence on which approaches are used has developed. Not only has there been an explosion in intellectual activity, there also are clear improvements in TTS by any quality measure. Finally, although the global market is still small (less than $250 million annually), commercial application of TTS has also increased.

We feel it is nevertheless important to point out that the ultimate goal — that of accurately mimicking a human speaker — is as elusive as ever, and that the reason for this is no secret. After all, for a system to sound natural, the system has to have real-world knowledge; know rules and exceptions of the language; appropriately convey emotive states; and accurately reproduce the acoustic correlates of intricate motor processes and turbulence phenomena involving the vocal cord, jaws, and tongue. What is known about these processes and phenomena is extremely incomplete, and much may remain beyond the grasp of science for many years to come.

In view of this, the convergence in current work on TTS is perhaps somewhat disturbing. For example, a larger percentage of current systems use concatenative synthesis (e.g., the Bell Labs system) than parametric/articulatory synthesis (e.g., The MITalk System). We believe that this is not for theoretical reasons but for practical reasons: the quality levels that are the current norm are easier to attain with a concatenative system than with a parametric/articulatory system. However, we feel that the complex forms of coarticulation found in human speech ultimately can only be mimicked by accurate articulatory models, because concatenative systems would require too many units to achieve these — significantly higher — levels of quality. In other words, the convergence may reflect not so much agreement on major theoretical issues as similarity of the constraints under which researchers work.

We make these remarks to express our belief that TTS is not a mature technology. There may exist standard solutions to many TTS problems, but we view these solutions only as temporary, in need of constant scrutiny and if necessary rejection.

1.1 What is Multilingual TTS?

By "multilingual" we mean a system that uses common algorithms for multiple languages. Thus, a collection of language-specific synthesizers does not qualify as a multilingual system. Ideally, all language-specific information should be stored in data tables, and all algorithms should be shared by all languages.

Needless to say, this ideal state is in practice hard to achieve. One reason is that most TTS researchers develop their ideas primarily with one particular language or a handful of particular languages in mind. As a result, their algorithms often contain parameters that are sufficient to cover the languages they have dealt with, but do not extend easily (or at all) to new languages. Thus, rather than being an actuality, multilinguality in text to speech synthesis is a goal, a guiding principle that orients research away from the language-specific questions and towards language-independent questions.

Our reading of the literature leads us to conclude that we have proceeded further towards the ideal than other research groups. While most if not all TTS systems contain some components that are arguably language-independent, it is unusual to find systems that are completely multilingual (in our restricted sense) in every single component — i.e. from text-analysis to synthesis. Many systems that claim to be such have only been applied to a handful of related languages (e.g., a few Western European Indo-European languages), so it is rather hard to evaluate the claim that they are truly multilingual.

1.2 Guiding Assumptions and Themes

Our work on synthesis, and as a result this book, makes basic assumptions about TTS that we want to make explicit at the outset.

1.2.1 Modularity

Like almost all currently existing TTS systems, the Bell Labs TTS system architecture is modular. Text analysis modules are followed by prosodic modules, which are then followed by synthesis modules. Within each group, there is further modularity, such as morphology vs. intonational phonology. Although we admit to the possibility of many alternative modular architectures, we are less flexible on the fundamental assumption of modularity itself.

1.2.2 Component diversity

We believe that different modules may have to use different methods to accomplish their respective tasks. For example, we use finite state transducer based methods for certain text analysis modules, timewarping methods for intonation, and quasi-linear regression methods for timing. But we do not use a single method such as classification and regression trees throughout the system, even when this is theoretically possible and would lead to significant system simplification. In other words, we prefer appropriateness-to-task over uniformity of method.

1.2.3 Knowledge *and* data

One of the largest changes since 1987 is the availability of large text and speech data bases. These data bases, together with the development of general purpose prediction engines such as classification and regression trees and neural nets have led to systems that were "trained" to convert textual input into speech with little or no manual intervention, and with little or no structure imposed on the estimated parameters. On the other hand, the MITalk system was constructed largely by hand, including both principled rules ("knowledge") and "tweaked" parameter values.

These changes have led some to believe that corpus-based methods with minimal use of knowledge are the wave of the future. We believe, however, that the size of current corpora, while impressive, is still infinitesimal compared to the combinatorial complexity of a natural language. In addition, we believe that the resulting poor coverage of a language by a training corpus can only be overcome using knowledge-based models. Once these models have been proposed, data are used to refine them and generate parameter estimates.

1.2.4 What works vs. what we would like to work

Finally, our approach tends to be pragmatic. For example, we use concatenative synthesis because that is currently the best available method to produce synthetic speech of consistently high quality. As pointed out earlier, however, at the same time we also believe that in the long run concatenative synthesis is not the answer.

1.3 Role of the Bell Labs System in this Book

The topic of this book is multilingual text-to-speech speech synthesis in general: what methods seem important in designing a truly multilingual text-to-speech system, and some of the mechanical details of building such systems. Naturally our views on these topics are heavily influenced by our own experience so that, inevitably, the Bell Labs TTS system plays a central role in our discussion throughout.

There are several reasons why this seemingly parochial mode of discussion is in fact reasonable. First, in any discussion of text-to-speech technology, it is useful to get a clear sense of how the scientific issues and methodologies fit into the context of an entire end-to-end system. Since the Bell Labs system is naturally the one with which we are most familiar, this purpose is most adequately served by our centering the discussion around our own work.

A second, related, point is the following: the speech synthesis literature is replete with promising proposals for new approaches to TTS that for a number of reasons are never fully tested. One reason is that due to dependencies between components an individual component cannot be seriously tested without building an entire TTS system. For example, a new component for computing intonational phonological structure from text cannot be tested without a component that computes the acoustic correlates (timing, F_0). Insofar as it is a complete working system, the Bell Labs TTS system has functioned as an excellent testbed for the exploration of new ideas.

Third, by discussing a system that has been in existence in some form for at least two decades, we gain insight in the *process* of system building, and in trends in TTS in general — to the extent that our system reflects these trends.

And finally, the (fairly conventional) Bell Labs TTS architecture presents a modular framework that allows us to organize chapters around these modules.

1.4 A Historical Note on the Bell Labs System

The Bell Labs TTS system reflects the interests of the people who have contributed to the system over the years. There is no doubt that a different group functioning in otherwise identical circumstances would have produced a very different system. One feature we pride ourselves on is the diversity of these interests, including statistical mechanics, linguistics (e.g., morphology, intonational phonology, phonetics), electrical engineering, statistics, mathematical psychology, and computer science (e.g., language modeling, discourse analysis, algorithms).

Our work on text-to-speech has depended heavily on the contributions of many people over many years, of which the present authors represent only a small fraction. Significant contributions to the predecessor of our current American English system were made by Mark Anderson, Adam Buchsbaum, Ken Church, Cecil Coker, Mark Liberman, Doug McIlroy, Peter Pavlovcik,

Janet Pierrehumbert, and David Schulz. Mark Liberman made significant contributions to the overall design of the system, which Adam Buchsbaum helped implement. Liberman and Pierrehumbert designed the original model of intonation used in the system, and they along with Mark Anderson produced an implementation of it. Doug McIlroy produced the first letter-to-sound rules; subsequently Cecil Coker implemented the first version of dictionary-based pronunciation methods. Ken Church contributed his part of speech disambiguation algorithm and various algorithms for the pronunciation of proper names, which were ported to C by Adam Buchsbaum. Pavlovcik and Schulz, along with various others at (then AT&T) Network Systems, were responsible for the "front end" preprocessing module.

The current version of the American English system was, of course, based heavily on this earlier work, but benefited from the contributions of an equally large number of people, including: Mark Beutnagel, Julia Hirschberg, Luis Oliveira, Michael Riley, Scott Rosenberg, James Rowley, Juergen Schroeter, Kim Silverman, David Talkin, and David Yarowsky. New synthesis methods were provided by David Talkin and Luis Oliveira. Julia Hirschberg provided models of accentuation and prosodic phrasing. David Yarowsky applied models for sense disambiguation to the problem of homograph disambiguation, and also developed and implemented a model of lemmatization. Contributions to the implementation and design of the modular architecture described in Section 2.6 were made by Rowley and Beutnagel, and important design suggestions were made by Riley. Riley was also responsible for the decision tree software which Hirschberg used in her work on accenting and phrase boundary prediction.

Our synthesis work depends upon large amounts of labeled speech and text data, and this in turn has depended mostly on human labelers, including Jill Burstein, Amy Delcourt, Alice Greenwood, Nava Shaked and Susanne Wolff. (Jill Burstein also performed the somewhat thankless task of editing and correcting our English pronunciation dictionaries.) In this regard we would like to give special recognition to Rob French, who has spent thousands of hours segmenting speech in English and other languages.

Our multilingual work has depended, naturally enough, on the input of many specialists in the languages concerned. Alejandro Macarrón constructed the concatenative unit inventory for our earlier Castilian Spanish system. Pilar Prieto did most of the work for Mexican Spanish, and John Coleman made some initial contributions to the construction of the unit inventory; Holly Nibert worked on a model of Spanish intonation. Our Italian system was constructed with much help from Cinzia Avesani and Ana Pereira; Mariapaola D'Imperio has worked on improving the duration models. (Julia Hirschberg was also instrumental in initiating the work on Italian, and in addition worked with Prieto on the modeling of Spanish prosodic phrasing.) Olivier Soumoy contributed to the modeling of French duration. Stefanie Jannedy worked on the pronunciation of proper names in German. Russian was done with the help of Elena Pavlova and Yuŗiy Pavlov. Benjamin Ao worked with us on duration modeling for Mandarin. Nancy Chang investigated morphological productivity in Mandarin, and the results of her work are incorporated in the word segmentation model in our Mandarin system; William Gale also collaborated in this research. Japanese has benefited from the input of Mary Beckman and Jennifer Venditti. Finally, two of our systems — Romanian and Navajo — were built by summer students, Karen Livescu, and Bob Whitman, respectively.

The finite-state technology underlying our multilingual text analysis work is due to Michael Riley, Fernando Pereira and Mehryar Mohri.

Recently we have branched out into a new area, visual synthesis. We would like to acknowledge the work of Jialin Zhong in this area.

Finally, most of our programs have at one time or another passed through the hands of Michael Tanenblatt, whom we would like to thank for many contributions to our suite of TTS software. Michael is also the maintainer of our WWW page (`http://www.bell-labs.com/project/tts`).

1.5 Organization of the Book

The remainder of the book is organized as follows. Chapter 2 discusses general methods and the modular structure of the Bell Labs TTS system. This modular structure serves to organize the remaining chapters in the book. In particular there are two chapters — Chapters 3 and 4 — devoted to issues in text analysis; one Chapter (5) on timing; one on intonation (6); and one on synthesis (7) including the issues involved in concatenative inventory construction. Chapter 8 discusses issues in TTS evaluation. Finally, Chapter 9 discusses a few of the issues in TTS that we expect to become increasingly important in the years to come.

Chapter 2

Methods and Tools

Jan van Santen, Richard Sproat

2.1 Introduction

In our work on text-to-speech, there are several fundamental techniques that have played an important role — and will continue to play such a role — in our work on various components of the system. We have decided to start with a description of these techniques. Specifically, we will present:

- The role of TTS text-analysis components in the construction or testing of later TTS components.
- Various multivariate statistical methods that we have used in the analysis of the typically (highly) multidimensional data that are characteristic of speech and natural language.
- Greedy methods for optimal text selection.
- An introduction to finite-state automata and finite-state transducers.

We end the chapter with a discussion of the modular architecture of the Bell Labs TTS system.

2.2 TTS Components

We often use text-analysis components of TTS to process text for either construction or tests of other TTS components. For example, to construct our phrasing component (which inserts phrase boundaries between words not separated by punctuation), we must also automatically tag text in terms of parts of speech; and to generate text allowing us to evaluate all concatenative units (which are specified as sequences of phone labels), we automatically transcribe a large on-line text corpus.

This has implications for the order in which TTS components are constructed for a new language. In the ideal case, we postpone construction of components that rely most heavily on automatic text analysis, such as the timing and intonation components. We prefer a process where work on text analysis proceeds in parallel with acoustic inventory construction. This has the added advantage that the prosodic modules can be fine-tuned on the basis of actual listening experiments.[1]

2.3 Statistical Analysis of Multivariate Effects

Several TTS components contain quantitative models whose structure and parameters are determined on the basis of text or speech corpora. The purpose of this section is to discuss methods where some quantitative variable (e.g., F_0 peak height) depends on several factors (e.g., stress, phrasal location, vowel identity). Many standard textbooks exist that discuss some of these methods in great detail (e.g., general statistics: (Hays, 1981), multivariate linear statistics: (Morrison, 1967; Rao, 1965), generalized additivity: (Krantz et al., 1971), isotonic regression: (Barlow et al., 1972), and assorted exploratory analysis techniques: (Hoaglin, Mosteller, and Tukey, 1983).)

The key questions addressed by these methods are:

1. Which factors are relevant for a given variable?
2. How do the factors in combination affect the variable?

2.3.1 General issues in modeling multivariate effects

Factors, levels, parameters, interactions

The inputs to TTS components can usually be described as discrete vectors. For example, the input to the timing component consists of vectors of the type

$$\vec{\mathbf{f}} = \; < /e/, \text{ 1 stressed, accented}, \; \cdots, \; \text{phrase medial} > . \qquad (2.1)$$

[1] Of course, in addition to depending upon text-analysis modules, our work also depends upon large text and speech corpora; we shall see many instances of this dependence throughout this book. Due to initiatives by Ken Church and Mark Liberman, we started collecting text corpora in the early 80's. Initially, these were used for training statistical methods for part-of-speech tagging and name pronunciation. But soon enough, these corpora were also used for training or testing of other TTS components.

Currently, large text and speech corpora are widely available via various European organizations (for European languages); organization such as the Republic of China Computational Linguistics Society (for Chinese); and the Linguistic Data Consortium (for assorted languages).

Each component of such a vector is a *level* on a *factor*, where we identify a factor as a *set* such as

$$\text{Stress} = \{1 \text{ stress}, 2 \text{ stress}, \text{unstressed}\}. \tag{2.2}$$

Denoting factors as $\mathcal{F}_1, \cdots \mathcal{F}_N$, the set of all vectors $\vec{\mathbf{f}}$ forms the *factorial space*

$$\mathbf{F} \ = \ \mathcal{F}_1 \times \cdots \times \mathcal{F}_N. \tag{2.3}$$

The task of a component is to assign some output to each input vector. For example, the timing component assigns segmental durations to inputs of the type in Equation 2.1. In other words, the timing component *maps* vectors onto the real numbers, $\mathbf{R}$:

$$\text{DUR} : \mathbf{F} \rightarrow \mathbf{R}. \tag{2.4}$$

Usually, this mapping is computed in multiple stages. For example, for each i we first map factor $\mathcal{F}_i$ on $\mathbf{R}$, and then map all combinations of the outputs onto the final output value by adding these numbers (*additive model*). Denoting the i-th component of $\vec{\mathbf{f}}$ as $\vec{\mathbf{f}}_i$, for each i we have

$$D_i : \mathcal{F}_i \rightarrow \mathbf{R}, \tag{2.5}$$

and we combine the resulting values via

$$\text{DUR}(\vec{\mathbf{f}}) = \sum_{i=1}^{N} D_i(\vec{\mathbf{f}}_i). \tag{2.6}$$

In the additive model (Equation 2.6), each D_i is a *parameter vector*, whose number of components is equal to the number of levels of factor $\mathcal{F}_i$. For example, if $\mathcal{F}_1$ is the *Stress* factor, then, say, $D_1(1\ stressed) = 20\ ms$, $D_1(2\ stressed) = 5\ ms$, and $D_1(unstressed) = -25\ ms$. These parameter values can be estimated from data (see below), and represent the *effects* of the *Stress* factor; in this case, the effect is that 1 stressed vowels are 45 ms. longer than unstressed vowels.

In a different type of model, there could be more than one per-factor mapping. For example, for $i = 1, 2, 3$ we would again have $D_i : \mathcal{F}_i \rightarrow \mathbf{R}$, but in addition we have $E_1 : \mathcal{F}_1 \rightarrow \mathbf{R}$; and the combination rule is given by:

$$\text{DUR}(\vec{\mathbf{f}}) = D_1(\vec{\mathbf{f}}_1) \ + \ D_2(\vec{\mathbf{f}}_2) \ + \ E_1(\vec{\mathbf{f}}_1) \times D_3(\vec{\mathbf{f}}_3). \tag{2.7}$$

In this equation, determining what the magnitudes of the effects of factors 1 and 3 are is not straightforward any more, because they *interact*. Two factors are said to *interact in the additive sense* if the effects of one factor are modified by the other factor. For example, if the third factor is phone identity, then the effects of stress are given by:

$$\begin{aligned} &\text{StressEffect}(/\text{e}/) \ = \\ &\quad [D_1(1\ stressed) \ + \ E_1(1\ stressed) \times D_3(/\text{e}/)] \ - \\ &\quad\quad [D_1(unstressed) \ + \ E_1(unstressed) \times D_3(/\text{e}/)] \end{aligned} \tag{2.8}$$

for vowel /e/ and by

$$\begin{aligned}\text{StressEffect}(/\text{i}/) &= \\ &[D_1(1\ \textit{stressed}) + E_1(1\ \textit{stressed}) \times D_3(/\text{i}/)] - \\ &\quad [D_1(\textit{unstressed}) + E_1(\textit{unstressed}) \times D_3(/\text{i}/)]\end{aligned} \tag{2.9}$$

for vowel /i/. Unless $D_3(/\text{i}/) = D_3(/\text{e}/)$, the stress effects will be different for these two vowels, or, equivalently, the stress factor and the vowel identity factor interact.

For interactions in the *multiplicative sense*, we replace in the above "+" and "−" by "×" and "÷", respectively. In multiplicative interactions, we measure effects as fractions rather than millisecond amounts.

To summarize, many TTS components produce quantitative output based on discrete factorial input, by first mapping the individual factors on quantitative parameters, and then combining these parameters with some combination rule. Factors interact in the additive (multiplicative) sense if and only if the effects of one factor measured in milliseconds (as a fraction) are modulated by other factors. Of course, the additive and multiplicative models are just two of an infinite set of combination rules so that the statement that factors interact in either sense is not terribly informative; the key challenge is to find models that are simple and principled and capture the specific interaction patterns presented by the data. The main reason for the use of the additive and multiplicative models is that parameter estimation is quite simple (see below).

Factor confounding

Before we continue our discussion on models, a brief word on *factor confounding*. In some specially designed *balanced* experiments, all factor level combinations occur equally often. For example, in Port's experiments on the effects of postvocalic voicing on timing (1981), all combinations of postvocalic consonant (/p/ vs. /b/), word length (1, 2, and 3 syllables), and vowel (/i/ vs. /ɪ/) occurred equally often. However, in any naturally occurring text or speech, factors are not balanced — they are *confounded* or, equivalently, *correlated*. In the presence of confounding, separately analyzing the effects of factors can be deceptive. For example, in English two-syllable words, stress is likely to be on the first syllable. As a result, we are likely to find longer syllable durations for word-initial than for word-final positions. However, when we only look at stressed syllables, the opposite is found. The reason is, of course, that the lengthening effects of stress are stronger than the those of word-finality, and that these two factors are seriously confounded.

Factor confounding is not to be confused with interactions. To make this really clear: confounding refers to relations between factors and is independent of the effects factors have on some variable, whereas interaction refers to how the effects of factors in combination affect a variable.

Generalized interactions

We now generalize the concept of interaction. First consider (for $N = 2$ factors) the model

$$\mathrm{DUR}(\vec{\mathbf{f}}) = \frac{1}{1 + e^{-\left[D_1(\vec{\mathbf{f}}_1)^{D_2(\vec{\mathbf{f}}_2)}\right]}}. \tag{2.10}$$

When we let

$$\mathrm{dur}(\vec{\mathbf{f}}) = log\ log\ log\left[\frac{\mathrm{DUR}(\vec{\mathbf{f}})}{1 - \mathrm{DUR}(\vec{\mathbf{f}})}\right], \tag{2.11}$$

then

$$\mathrm{dur}(\vec{\mathbf{f}}) = d_1(\vec{\mathbf{f}}_1) + d_2(\vec{\mathbf{f}}_2), \tag{2.12}$$

where $d_1(x) = log(log(D_1(x)))$ and $d_2(x) = log(D_2(x))$. In other words, when we transform DUR according to Equation 2.11, then the model in Equation 2.10 becomes additive. Likewise, the multiplicative model can be changed into an additive model by logarithmic transformation of DUR. Noting that both the logarithm and the transformation in Equation 2.11 are strictly increasing, we define the following form of generalized additivity:

> A mapping $\mathrm{H} : \mathbf{F} \rightarrow \mathbf{R}$ is *generalized additive* if there exists a strictly increasing transformation $\mathcal{T}$ and per-component mappings h_i such that
>
> $$\mathcal{T}\left[\mathrm{H}(\vec{\mathbf{f}})\right] = \sum_{i=1}^{n} h_i(\vec{\mathbf{f}}_i). \tag{2.13}$$

An important implication — but by no means the only implication (see (Krantz et al., 1971)) — of generalized additivity is *directional invariance*, by which the following is meant:

$$\begin{aligned} \text{If } \mathrm{H}(\mathbf{f}_1, \mathbf{f}_2, \cdots, \mathbf{f}_N) &> \mathrm{H}(\mathbf{f}'_1, \mathbf{f}_2, \cdots, \mathbf{f}_N) \\ \text{then } \mathrm{H}(\mathbf{f}_1, \mathbf{f}'_2, \cdots, \mathbf{f}'_N) &> \mathrm{H}(\mathbf{f}'_1, \mathbf{f}'_2, \cdots, \mathbf{f}'_N). \end{aligned} \tag{2.14}$$

To illustrate, consider two vectors describing the context of the same vowel that are identical except for stress (1-stressed ($\mathbf{f}_1$) vs. unstressed ($\mathbf{f}'_1$)). The definition implies that the stressed case will yield the longer duration for all vowels, phrasal positions, etc., as long as the two vectors in these minimal pairs are matched on everything except for stress.

This provides a much broader definition of interaction — as direction reversals. Of course, far fewer factors interact in this broader sense. Yet, stress certainly interacts in the additive/multiplicative sense: in English longer vowels tend to be stretched by larger millisecond amounts than shorter vowels, and in German the percentage increase in duration due to stress is larger for utterance-final syllables.

We conclude this section on generalized interactions with two remarks. First, analyzing data for the presence of generalized interactions is a useful heuristic for determining which class of non-additive models might be appropriate or which nonlinear strictly increasing transformation $\mathcal{T}$ may produce additivity. Concerning the latter, there already exist several standard transformations of the frequency scale such as semi-tones, mel, the logarithm, and the identity transformation (linear scale); many more could be explored in the search for additivity. This may be useful for the modeling of fundamental frequency.

The second remark is that the prevalence of directional invariance is a key factor for why additive and multiplicative models perform as well as they do, despite their blatant over-simplicity: directional invariance virtually guarantees a decent fit of the data to the additive or multiplicative model. Because, in addition, these models have excellent parameter estimation properties, they are often surprisingly difficult to best in actual data analysis situations, in particular when one is dealing with missing or noisy data.

2.3.2 Some standard multi-variate models

Linear regression

Linear, or multiple, regression analysis is a close cousin of the additive model, the difference being that the factors themselves are quantitative and not discrete, and the functions D_i multiply their arguments with constants:

$$D_i(\vec{\mathbf{f}}_i) \;=\; \alpha_i \times \vec{\mathbf{f}}_i. \tag{2.15}$$

For example, when $\vec{\mathbf{f}} = <sdur,\ spos>$, where *sdur* is syllable duration and *spos* is the number of syllables from the end of the phrase, then we could model the dependency of F_0 peak height on these two variables by

$$\text{PeakHeight}(sdur, spos) = \alpha_1 \times sdur \;+\; \alpha_2 \times spos + \mu. \tag{2.16}$$

After proper normalization, the magnitudes of the estimated values of the α weights give a measure of the importance of a variable. There are two main drawbacks of multiple regression. First, when the factors are correlated (e.g., utterance-final syllables tend to be longer) then the estimates of the weights can be quite unreliable (the same is true, of course, for the additive and multiplicative models when there is heavy factor confounding). To see this, observe that when the factors are almost perfectly correlated, one can substitute one factor for the other, so that the weights have become arbitrary. Second, the model does not allow for interactions.

Analysis-of-variance model

The analysis of variance customarily is used not for modeling purposes but for hypothesis testing, in particular testing for the existence of main effects and (additive) interactions. However, underlying this statistical technique is a model in which some observed variable (*Obs*) is the sum of a set of *interaction terms*:

$$\text{Obs}(\vec{\mathbf{f}}) = \sum_{I \in K} D_I(\vec{\mathbf{f}}_{[I]}). \tag{2.17}$$

Here, K is some collection of subsets of the set of factors, $\{1, \cdots, N\}$, I is one of these subsets, and $\vec{\mathbf{f}}_{[I]}$ is the sub-vector of $\vec{\mathbf{f}}$ corresponding to subset I. The interaction terms $D_I(\vec{\mathbf{f}}_{[I]})$ are constrained to have zero sums.

To illustrate, for $I = \{1, 2, 4\}$, $\vec{\mathbf{f}}_{[I]}$ is the vector $< \vec{\mathbf{f}}_1, \vec{\mathbf{f}}_2, \vec{\mathbf{f}}_4 >$. The additive model corresponds to the special case where K consists of the singleton sets $\{1\}, \cdots, \{N\}$.

One reason that the analysis of variance model is rarely used for predictive modeling purposes is that the interaction terms are, except for the zero-sum assumption, completely unconstrained, and hence can "model" any interaction pattern. That is why the only question asked about the estimated of D_I is whether they are statistically zero or not.

Sum-of-products model

The sums-of-products model (van Santen, 1993a) attempts to make interaction terms more meaningful by dropping the zero sum assumption (which is made only for reasons of mathematical convenience), and replacing it with the assumption that each term is a *product of single-factor parameters*. Thus, the interaction terms have the form:

$$D_I(\vec{\mathbf{f}}_{[I]}) = \prod_{i \in I} s_{Ii}(\vec{\mathbf{f}}_i). \tag{2.18}$$

Again, when K consists of the singleton sets $\{1\}, \cdots, \{N\}$, the additive model emerges as a special case. However, Equation 2.18 also generalizes the multiplicative model, which can be obtained by letting $I = \{1, \cdots, N\}$ and $K = \{I\}$.

The key reason that this model can be used for predictive purposes is that it captures *amplificatory* or *attenuative* interactions in a natural way. For example, in Equation 2.7, the multiplicative term allows $\mathcal{F}_1$ to *attenuate* or *amplify* the effects of $\mathcal{F}_3$, *but without reversing the direction* provided that the parameters are all positive. There are many interactions of this type in speech. For example, utterance finality amplifies the effects of the postvocalic consonant on vowel duration (see Chapter 5), and the effects of the preceding consonant on F_2 measured in the vowel center is stronger for some vowels than for others but shows no reversals (see Chapter 7).

In other words, the sum-of-products model provides exactly what is needed for situations with additive and multiplicative interactions that, however, do not involve direction reversals.

Corrected means

Deciding what the effects of a factor are is difficult when factors are confounded. However, if one is willing to make assumptions about how the factors are combined, inferences can be made even in the presence of severe imbalances. We now discuss three techniques for accomplishing this task: *corrected means*, *quasi-minimal sets*, and *multivariate isotonic regression*.

The idea behind corrected means is quite simple: if $\mathcal{F}_1$ is the factor of interest (i.e., the factor whose effect we are interested in), then we can define a new factor $\mathcal{F}_{\text{remainder}}$ which is the product of all remaining factors, $\mathcal{F}_2 \times \cdots \times \mathcal{F}_N$. We can then apply the 2-factor additive or multiplicative model to this new factorization, and estimate the parameters D_1. These parameter estimates can be interpreted as means of the levels on the factor of interest that *have been corrected for the effects of the remaining factors*.

The assumption made here is much weaker than assuming additivity or multiplicativity, because nothing is assumed about how the factors $\mathcal{F}_2, \cdots, \mathcal{F}_N$ combine with each other; it is only assumed that their joint effects combine additively or multiplicatively with the factor of interest.

A problem is that the number of levels on these combined factors is often quite large, and produces a design matrix (see below) of less than full column rank, so that the parameters (i.e., D_1 and $D_{\text{remainder}}$) cannot be estimated. However, since the only goal is to estimate D_1, it is not necessary that the $D_{\text{remainder}}$ parameters also be estimable. One way to analyze these data is to set up a matrix $M(D_1, D_{\text{remainder}})$ whose rows correspond to factor $\mathcal{F}_1$ and whose columns correspond to $\mathcal{F}_{\text{remainder}}$, and whose cells contain the observed values; most cells are empty, of course. Dodge (Dodge, 1981) describes a method for eliminating from this matrix columns, and hence levels on $\mathcal{F}_{\text{remainder}}$, such that the parameters for the factor levels that remain are estimable.

This method can obviously be generalized to *sets of factors of interest*, instead of only one such factor.

Quasi-minimal sets

A method that serves the same purpose as the above correction method consists of constructing the same matrix $M(D_1, D_{\text{remainder}})$, and eliminating all columns that are not fully filled. This leads naturally to far more columns being eliminated than in Dodge's method, and often all columns end up being removed. In the resulting matrix, we assign rank numbers to each value in a column, and then add these rank numbers across columns. (There are many other ways of combining the columns.) These sums reflect the effects of the factor of interest, modulo a scale factor.

This method is still weaker in its assumptions than the corrected means method, but is also less likely to be feasible when many cells are empty to begin with.

Multivariate isotonic regression

Multivariate isotonic regression (Dykstra and Robertson, 1982; Bril et al., 1984), like quasi-minimal sets analysis, is an estimation method that only assumes directional invariance, and not additivity. It is based on *isotonic regression* (Barlow et al., 1972), which is useful for speech applications all by itself.

In isotonic regression, we consider a sequence of numbers $x_1, x_2, \cdots, x_m$, which are taken to be somewhat noisy measurements of an underlying variable whose values one has good reasons to believe are strictly increasing. Isotonic regression finds numbers $\hat{x}_1, \hat{x}_2, \cdots, \hat{x}_m$ that have the property that $\hat{x}_1 \leq \hat{x}_2 \leq \cdots \leq \hat{x}_m$ and the sum of squares $\sum_{i=1}^{n} w_i \times (x_i - \hat{x}_i)^2$ is minimized (the w_i's are optional weights).

Isotonization (i.e., replacement of x_i by $\hat{x}_i$) can be used as a smoothing method, but also as a way to fill in missing data. For example, if no value has been observed for x_3, the assumption that $x_2 \leq x_3 \leq x_4$ will produce an interval estimate (or a point estimate if $\hat{x}_2 = \hat{x}_4$) for $\hat{x}_3$.

To illustrate, observed F_0 curves are often quite rough in low-energy regions; by judiciously defining in which region one expects the "true" F_0 to be strictly increasing (or decreasing), one can apply isotonic regression weighted by the product of local energy and voicing probability to fill in these rough areas.

Turning now from univariate to multivariate applications, Dykstra (Dykstra and Robertson, 1982; Bril et al., 1984) has shown that the same method can be applied to multivariate problems. Here, one first fits the additive model to obtain values for the empty cells, but assigns extremely small weights to these cells. Next, an iterative process follows which proceeds by matrix dimension (rows, columns, hyper-rows, etc); within each dimension, the process isotonizes each individual row (column, etc). This process converges, and the initial values assigned to the missing cells have little impact on the final result. Of course, this process has to be repeated for several permutations of dimensions, because the within-dimension order is not given a priori. We have found, however, that one can limit the search to a small neighborhood of the permutations defined by the corrected means.

This process produces a complete (i.e., without empty cells) N-dimensional matrix. This matrix can then be inspected for the per-dimension permutations, which provide information about the individual effects of the factors. In addition, the matrix as a whole can be used as a lookup table, without having to model its entries.

There are several drawbacks, however. First, this is a computationally intensive process. Second, the accuracy of the results depends critically on the distribution and reliability of the non-empty cells. For example, if few cells are occupied in the extreme corners of the matrix, then the method is required to extrapolate to these corners, which can produce unreliable or even indeterminate results.

2.3.3 Least squares estimation

We briefly discuss here the dominant approach to parameter estimation, *least squares estimation*. We also discuss this method here because of its relevance for a text selection method presented below in Section 2.4.5.

Consider the additive model, and let the observations be denoted $Obs(\vec{\mathbf{f}})$, for $\vec{\mathbf{f}} \in \mathbf{F}'$, where $\mathbf{F}'$ is some subset of the complete factorial space $\mathbf{F}$. We can list the vectors $\vec{\mathbf{f}}$ as $\vec{\mathbf{f}}^{[1]}$, $\vec{\mathbf{f}}^{[2]}$, ..., $\vec{\mathbf{f}}^{[p]}$, where p is the number of observations in $\mathbf{F}'$.

Least-squares estimation finds parameters D_i that minimize

$$\sum_{j=1}^{p} \left[Obs(\vec{\mathbf{f}}^{[j]}) - \sum_{i=1}^{n} D_i(\vec{\mathbf{f}}_i^{[j]}) \right]^2 . \tag{2.19}$$

This can be simplified using matrix notation. We can form the $p \times 1$ column vector $\vec{O}$ whose elements are $Obs(\vec{\mathbf{f}}^{[1]})$, $Obs(\vec{\mathbf{f}}^{[2]})$, ..., $Obs(\vec{\mathbf{f}}^{[p]})$. Let $\vec{D}$ be a vertical stack of the vectors defined by the mappings D_1, ..., D_N.

For each factor $\mathcal{F}_i$ having $\mathcal{L}_i$ levels, associate the first $\mathcal{L}_i - 1$ levels with $\mathcal{L}_i - 1$-component row vectors having 0 everywhere but in the k-th position (where k is level number), and associate the $\mathcal{L}_i$-th level with a row vector of -1's. A given vector $\vec{\mathbf{f}}^{[j]}$ can then be associated by N such row vectors, which can be abutted horizontally to form a single M-component row vector, $\vec{x}(\vec{\mathbf{f}}^{[j]})$; here, $M = \sum_{i=1}^{n} \mathcal{L}_i$. Let X be the $p \times M$ matrix obtained by piling up these row vectors (the *design matrix*). Then we can write:

$$\vec{Obs} \approx X\vec{D}, \tag{2.20}$$

where "$\approx$" indicates that we want to approximate observations $\vec{Obs}$ with model $X\vec{D}$. The corresponding least-squares formulation is:

$$|| \vec{Obs} - X\vec{D} ||, \tag{2.21}$$

where "$|| \vec{y} ||$" indicates the norm, or sum of squares, of a vector $\vec{y}$. If the matrix X has full column rank, then the standard solution is given by:

$$\hat{\vec{D}} = (X^t X)^{-1} X^t \vec{Obs}, \tag{2.22}$$

where superscript "t" indicates the matrix transpose and where superscript "-1" the matrix inverse.

A similar closed-form estimation is possible for linear regression and for analysis of variance. However, this is not the case for sum-of-products models, where parameters must be estimated with iterative methods, and certainly not for multivariate isotonic regression.

2.4 Optimal Text Selection

Not only do TTS components containing quantitative models require training data bases, but also other components such as those for accenting and phrasing; the latter components produce discrete, not quantitative output. The quality of a component depends both on how clever our models are and on how well the training data base covers the TTS input domain. In this section we discuss some methods for optimal coverage.

We can think of the input domains of most component sets of *units*. For example, the accenting component, which assigned a pitch accent type to a word, processes vectors describing the lexical identity of the word in question, its location in the phrase, and part-of-speech tags of surrounding words. The intonation component uses vectors describing features associated with a syllable, including its segmental makeup, lexical stress, and pitch accent type of its associated word. In each of these three cases, the input domains are discrete, but extremely large. For example, using parts-of-speech tags in a window of five words in combination with five location positions already creates a space of potentially one hundred million units.

There are two complementary approaches to deal with the fact that training data bases sample only an infinitesimal subset of the domain. One is to carefully select training data — which is the topic of this section. The other is to use models and rules that are based on general regularities or invariances, and that one can trust to accurately generalize from training data to the domain at large. This very broad topic was touched on above in our discussion of statistical methods, but also implicitly pervades our discussion of text analysis, with its emphasis on the discovery of *principled rules*.

2.4.1 Greedy algorithm

Consider a simple problem: constructing a text data base in which each diphone occurs at least once. This can be done manually, by perusing a pronouncing dictionary and finding one word for each diphone. However, there is an alternative method, which is faster because it is completely automatic, and produces a short word list because it attempts to optimize the number of diphones per word. The method works as follows. First, the TTS pronunciation component is applied to a large on-line text corpus to produce two lists, a sentence list and a diphone list. The sentence list contains individual sentences on each line, and the diphone list contains the corresponding diphones with duplicates removed. Second, we use the well-known *greedy algorithm* (Cormen, Leiserson, and Rivest, 1990) to select sentences. This algorithm starts by finding the sentence that has the largest number of distinct diphones, and then removes the names of these diphones from the diphone list. Once N sentences have been selected, the $N+1$st selection is based on which sentence has the most remaining diphones. This algorithm provides an approximate solution to the equally well-known set-covering problem: given a collection of sets, find the smallest number of sets whose union contains all elements in the union of the original collection. This problem is known not to have a solution in polynomial time, but it is also known that the greedy solution is off by on the average only a factor of $log_2(k)$, where k is the true minimum number of covering sets. Thus, when k is 2 or less the greedy algorithm is optimal.

In the remainder of this section, we discuss applications of the greedy algorithm and its variants.

2.4.2 Standard application

Our first application of greedy algorithms was for a perceptual experiment, in which we wanted listeners to process the smallest number of sentences that yet contained each acoustic inventory element at least once (van Santen, 1993b). Starting with a corpus of 67,440 sentences, the greedy algorithm found a subset of 650 sentences with complete coverage of all (2533) elements in the larger list. This represents a *reduction percentage* of 99%.

In other cases, the reduction percentage has been less favorable. For example, the reduction percentage was less than 90% with a corpus of 169,328 personal names containing 16,422 triphones. Obviously, this reduction percentage increases with the ratio of corpus size to the set of units (in these two samples, the values of this ratio were 26.6 and 10.3, respectively), because the greedy algorithm guarantees that each newly selected sentence contains at least one new unit so that the number of greedily selected sentences can never exceed the number of units. However, the reduction percentage also depends on co-occurrence frequencies, or correlations, between units. Favorable correlational patterns can produce selected lists much shorter than the number of units, as we found in the acoustic inventory element application (650 vs. 2533).

2.4.3 Frequency weighted greedy coverage

In many cases, we can meaningfully associate weights with the units. The greedy algorithm can be generalized to select sentences based on the sum of the weights of the (unique) units contained in a sentence. The basic algorithm is a special case where these weights are all equal to 1.

We have found weighted greedy coverage useful in two circumstances. First, there are cases where it proves impossible to cover all units, so that it becomes attractive to at least attempt to cover the most frequently occurring units. In this case, we use frequencies of occurrence as weights. This happens, for example, when we try to cover all distinct triphones that occur in the full names of all US citizens, but for practical reasons have to accomplish this with fewer than 10,000 names; there are more than 70,000 such triphones, assuming a 43-phone alphabet. The standard greedy algorithm failed to accomplish this. However, using fewer than 10,000 names, we were able to obtain frequency-weighted coverage in excess of 99%; unweighted coverage was less than 70%.

A second case where weights can be useful is in the standard situation, where complete coverage is feasible. We have found that assigning weights equal to one divided by the frequency of occurrence has the effect of forcing the algorithm to select sentences with "hard-to-get" units first. In the process, more frequent units are encountered as a by-product, so that the algorithm does not have to go out of its way to capture those units. Often this produces shorter sentence lists than can be generated by the standard greedy algorithm

2.4.4 Inside units: sub-vectorization

So far, we have treated units as atomic entities. However, often units are complex and consist of feature vectors, such as the input units to the phrasing, intonation, and duration components.

We have seen in Section 2.3 that we often model vector input by assigning parameters to each vector component, via explicit arithmetic equations that specify which features interact and which do not. We also saw that the number of parameters in these cases is typically small, and depends linearly on the number of levels on each factor, even when there are several interactions. By contrast, the feature space depends multiplicatively on these numbers. Hence, attempting to cover the entire feature space might be unnecessary if the only goal is to estimate these parameters, because the number of judiciously chosen training units need not exceed the number of parameters.

The next subsection discusses a highly efficient algorithm that can be used if we assume a specific model. i.e., list of factors, their levels, and an equation modeling the effects of these factors. However, there often are situations where one is reluctant to go that far, even though one does have a good idea about what interactions might occur.

For this type of situation, we propose using a method called *sub-vectorization* in which we construct from each feature vector a set of sub-vectors corresponding to precisely those components that we expect to interact. For example, if we expect postvocalic consonant identity and phrasal location to interact with each other but not with vowel identity, then a vector such as </i/, phrase-final, voiced stop> would generate the sub-vectors </i/> and <phrase-final, voiced stop>. We then apply standard greedy algorithms to a list in which each sentence is represented as a set of sub-vectors instead of full vectors.

We applied this idea to Mandarin Chinese (Shih and Ao, 1997), and found that the 8,233 distinct sub-vectors occurring in a data base of 15,630 sentences could be covered completely in just 427 sentences. Coverage of the original feature space would have required thousands of sentences. Yet, indications are that the resulting set of sentences was prosodically and segmentally quite rich, producing both robust research results and a good timing component.

2.4.5 Inside units: model-based greedy selection

As just mentioned, the idea underlying sub-vectorization can be taken a step further. If, instead of only knowing which factors might interact, one has a complete equation for predicting an acoustic variable from the feature space, it should be possible to derive from this equation how many parameters are to be estimated and what the combinatorial structure of the training data base must be to allow unique parameter estimation.

We analyzed this idea for the class of sums of products models. In the simplest case (the additive model) the total number of parameters is roughly equal to the total sum of the numbers of distinctions made on each factor. In the typical case (e.g., vowel duration), this is about 30. Because a long sentence could contain at least thirty vowels and hence data points, it should in principle be possible to estimate all parameters *using exactly one spoken sentence.*

However, estimability requires not only that the number of data points exceeds the number of parameters, but also that the combinational structure of the set of feature vectors in selected set of sentences is appropriate. We solved this problem by first computing for each sentence the corresponding design matrix. Parameter estimability is equivalent to having the vertical pile of the design matrices of the selected sentences be of full column rank. We developed a fast algorithm that, using a generalized concept of greedy coverage and using basic linear algebraic results on orthogonality, finds the smallest set of sentences that allows parameter estimation (Buchsbaum and van Santen, 1996). Applied to the Mandarin Chinese data base, we found that two sentences are adequate.

When one is hesitant to base an entire duration component on only two recorded sentences, one can repeat the procedure and generate a set of 20 sentences. Because each subset of 2 sentences provides complete coverage (i.e., estimability) of all parameters, these twenty sentences may have very favorable properties in terms of the maximal standard error of estimate of the parameters. In other words, we conjecture that this procedure provides more reliable parameter estimates with fewer sentences than any other method we are aware of.

Of course, this method only makes sense when we are willing to make strong assumptions about the model. In cases where we want to adapt duration parameters to a new speaker of a dialect whose durational patterns already have been successfully modeled, this assumption may very well be justified.

2.4.6 Coverage

We end this section on optimal text selection by pointing out that a text corpus, no matter how large, is only an infinitesimal sample of the input domain of unrestricted text-to-speech. No matter how carefully we select training text and how well it covers the text corpus, this by itself is no basis for asserting that the training corpus covers the input domain.

There are some exceptions. For example, since it is known which diphones exist in the language, the input domain can indeed be covered by a training corpus — as far as these units are concerned. However, it is not known how many distinct vectors relevant for F_0 prediction exist in the language. This latter situation is probably the rule rather than the exception.

Another important fact to keep in mind is that there are surprisingly large statistical differences between different text "genres", such as newspaper text, technical manuals, movie reviews, and dialogue in various telephone transactions. Even within a certain genre, unit frequency distributions differ markedly, regardless of what type of unit one looks at — words, triphones, parts-of-speech pairs. For example, we analyzed different text corpora by looking at frequency distributions of shared words, and found correlations of less than 0.50 between the log frequencies in the Associated Press from the years 1989 vs. 1990 (van Santen, 1997).

2.5 Finite-State Methods and Tools

Our work on multilingual text-analysis is based nearly entirely upon a single computational device, the *weighted finite-state transducer*, or WFST. In this section we give an overview of WFST's and the *(weighted) regular relations* that they compute.[2] This is by no means intended to be a complete introduction to this subject. For further and more complete discussion the reader is urged to consult other sources. For ordinary (unweighted) finite-state *acceptors* (FSA's), and the corresponding *regular languages*, any decent introduction to automata theory (Harrison, 1978; Hopcroft and Ullman, 1979; Lewis and Papadimitriou, 1981, for example) will suffice. For transducers and weighted finite-state machines, there is lamentably less material accessible to the non-specialist, but there are a few papers one might recommend. One is (Kaplan and Kay, 1994), whose discussion we draw on to some extent here. More recent is (Mohri, 1997), as well as a few of the references cited therein. Mohri focuses on applications of WFST's to language and speech processing, and efficient algorithms to support these kinds of applications: some of the algorithms that he discusses are the same as are used in our own finite-state toolkit. Finally, one might consult (Sproat, 1992) for a fairly non-technical introduction to FST's as applied in computational morphology and phonology.

2.5.1 Regular languages and finite-state automata

We start with a notion of an alphabet of symbols, where the entire alphabet is conventionally denoted Σ. Also conventional is the use of the symbol ϵ to denote the *empty string*, which is *not* an element of Σ and which is distinct from the *empty set* $\emptyset$. Regular languages are usually defined with a recursive definition along the following lines:[3]

[2] One also sees the term *weighted rational transductions*; cf. (Pereira, Riley, and Sproat, 1994).

[3] This definition is based on the definition given in (Kaplan and Kay, 1994, page 338).

1. $\emptyset$ is a regular language
2. For all symbols $a \in \Sigma \cup \epsilon$, $\{a\}$ is a regular language
3. If L_1, L_2 and L are regular languages, then so are
   (a) $L_1 \cdot L_2$, the *concatenation* of L_1 and L_2: for every $w_1 \in L_1$ and $w_2 \in L_2$, $w_1 w_2 \in L_1 \cdot L_2$
   (b) $L_1 \cup L_2$, the *union* of L_1 and L_2
   (c) L^*, the *Kleene closure* of L. Using L^i to denote L concatenated with itself i times, $L^* = \cup_{i=0}^{\infty} L^i$.

The above definition defines all and only the regular languages. However, there are additional *closure* properties which regular languages observe. We will make use here of the term Σ^*, which denotes the set of all strings (including the empty string) over Σ.[4]

- *Intersection*: if L_1 and L_2 are regular languages then so is $L_1 \cap L_2$.
- *Difference*: if L_1 and L_2 are regular languages then so is $L_1 - L_2$, the set of strings in L_1 that are not in L_2.
- *Complementation*: if L is a regular language, then so is $\Sigma^* - L$, the set of all strings over Σ that are *not* in L. (Of course, complementation is merely a special case of difference.)
- *Reversal*: if L is a regular language, then so is $Rev(L)$, the set of reversals of all strings in L.

Regular languages are defined as sets of strings. They are commonly *notated* as *regular expressions*. And they can be *recognized* by *finite-state automata*. The equivalence of these three objects is a well-known result of automata theory known as Kleene's theorem(s). A finite-state automaton can be defined as follows (see (Harrison, 1978; Hopcroft and Ullman, 1979; Lewis and Papadimitriou, 1981)):

> A finite-state automaton is a quintuple $M = (K, s, F, \Sigma, \delta)$ where:
>
> 1. K is a finite set of states
> 2. s is a designated initial state
> 3. F is a designated set of final states
> 4. Σ is an alphabet of symbols, and
> 5. δ is a transition relation from $K \times \Sigma$ to K

[4] It should be stressed that it is not necessary to explicitly state all of these properties: for example, given union and complementation, closure for intersection follows from De Morgan's Laws. We merely list all of these for completeness.

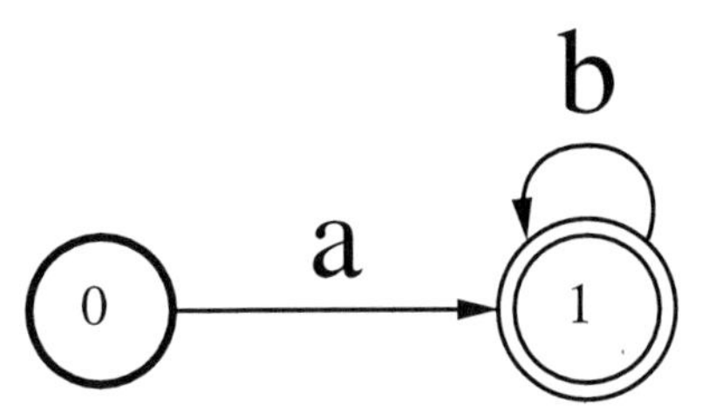

Figure 2.1: An automaton that computes ab^*. By convention, the heavy-circled state (0) is the initial state, and the double-circled state is the final state.

To illustrate with a simple example, consider the following infinite set of strings: $\{a, ab, abb, abbb \ldots\}$. In other words, the set consisting of a followed by zero or more bs. This can be represented, using fairly standard regular-expression notation, as the compact expression ab^*. And it can be recognized by the finite-state machine depicted in Figure 2.1. This machine is *deterministic*, meaning that at any state, for a given input symbol, there is exactly one state to which the machine may move. For example, on the input abb, the machine will start in state 0, move on a to state 1, move on b to state 1 again, and repeat this process once more to end in state 1 which, since this state is final, means that the machine has accepted the input. The definition of finite-state automaton above also allows for *non-deterministic* machines, meaning that at a state one may in principle visit more than one next state on a given symbol. As it happens non-deterministic *unweighted automata* can always be determinized — that is, converted into an equivalent deterministic automaton. As a practical matter, an equivalent deterministic automaton may be (much) larger than a non-deterministic automaton, but it will generally be more efficient to operate simply because there is never any need to search more than one path in the machine.[5]

2.5.2 Regular relations and finite-state transducers

One can extend the notion of regular languages to *regular n-relations*, which are sets of n-tuples that can be recursively defined in a parallel manner to the definition of regular languages:

1. $\emptyset$ is a regular n-relation

[5] Strictly speaking, it is not even true that the deterministic machine needs to be larger. See (Mohri, 1994) for discussion of space-efficient representations of deterministic automata, and see also Section 4.2 for an application of such representations.

2. For all symbols $a \in [(\Sigma \cup \epsilon) \times \ldots \times (\Sigma \cup \epsilon)]$, $\{a\}$ is a regular n-relation

3. If R_1, R_2 and R are regular n-relations, then so are

   (a) $R_1 \cdot R_2$, the *(n-way)concatenation* of R_1 and R_2: for every $r_1 \in R_1$ and $r_2 \in R_2$, $r_1 r_2 \in R_1 \cdot R_2$

   (b) $R_1 \cup R_2$

   (c) R^*, the *n-way Kleene closure* of R.

Regular n-relations can be thought of as *accepting* strings of a relation stated over an m-tuple of symbols, and mapping them to strings of a relation stated over a k-tuple of symbols, where $m + k = n$. We can therefore speak more specifically of $m \times k$-relations. As with regular languages, regular n-relations obey further closure properties:

- *Composition*: if R_1 is a regular $k \times m$-relation and R_2 is a regular $m \times p$-relation, then $R_1 \circ R_2$ is a regular $k \times p$-relation.
- *Reversal*: if R is a regular n-relation, then so is $Rev(R)$.
- *Inversion*: if R is a regular $m \times n$-relation, then R^{-1}, the *inverse* of R, is a regular $n \times m$-relation.

Composition will be explained below.

In most practical applications of n-relations in natural language and speech processing, $n = 2$, with (obviously) $k = m = 1$; in this specific case, we can speak of a relation mapping from strings of one regular language into strings of another. (A notable exception is the work of Kiraz (1996).) For this reason we will speak henceforth simply of *relations*, and have it unambiguously mean *2-relations*. Note that difference, complementation and intersection are omitted from the set of closure properties, since in general regular relations are *not* closed under these operations, though particular subsets of regular relations, including length-preserving (or "same-length") relations, *are* closed under these operations; see (Kaplan and Kay, 1994).

Corresponding to regular relations are n-way regular expressions and *finite-state transducers* (FST's). Rather than give a formal definition of an FST (which can in any event be modeled on the definition of FSA given above), we will illustrate by example. Consider an alphabet $\Sigma = \{a, b, c, d\}$ and a regular relation over that alphabet expressed by the following set: $\{(a, c)$, (ab, cd), (abb, cdd), $(abbb, cddd) \ldots \}$. In other words, the relation consisting of a mapping to c followed by zero or more b's mapping to d. We can represent this compactly by the two-way regular expression *a:c (b:d)**. A finite-state transducer that computes this relation is depicted in Figure 2.2. By convention, we will refer to the expressions on the lefthand side of the ':' as the input side, and the expressions on the righthand side as the output side. So in Figure 2.2, the input side is characterizable by the regular expression ab^*, and the output side by the expression cd^*.

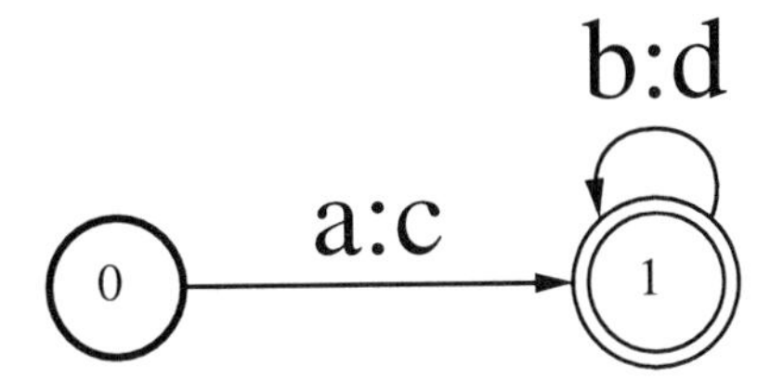

Figure 2.2: A transducer that computes *a:c (b:d)**.

As it happens, the transducer in Figure 2.2 is deterministic, both on the input side and the output side: for any state and any input or output symbol, there is at most one destination state corresponding to that symbol-state pair. But it is important to bear in mind that not all FST's can be determinized. All *acyclic* FST's[6] are certainly determinizable, but for cyclic transducers the issue is subtle and complex: the reader is referred to (Mohri, 1997) for extensive discussion of this issue.

The term *composition* has its normal algebraic interpretation: if R_1 and R_2 are regular relations, then the result of applying $R_1 \circ R_2$ to an input expression I is the same as applying first R_1 to I and then applying R_2 to the output of this first application. Consider the two transducers in Figure 2.3 labeled T_1 and T_2. T_1 computes the relation expressable as *(a:c (b:d)*) | ((e:g)* f:h)*. T_2 computes *g:i* ϵ*:j h:k*, where the ϵ*:j* term inserts a *j*.[7] $T_1 \circ T_2$ computes the trivial relation, *e:i* ϵ*:j f:k*. In this particular case, though both T_1 and T_2 express relations with infinite domains and ranges, the result of composition only maps the string *ef* to *ijk*.

The inverse of a transducer is computed by simply switching the input and output labels. The fact that regular relations are closed under inversion has an important practical consequence for systems based on finite-state transducers, namely that they are fully bidirectional (Kaplan and Kay, 1994). One can for example construct a set of rewrite rules (see Section 2.5.4) that maps from lexical to surface form; and then invert the resulting transducer to use it to compute possible lexical forms from surface forms.

One notion that we will make use of from time to time is the notion of *projection* onto one dimension of a relation. For example, for a 2-way relation R, $\pi_1(R)$ projects R onto the first dimension and $\pi_2(R)$ projects onto the second dimension. (π_1 and π_2 are also said to compute, respectively, the domain and

[6] An acyclic automaton or transducer is one in which there are no cycles, meaning that once one leaves any state, there is no path that will lead back to that state.

[7] Here, and elsewhere, we adopt the Unix regular-expression convention whereby '|' denotes disjunction.

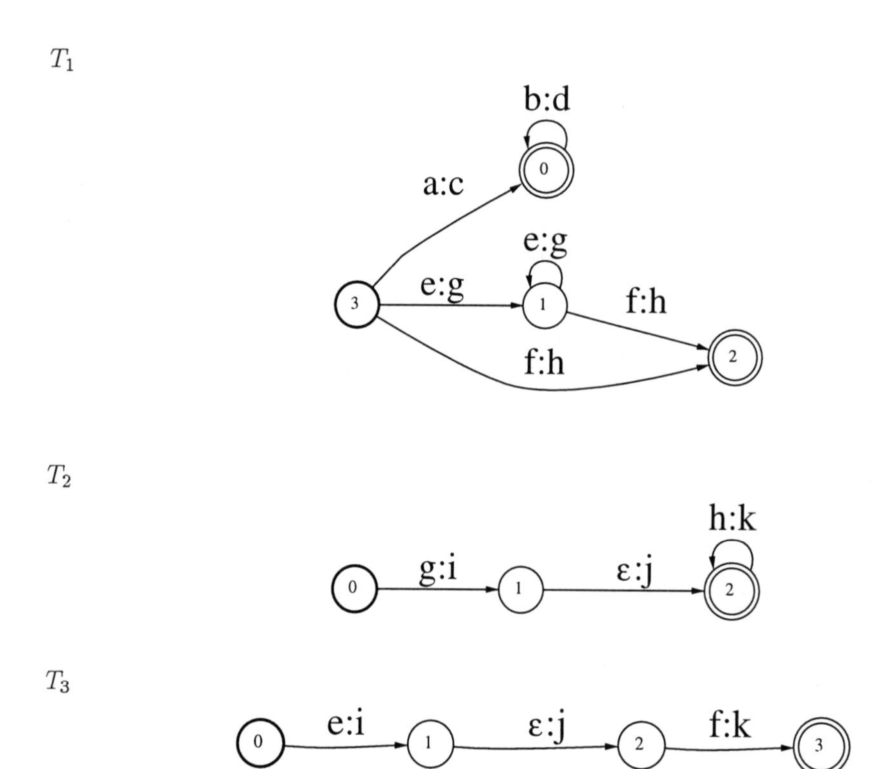

Figure 2.3: Three transducers, where $T_3 = T_1 \circ T_2$

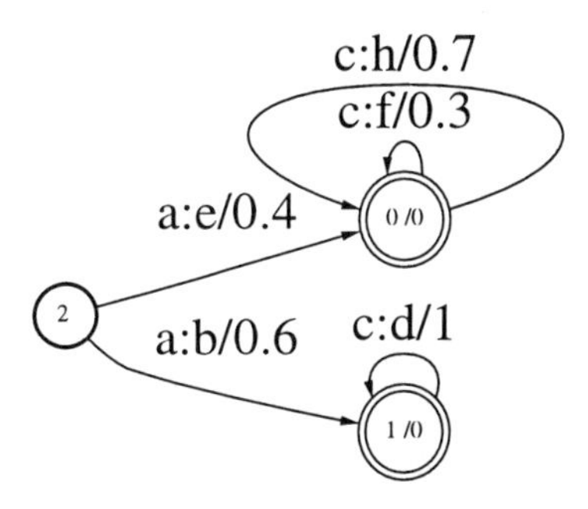

Figure 2.4: An example of a weighted finite-state transducer. Here, and elsewhere, the number after the '/' is the weight.

range of the relation.) The analog of the projection operation for 2-way FST's produces an FSA corresponding to one side of the transducer. So the first projection (π_1) of the transducer in Figure 2.2 is the acceptor in Figure 2.1.

2.5.3 Weighted regular X and weighted finite-state X

We have so far spoken of automata and transducers where the arcs are labeled with elements from an alphabet. It is also possible to add to these arcs *weights* or *costs*, which may in general be taken from the set of real numbers. Let us consider here the most general case, namely that of a *weighted finite-state transducer* or WFST. An example WFST is given in Figure 2.4. This WFST accepts any string that matches ac^*, and transduces it to several possible outputs. Consider as a concrete case the input *acc*. The possible outputs for this input are: *bdd*, *eff*, *efh*, *ehf*, *ehh*. The weights allow us to rank these outputs, but the ranking will depend upon how the weights are interpreted. The most common interpretations of the weights in practical applications are:

- As *probabilities*: in this case the weights along a path are *multiplied*, the cost of a particular analysis is the *sum* of the weight of all paths having that analysis, and the best analysis has the *highest* weight.

- As *costs*, or negative log probabilities: in this case the weights along a path are *summed*, the cost of a particular analysis is the *minimum* of the weight of all paths having that analysis, and the best analysis has the *lowest* weight.

Of these two, the cost interpretation is by far the more common (since using real probabilities will lead to underflow problems on large examples), and it is the one that will be assumed here; for the purposes of this discussion, then, *weight* and *cost* have the same meaning, and we will freely interchange these terms. On the cost interpretation, the weights of the possible outputs of *acc* are as follows: *bdd<2.6>*, *eff<1.0>*, *efh<1.4>*, *ehf<1.4>*, *ehh<1.8>*. Here, and elsewhere we will denote a cost on an expression with a number in angle brackets. Of this set, *eff* has the lowest cost, and is therefore the best analysis. In subsequent discussion, when we refer to the best path, we will always mean the path with the lowest cost, and when we refer to computing the best path, we will presume an algorithm for finding this lowest-cost path.

The closure properties of WFST's are identical to those of FST's. For WFSA's, the closure properties are identical to those of FSA's, except that they are not closed under difference or complementation. On properties for determinizable WFSA's, and algorithms for computing deterministic equivalents see (Mohri, 1997).

2.5.4 A sketch of some lexical tools

The operations on WFST's that we have described in the preceding sections are provided by our underlying *fsm* toolkit. But in order to make use of WFST's in linguistic description, it is desirable to have a set of tools that enable one to convert from a high-level description understandable to a human expert, into an internal WFST representation. Such tools are now fairly widespread: a particularly well-known set is that developed at Xerox and described in (Karttunen, 1993; Karttunen and Beesley, 1992; Karttunen, 1995; Karttunen, 1996) and elsewhere. We make use of our own toolkit, *lextools*, which (it should be stressed) is currently still under development. The tools include:

compwl: A general regular expression and word-list compiler. This is mostly used in the construction of lexicons (which can be represented as lists of regular expressions), and grammatical constraints (which can be represented as regular expressions).

arclist: A finite-state grammar compiler, the name being taken from (Tzoukermann and Liberman, 1990). An example of the use of this tool can be found in Section 3.6.4.

paradigm: A compiler for morphological paradigms. See Section 3.4.1.

rulecomp: A rewrite rule compiler for compiling rewrite rules into WFST's. The rules are of the form $\phi \rightarrow \psi/\lambda \_\_ \rho$. Here ϕ, λ and ρ are regular expressions defining the input, left context, and right context, respectively; and ψ is a *weighted* regular expression defining the output. The algorithm is described in (Mohri and Sproat, 1996).

numbuilder: A compiler for digit-string-to-number-name conversion. Applications of this tool are described in Sections 3.4.1 and 3.6.3.

2.6 TTS Architecture

In designing TTS systems it is important to have powerful and flexible methodologies, as well as tools for implementing those methodologies. It is also important that the actual TTS software be designed in a way that makes further development relatively easy and efficient.

The Bell Labs Multilingual TTS engine is a modular system consisting of ten modules, which are pipelined together. Each module is responsible for one portion of the problem of converting from text into speech. Each module is language-independent in the sense that it contains no language particular code.[8] To switch to a new language, one merely has to provide an appropriate set of language-specific tables for text-analysis, duration, intonation and concatenative units. The modules, in order of application in the pipeline, are as follows:

gentex: The text-analysis module.

phrase: The phrasing module, which simply builds phrases as specified by gentex.

phrasal phonology: The phrasal phonology module, which sets up phrasal accents.

duration: The segmental timing module.

intonation: The intonation module.

amplitude: The amplitude module.

source: The glottal source module.

getdy: The unit selection module, which selects the concatenative units from a table, given the phonetic transcription of the words.

catdy: The diphone concatenation module, which performs the actual concatenation of the units.

synthesis: The synthesis module, which synthesizes the final waveform given the glottal source parameters, the F_0 values, and the LPC coefficients for the diphones.

The description of most of these modules will be the topic of subsequent chapters: *gentex* is described in Chapters 3 and 4; *duration* is described in Chapter 5; *intonation* and *phrasal phonology* are the topic of Chapter 6;[9] aspects of *source*,

[8] In our American English system, which shares the same general architecture with our multilingual system (Sproat and Olive, 1997), most of the modules are language-particular.

[9] Actually there are currently two intonation models available. One is based on the Pierrehumbert-Liberman-style tonal target model; currently this is used for Mandarin, Japanese, Taiwanese and Navajo. The other one, which is more similar to the Fujisaki model, is used for our other languages. Models of both kinds are available for American English.

amplitude, *getdy*, *catdy* and *synthesis* are described in Chapter 7. We will have nothing to say about the *phrase* module, which is very simple. Amplitude will be discussed briefly in Chapter 7.

Each module in the pipeline communicates with the next via a uniform set of data structures: the only exception to this is the unit-concatenation module, which outputs a stream of LPC parameters and source state information for the synthesis module. Each set of data structures corresponds to a single sentence, as determined by the text analysis module. The data structures consist of a set of arrays of linguistic structures. Each array is preceded by the following information:

1. The structure's type T: e.g., word, syllable, phrase, phoneme, concatenative unit ...

2. N, the number of structures of type T for this sentence

3. S, the size of an individual structure of type T

After each of these headers is an $N \times S$ byte array consisting of N structures of type T.

This modular pipeline design confers a number of advantages. The first of these is a standard observation: if the system's modules share a common interface then it is easy for different workers to work independently of one another on different modules, as long as they agree on the input-output behavior of each module, and the manner in which the modules should relate to one another. Secondly, it is easy to terminate the processing at any point in the pipeline. For example, when one is testing a segmental duration module, one can test the system by running it over a large corpus of text, and sampling the segmental durations that are computed. Since modules subsequent to the duration module are irrelevant for this test, one can simply terminate the pipeline after this module with a program that prints out information on segments and their duration. Other useful properties include the ability to insert into the pipeline programs that interactively modify TTS parameters in various ways; see (Sproat and Olive, 1997) for further discussion.

Chapter 3

Multilingual Text Analysis

Richard Sproat, Bernd Möbius, Kazuaki Maeda, Evelyne Tzoukermann

3.1 Introduction

The first task faced by any text-to-speech system is the conversion of input text into an internal linguistic representation. By "input text" we normally mean a string of characters from some character set (e.g. ascii).[1] By "internal linguistic representation", we mean a set of structures that represent the words that occur in the text, their grammatical categories, morphological analyses, phonological transcriptions, accentual or tonal properties, and the placement of prosodic phrase boundaries. We use the phrase "text analysis" as a cover term for all of the computations involved in this conversion. That is, it comprises such tasks as tokenization, end-of-sentence detection, abbreviation and number expansion, morphological analysis and phonological modeling, accentuation and phrasing.

It is worth noting that our use of the term "text analysis" is a far broader use of the term than one typically finds in the TTS literature. For example, van Holsteijn (van Holsteijn, 1993, page 29) uses the term to denote "the first submodule [of the text preprocessor which] deals with the segmentation and labeling of text". There is, however, good justification for our unorthodox use of the term. The more familiar usage invariably presumes a model of TTS conversion where text "preprocessing" or "normalization" is logically separate from and prior to "linguistic analysis". We shall argue that this separation is

[1] See Appendix A for some discussion of character encodings and their role in text analysis for TTS.

neither desirable nor necessary: "preprocessing" functions such as abbreviation expansion can and should be performed in parallel with "linguistic analysis" functions such as morphological analysis. Unfortunately there is no standard term for such an integrated model; "text normalization" and "preprocessing" are obviously too narrow in scope, and linguists (at least) would consider it bizarre to consider (for example) number expansion to be part of linguistic analysis. The term "text analysis" seems a reasonable compromise.

One of the main reasons that text analysis is such a challenging problem is that the standard written form of (any) language gives an imperfect and often distant rendition of the corresponding spoken forms. Among the problems that one faces in handling ordinary text are the following:

1. While a large number of languages delimit words using whitespace or some other device, some languages, such as Chinese and Japanese do not. One is therefore required to "reconstruct" word boundaries in TTS systems for these languages.

2. Digit sequences need to be expanded into words, and more generally into well-formed number names: so *243* in English would generally be expanded as *two hundred and forty three*.

3. Abbreviations need to be expanded into full words. In general this can involve some amount of contextual disambiguation: so *kg.* can be either *kilogram* or *kilograms*, depending upon the context.

4. Ordinary words and names need to pronounced. In many languages, this requires morphological analysis: even in languages with fairly "regular" spelling, morphological structure is often crucial in determining the pronunciation of a word.

5. Prosodic phrasing is only sporadically indicated (by punctuation marks) in the input, and phrasal accentuation is almost never indicated. At a minimum, some amount of lexical analysis (in order to determine, e.g. grammatical part of speech) is necessary in order to predict which words to make prominent, and where to place prosodic boundaries.

3.2 The "Two-Stage" Traditional Approach

In many TTS systems the first three tasks — word segmentation, and digit and abbreviation expansion — would be classed under the rubric of *text normalization* and would generally be handled prior to, and often in a quite different fashion from the last two problems, which fall more squarely within the domain of linguistic analysis.[2]

[2]But see Traber (1995), who treats numeral expansion as an instance of morphological analysis, and also the work of van Leeuwen (van Leeuwen, 1989b; van Leeuwen, 1989a), which uses cascaded rewrite rules within his TooLiP system towards the same ends.

As a recent example of this view, consider the *TextScan* system described in (van Holsteijn, 1993), which serves as the preprocessing module for the *Speechmaker* (Dutch *Spraakmaker*) text-to-speech system. The purpose of this module is to "perform a segmentation of the input text, and to convert anomalous symbol strings such as numerals and abbreviations into a lexical format" (page 27). Also performed are such tasks as period disambiguation, for the purpose of end-of-sentence detection. These "normalized" strings are then linguistically analyzed by later components. Any reasonable preprocessing module must of course perform some disambiguation of the input that it is expanding; for example, strings such as *DFl 2,50* (2.50 Dutch guilders) contain information that what is being expanded is a money amount, which should therefore not be expanded in the same way that *2,50* (= *2.50*) would be expanded. But such disambiguation is, in van Holsteijn's terminology, entirely *form-based*, meaning that there is no access to information derived from (subsequent stages of) linguistic analysis, including morphological and syntactic information. Still, according to van Holsteijn (page 32): "In some cases, disambiguation can only be achieved with help from syntactical and sometimes even semantic analysis. It should be said, however, that ambiguities with which a form-based preprocessing module cannot cope are relatively few." While this may be true for Dutch, and even to a large extent for English, as we shall show directly there are languages where this approach cannot work.

3.3 Problems with the Traditional Approach

Even in English, one can find examples where purely form-based disambiguation is insufficient. Consider an example that is problematic for the Bell Laboratories American English TTS system, a system that treats text normalization prior to, and separately from, the rest of linguistic analysis. If one encounters the string *$5* in an English text, the normal expansion would be *five dollars*. But this expansion is not always correct: when functioning as a prenominal modifier, as in the phrase *$5 bill*, the correct expansion is *five dollar*, since in general plural noun forms cannot function as modifiers in English. The analysis of premodified nominals in the American English system (see (Sproat, 1994)) comes later than the preprocessing phase, and since a hard decision has been made in the earlier phase, the system produces an incorrect result.

An even more compelling example can be found in Russian. While in English the percentage symbol '%', when denoting a percentage, is always read as *percent*, in Russian selecting the correct form depends on complex contextual factors. The first decision that needs to be made is whether or not the number-percent string is modifying a following noun. Russian in general disallows noun-noun modification: in constructions equivalent to noun-noun compounds in English, the first noun must be converted into an adjective: thus *rog* (Cyrillic рог) 'rye', but *rzhanoj xleb* (ржаной хлеб) (rye+adj bread) 'rye bread'. This general constraint applies equally to *procent* 'percent', so that the correct rendition of *20% skidka* (20% **скидка**) 'twenty percent discount' is *dvadcati-procentnaja skidka* (**двадцати-процентная скидка**) ($\text{twenty}_{[\text{gen}]}$-$\text{percent+adj}_{[\text{nom,sg,fem}]}$ $\text{discount}_{[\text{nom,sg,fem}]}$). Note that not only does *procent* have to be in the adjectival form, but as with any Russian adjective it must also agree in number, case and gender with the following noun. Observe also that the word for 'twenty' must occur in the genitive case. With some well-defined exceptions, numbers that modify adjectives in Russian must occur in the genitive case: consider, for example, *etazh* (**этаж**) 'storey', and *dvux-etazhnyj* (**двухэтажный**) ($\text{two}_{[\text{gen}]}$-$\text{storey+adj}_{[\text{nom,sg,masc}]}$). If the percentage expression does not modify a following noun, then the nominal form *procent* is used. However this form appears in different cases depending upon the number it occurs with. With numbers ending in *one* (including compound numbers like *twenty one*), *procent* occurs in the nominative singular. After so-called paucal numbers — *two*, *three*, *four* and their compounds — the genitive singular *procenta* is used. After all other numbers one finds the genitive plural *procentov*. So we have *odin procent* (**один процент**) (one $\text{percent}_{[\text{nom,sg}]}$), *dva procenta* (**два процента**) (two $\text{percent}_{[\text{gen,sg}]}$), and *pyat' procentov* (**пять процентов**) (five $\text{percent}_{[\text{gen,pl}]}$). All of this, however, presumes that the percentage expression as a whole is in a non-oblique case. If the expression is in an oblique case, then both the number and *procent* show up in that case, with *procent* being in the singular if the number ends in *one*, and the plural otherwise: *s odnim procentom* (**с одним процентом**) (with $\text{one}_{[\text{instr,sg,masc}]}$ $\text{percent}_{[\text{instr,sg}]}$) 'with one percent'; *s pjat'ju procentami* (**с пятью процентами**) (with $\text{five}_{[\text{instr,pl}]}$ $\text{percent}_{[\text{instr,pl}]}$) 'with five percent'. As with the adjectival forms, there is nothing peculiar about the behavior of the noun *procent*: all nouns exhibit similar behavior in combination with numbers (see, for example, (Franks, 1994)). The complexity, of course, arises because the written form '*%*' gives no indication of what linguistic form it corresponds to. Furthermore — and crucially — there is no way to correctly expand this form without doing a substantial amount of analysis of the context, including some analysis of the morphological properties of the surrounding words, as well as an analysis of the relationship of the percentage expression to those words.

The obvious solution to these problems is to delay the decision on how exactly to transduce symbols like '$' in English or '%' in Russian until one has enough information to make the decision in an informed manner. This suggests a model where, say, an expression like '20%' in Russian is transduced into all possible renditions, and the correct form selected from the lattice of possibilities by filtering out the illegal forms.[3] An obvious computational mechanism for accomplishing this is the *finite-state transducer* (FST). Indeed, since it is well-known that FST's can also be used to model (most) morphology and phonology (Koskenniemi, 1983; Karttunen, Kaplan, and Zaenen, 1992; Sproat, 1992), as well as to segment words in Chinese text (Sproat et al., 1996), and (as we shall argue below) for performing other text-analysis operations such as numeral

[3] We will use the term *lattice* here and elsewhere to denote a finite-state machine, when we are particularly focusing on its ability to represent multiple analyses.

expansion, this suggests a model of text-analysis that is entirely based on regular relations. We present such a model below. More specifically we present a model of text analysis for TTS based on *weighted* FST's — WFST's — (Pereira, Riley, and Sproat, 1994; Pereira and Riley, 1996), which serves as the text-analysis module of the multilingual Bell Labs TTS system. We call our model *gentex* — pronounced /dž'εntεk/ — for "*gen*eralized *tex*t analysis". To date, gentex has been applied to eight languages: Spanish, Italian, Romanian, French, German, Russian, Mandarin and Japanese. Besides having a uniform computational treatment of a wide variety of text-analytic problems, the property of gentex that most distinguishes it from the majority of text-analyzers used in TTS systems is that there is no sense in which such tasks as numeral expansion or word-segmentation are logically prior to other aspects of linguistic analysis, and there is therefore no distinguished "text-normalization" phase.

3.4 Overall Architecture

The gentex model is best illustrated by a simple toy example that will show how various text-analysis problems — morphological analysis, abbreviation and numeral expansion, lexically dependent pronunciation — are handled using WFST's, but at the same time is small enough that we can display at least some of the transducers.

Consider an example from a fragment of English that includes the (admittedly silly) sentence:

(1) *Its top weighs 10 kg. now.*

A partial linguistic analysis of this sentence might be as in Figure 3.1. This figure includes some morphological analysis of the words, a (loose) syntactic analysis tree; some analysis of the prosodic structure of the sentence in terms of syllables, stress, phonological words, phonological phrases, and accentuation; and of course a segmental phonetic transcription of each of the words.[4] We will have little further to say about higher level syntactic analysis, since with the exception of the analysis of noun phrase structure (Sproat, 1994), we have done little syntactic analysis in our TTS work beyond the assignment of part of speech labels.

In order to be able to represent the mapping between the input string and the linguistic analysis in terms of WFST's, it is necessary to 'flatten' the linguistic representation in Figure 3.1 into a set of strings. Gentex assumes two linguistic analysis strings, representing roughly two levels of representation. The first, illustrated in (2a) represents the lexical and phrasal analysis, including the morphological category, grammatical part of speech, prosodic phrasing, and accentuation. The second, in (2b), represents the phonetic transcription, including word stress information. An explanation of some of the symbols is

[4] Although certain theoretical assumptions are clearly manifested in this diagram, it is actually intended to be as theory neutral as possible.

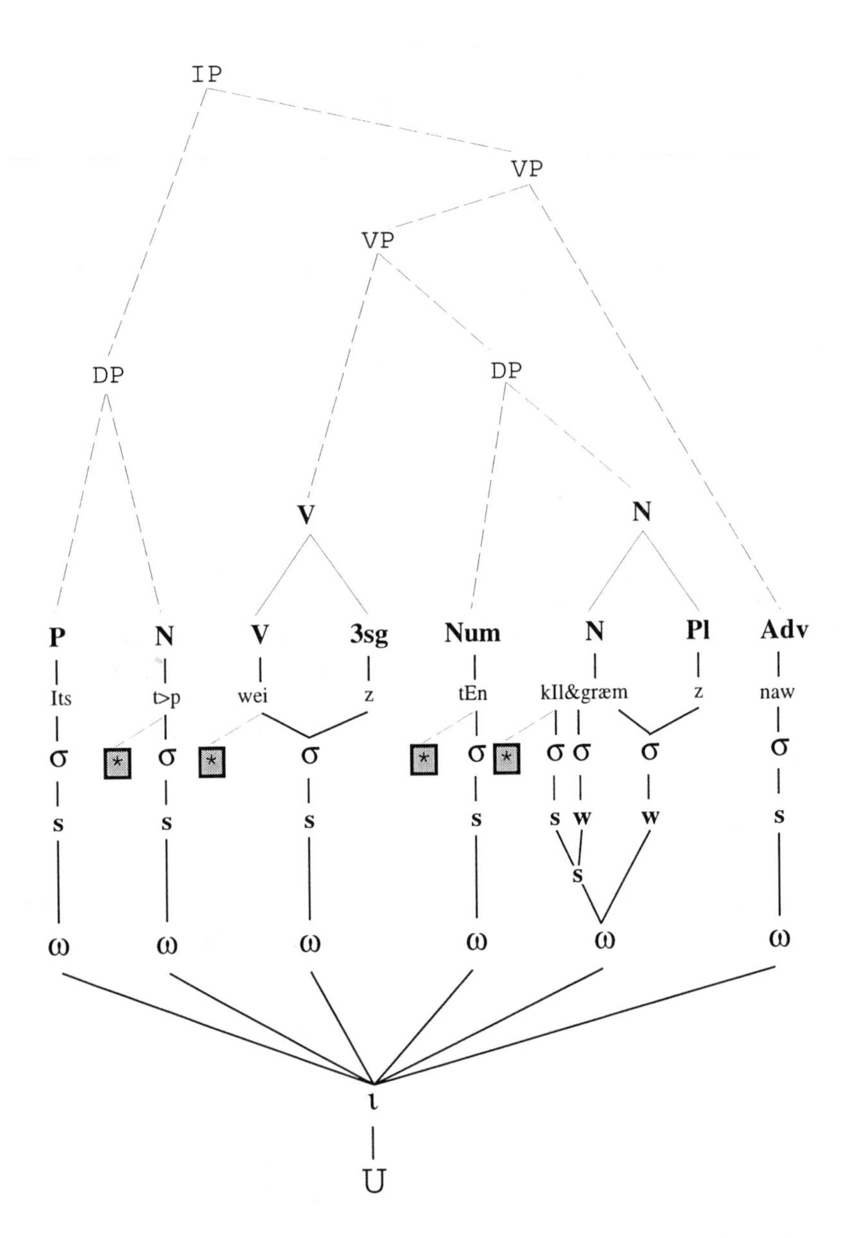

Figure 3.1: A partial linguistic analysis of the sentence *Its top weighs 10 kg. now.* Phonetic symbols are more or less IPA except that '&' represents /ə/, > represents /ɔ/, 'I' represents /ɪ/, 'E' represents /ɛ/, and 'aw' represents /aʊ/. Syllables are represented as σ; ω represents a phonological word; ι represents an intonational phrase and U represents an utterance. Stress (i.e., a strong, as opposed to a weak component) is indicated by s labels in the metrical tree, and word accent is represented with boxed asterisks.

given in Table 3.1, and see also Appendix B. For clarity, multi-character single symbols are enclosed in curly braces, and part of speech and grammatical tag symbols are enclosed in square brackets. Note that we will use the tag [Abbr] to mark expanded abbreviations: the reason for doing this will become clear later on. Finally, note that we omit a specific label for the intonational phrase ι, but for simplicity's sake merely tag this sentence as an utterance (U).

(2) (a) # '{i|ɪ}ts [PPos] # {Acc|*} t'{o|ɔ}p [N] # {Acc|*} w'{eigh|ei} [V] + {s|z} [3sg] # {Acc|*} t'{e|ɛ}n [Num][Pl] # {Acc|*} k'{i|ɪ}l{o|ə}gr{a|æ}m [N][Abbr] + {s|z} [Pl] # n'{ow:aʊ} [Adv] U

(b) # 'ɪts # t'ɔp # w'eiz # t'ɛn # k'ɪ$lə$græmz # n'aʊ U

The lexical representation in (2a) actually represents two levels of analysis, namely the (canonical) orthographic representation, and the morphophonological representation. Symbols like {eigh|ej} represent an element whose written representation is <eigh> and whose segmental representation is /ei/. The representation of *weighs* in (2a) can be viewed as a shorthand for a more structured representation such as the following, where the boxed numbers provide indices linking the orthographic symbols to the phonetic symbols:[5]

(3) [PHON : ⟨ [1] w [2] ei [3] z⟩
ORTH : ⟨ [1] w [2] eigh [3] s⟩]

This representation allows for a relatively straightforward mapping between orthography on the one hand — {eigh|ei} maps regularly to <eigh>, and phonology on the other — {eigh|ei} maps regularly to /ei/. It does *not* obviate the need for phonological rules or spelling rules that apply in complex constructions, but it does allow one to straightforwardly localize irregular or partially regular spellings to the lexicon, where they belong.

The mechanisms used in gentex are very simple. We sketch them here, and then provide a more detailed analysis immediately afterwards. The input *Its top weighs 10 kg. now.* is represented as a simple finite-state acceptor. Call this acceptor I. I is then composed with one or more lexical analysis transducers L to produce a mapping between I and a lattice of possible lexical representations.[6] This lattice is further composed with one or more language-model transducers Λ, which serve to eliminate or assign a low score to implausible analyses. The lowest-cost (best) path of the resulting transducer is then taken, and the right projection π_2 of this corresponds to the representation in (2a), which we shall term M. The derivation of M can be formally represented as

[5] See (Bird and Klein, 1994) for an example of a phonological theory cast in terms of typed feature structures.

[6] This use of the term *compose* may seem somewhat disingenuous since composition is defined as an operation on transducers rather than acceptors; see Section 2.5.2. However, observe that an acceptor can also be viewed as a transducer with identical input and output labels.

+	morpheme boundary
#	word boundary
U	utterance
$	syllable boundary
'	lexical stress
Acc|*	accented word
[PPos]	possessive pronoun
[Abbr]	abbreviation
[Num]	number word
[N]	noun
[V]	verb

Table 3.1: Symbols used in the analysis of the example *Its top weighs 10 kg. now.*

in (4a). From M we can derive P, the phonological representation in (2b), by applying one or more phonological transducers Φ to M, computing the best path, and taking the right projection π_2. This is represented formally in (4b):

(4) (a) $M = \pi_2[BestPath(I \circ L \circ \Lambda)]$
(b) $P = \pi_2[BestPath(M \circ \Phi)]$

We will have reason to refine this model somewhat as we proceed, but the account just sketched is almost a complete description of how gentex works. Once the representations in (4) have been computed, gentex reaps the information immanent in the transducers, parses them into a hierarchical structure more akin to the structure represented in Figure 3.1, and passes this structure to the next module of the system. The implementation of these structures, as well as the modular structure of the entire multilingual TTS system was discussed in Section 2.6.

3.4.1 Lexical analysis

So much for the general framework, but how are the transducers actually constructed? Let us start with the lexicon for ordinary words such as *its*,[7] *top*, *now* and *weighs*. For uninflected and underived words, it is sufficient to represent them in the lexicon with their lexical form and features:

(5) '{i|ɪ}ts [PPos]
t'{o|ɔ}p [N]
n'{ow|aʊ} [Adv]

[7] We ignore for the purposes of this discussion that *its* is actually morphologically complex.

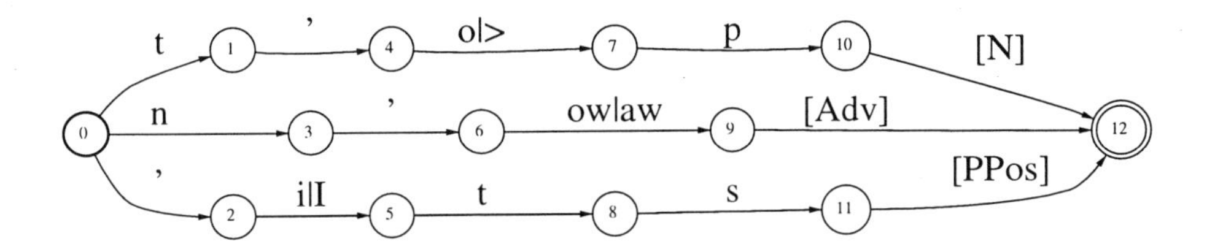

Figure 3.2: An acceptor for the uninflected words in (2).

For words that are inflected or derived via productive morphology it is generally more desirable to derive them than list them explicitly. For *weighs*, one would want to list the base as belonging to the morphological category *regular verb* (6), and list affixes of regular verbs separately (7):

(6) {regverb} w'{eigh|ei} [V]

(7) {regverb} {s|z} [3sg]

Lists of uninflected words can easily be compiled into a finite-state acceptor. Such an acceptor is given in Figure 3.2. For inflected words, we can first represent as a transducer the mapping between inflectional classes and the lexical items contained therein. Call this transducer *Base*; see Figure 3.3a. The inflectional affixes can be represented as an acceptor *Affix*, where the base is the inflectional class to which they attach; see Figure 3.3b. Finally, the set of inflections can be expanded out by composing *Affix* with the concatenation of *Base* and Σ^* (see Section 2.5.1), and then taking the righthand projection:

(8) $\pi_2[Affix \circ (Base \cdot \Sigma^*)]$

The resulting acceptor is shown in Figure 3.3c. This is of course only one of many possible finite-state models of productive morphology; for other approaches see, inter alia, (Koskenniemi, 1983; Karttunen, Kaplan, and Zaenen, 1992; Sproat, 1992; Sproat, 1997a).

The acceptors in Figures 3.2 and 3.3c can then be combined (unioned) into a single acceptor that will give a lexical representation for all ordinary words in the example in (1). This combined *Lexicon* acceptor is shown in Figure 3.4.

To go from the lexical acceptor in Figure 3.4 to the standard orthographic representation for these words requires a transducer that deletes such symbols as stress marks and morpheme boundaries, and at the same time translates symbols such as {eigh|ei} into their correct orthographic form. Such a transducer — let us call it *Lex2Surf*, where *Surf* denotes the surface orthographic representation — can be constructed by a variety of means, including rule compilers such as those reported in (Kaplan and Kay, 1994; Karttunen, 1995; Mohri

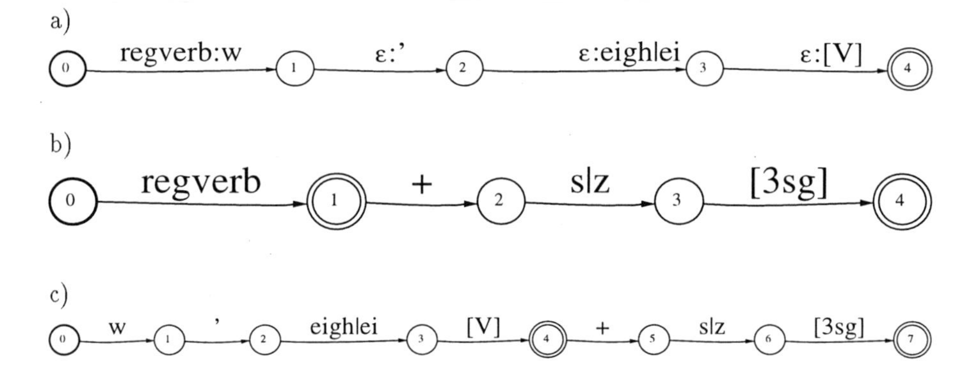

Figure 3.3: A transducer *Base* for the regular verb *weigh* (a); the acceptor *Affix* for a portion of the regular verbal paradigm (b), including both unaffixed forms and forms ending in third singular *-s*; and (c), $\pi_2[Affix \circ (Base \cdot \Sigma^*)]$. The acceptor in (c) will allow both *weigh* and *weighs*

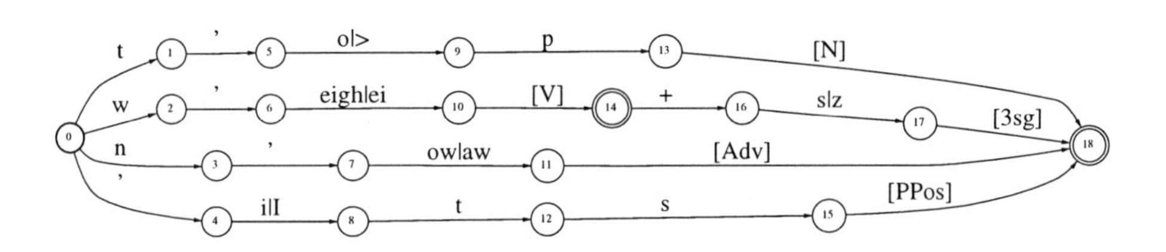

Figure 3.4: An acceptor for the lexicon containing both inflected and uninflected words.

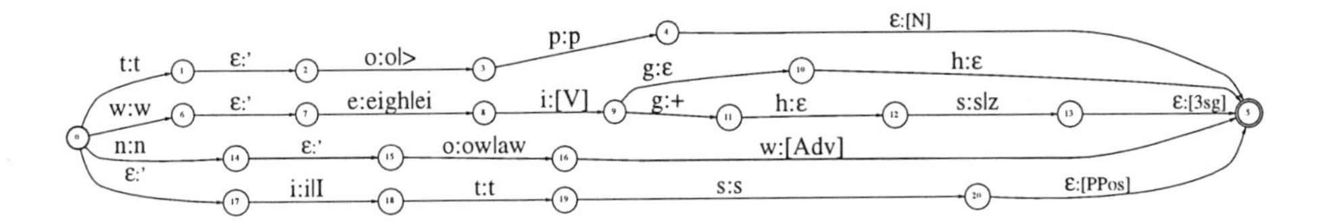

Figure 3.5: A transducer from the orthographic representation of ordinary words to their lexical representation.

and Sproat, 1996). For example, rules such as the following would be needed to define the mapping (where [Feat] is the union of all grammatical features):

$$
(9) \quad \begin{array}{lcl}
[\mathrm{Feat}] & \rightarrow & \epsilon \\
+ & \rightarrow & \epsilon \\
{}' & \rightarrow & \epsilon \\
\{\mathrm{i}|\mathrm{I}\} & \rightarrow & \mathrm{i} \\
\{\mathrm{o}|>\} & \rightarrow & \mathrm{o} \\
\{\mathrm{ow}|\mathrm{a}\upsilon\} & \rightarrow & \mathrm{ow} \\
\{\mathrm{eigh}|\mathrm{ei}\} & \rightarrow & \mathrm{eigh} \\
\{\mathrm{s}|\mathrm{z}\} & \rightarrow & \mathrm{s} \\
\{\mathrm{e}|\mathrm{E}\} & \rightarrow & \mathrm{e} \\
\{\mathrm{o}|\text{ə}\} & \rightarrow & \mathrm{o}
\end{array}
$$

The *Lex2Surf* transducer is already too complex (even on this small example!) to show. However, we *can* show $(Lexicon \circ Lex2Surf)^{-1}$, namely the result of composing *Lex2Surf* with the lexicon, and then inverting it so that it transduces from the orthographic representation to the lexical representation. This is shown in Figure 3.5.

Now consider abbreviations such as *kg*. In this case we want to construct directly the transducer that maps from orthographic representation into lexical form. But *kg* is ambiguous: depending upon the context it can represent either *kilogram* or *kilograms*. Our transducer must thus provide both options, and it must furthermore allow for the possibility that the period after *kg* may or may not be present. Such a transducer — call it *Abbr*, which again can be compiled from a simple linguistic description, is given in Figure 3.6.

The treatment of numerals such as *10* is somewhat more complex.[8] Obviously, since the set of numbers is infinite, it is in principle not possible to have

[8] We put to one side for the moment the question of determining whether a numeral sequence should be read as a number name or in some other way, i.e., as a money amount, or as a numerical label. For example, *747* could be *seven hundred (and) forty seven* or *seven four seven*, depending upon whether it is functioning as a real number or as a brand name (*Boeing 747*). This is an instance of *sense disambiguation* that can be handled by various techniques discussed in this Chapter and in Chapter 4.

Figure 3.6: A transducer mapping *kg* to its lexical renditions.

a general finite-state solution to the problem of converting strings of digits into appropriate number names. Indeed, even though no language has more than a finite number of simplex lexical items that name numbers,[9] the grammar of number names is arguably beyond context-free, let alone regular (Radzinski, 1991). But as a practical matter, there is a limit to how large a number a human listener will want to have read *as a number*, and for a sufficiently large number, any TTS system will cease to read it as a number name, but will instead revert to reading it as a string of digits, or in some other way.[10] Limited in this way, the problem becomes tractable for finite state transducers. The transducer *Num*, which converts strings of digits into number names is constructed in two phases.[11] The first phase involves expanding the numeral sequence into a representation in terms of sums of products of powers of the base — usually 10. Let us call this the *Factorization* transducer. Such a transducer that handles numbers up to multiples of 10^2 is shown in Figure 3.7. The factorization transducer is language- or more exactly language-area specific. English and other European languages have a word for 10^3, but not a separate word for 10^4; Chinese and many other East Asian languages, in contrast have a word for 10^4. Thus a number like 243,000 must be factored differently for English (10a) than it is for Chinese (10a):

(10) (a) $[2 \times 10^2 + 4 \times 10^1 + 3] \times 10^3$

(b) $[2 \times 10^1 + 4] \times 10^4 + 3 \times 10^3$

The second phase maps from this factored representation into number names, using a lexicon that describes the mapping between basic number words and their semantic value in terms of sums of products of powers of the base. Sample entries in such a lexicon for English, which we will call *NumberLexicon* might look as follows:

(11)

1 — '{o|wʌ}n{e|ϵ} [Num][Sg]
2 — t'{wo|u} [Num][Pl]
$1{\times}10^1 + 0$ — t'{e|ɛ}n [Num][Pl]
$1{\times}10^1 + 1$ — {e|ə}l'{e|ɛ}v{e|ə}n [Num][Pl]
$2{\times}10^1$ — tw'{e|ɛ}nt{y|i} [Num][Pl]
10^2 — h'{u|ʌ}ndr{e|ə}d [Num][Pl]

Num is then derived by composing the *Factorization* transducer with the Kleene closure of *NumberLexicon*:

[9] Classical Chinese has among the largest set of such names, with the highest number for which there is a monomorphemic number name being 10^{40} (Needham, 1959; Radzinski, 1991)

[10] Arguably, even human readers would have to give up at a certain point.

[11] The approach described here is related to the models of van Leeuwen (van Leeuwen, 1989b) and Traber (Traber, 1995). Traber, for example, treats number expansion as an instance of morphological analysis, and makes use of morphological analysis tools that are used more generally in the analysis of ordinary words.

For a linguistic treatment of number names see (Hurford, 1975). For an early set of computational treatments, see (Brandt Corstius, 1968).

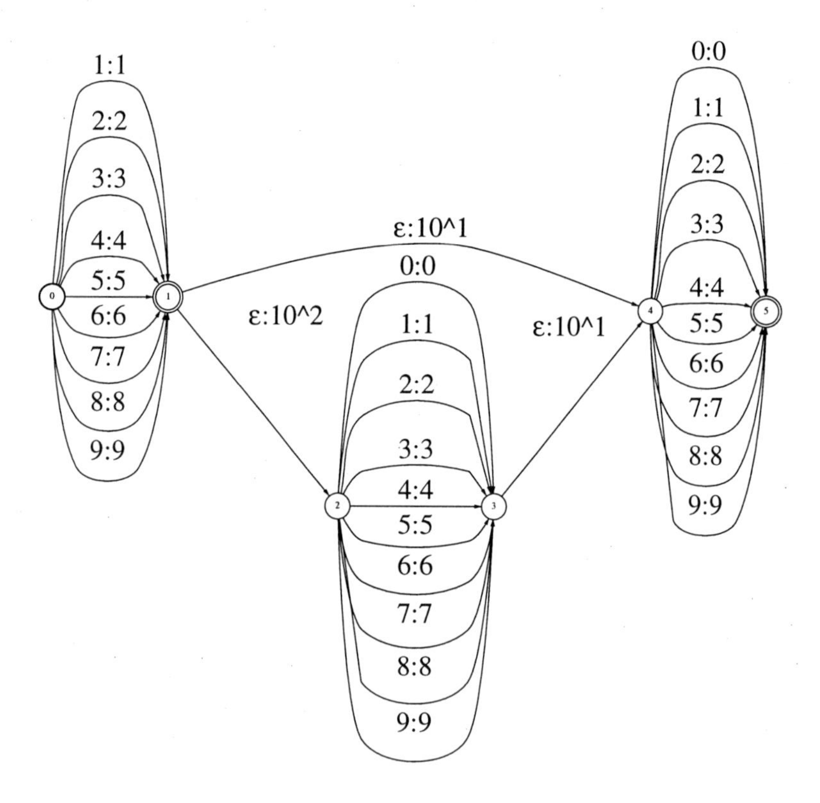

Figure 3.7: A numeral factorization transducer.

(12) $Num = Factorization \circ NumberLexicon^*$

Though $NumberLexicon^*$ allows words for numbers to occur in any order, *Factorization* acts as a filter, restricting the orders to well-formed sequences. Thus 345, for example would be factored as $3 \times 10^2 + 4 \times 10^1 + 5$. Given appropriately defined lexical correspondences we will arrive at the following lexical sequence (ignoring for the moment the full-blown lexical representations which we have heretofore used):[12]

(13)

3	three [Num][Pl]
$\times$	
10^2	hundred [Num][Pl]
+	
4×10^1	forty [Num][Pl]
+	
5	five [Num][Pl]

The model for numeral expansion just described cannot handle everything. Various "clean up" rules are necessary in order, for example, to allow for optional insertion of *and* between hundreds and decades. Such rules are also necessary to handle the highly complex agreement phenomena found in the expression of numbers in some languages, such as Russian; see Section 3.6.5. More exotic phenomena such as the "decade flop" found in various Germanic languages (including archaic English) require additional filters (see Section 3.6.3), and even more extreme reorderings can be found (see Section 3.6.8) which suggest the limitations of the model.

Returning to our toy example, we can construct a transducer that will map from the orthographic representation of any word — be it ordinary word, abbreviation or number — simply by constructing the union of $(Lexicon \circ Lex2Surf)^{-1}$, $Abbr$ and Num:

(14) $Word = (Lexicon \circ Lex2Surf)^{-1} \cup Abbr \cup Num$

We now need to build a model of the material that can come between words, items like space and punctuation. We will represent (strings of) spaces in the input using the word-boundary symbol '#'; punctuation is represented as itself. For the sake of the present discussion, we will assume that spaces always map to word boundaries, and that punctuation marks always correspond to phrase boundaries; in this case '.' corresponds to the utterance boundary 'U'. This is of course not always correct, and methods for doing a more accurate analysis are presented in Chapter 4. For the time being let us merely assume that we have the following correspondences between surface orthographic elements and their internal representation:

(15)

Surface	Lexical
#	#
.	U< 1.0 >

[12] The *Num* transducer, even for the small example we have examined, is too large to effectively show here.

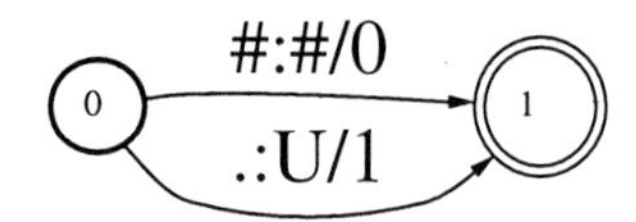

Figure 3.8: A very simple space and punctuation transducer.

The bracketed 1.0 is a cost: its function will be explained later on. The transducer modeling this (*Punc*) is pretty simple. See Figure 3.8. Given *Word* and *Punc*, we can define the lexical analysis transducer for the example in (1) as in (16). Note that at least one *Punc* must occur between words, but that more than one (e.g. a punctuation mark followed by a space) may occur.

(16) $L = Punc^* \cdot (Word \cdot Punc^+)^*$

The result of composing I with L is given in Figure 3.9. The path is unambiguous, except for the expansion of *kg.*, which has four possible analyses: the expansion may be either as *kilogram* or *kilograms*, and the period may count as part of the abbreviation (in which case it is deleted) or as a real end-of-sentence marker (in which case it is mapped to U). Some particular methods for disambiguating these alternatives are given in the next section. Of course, in any real application, there will in general be many more than four possible analyses for a sentence.

3.4.2 "Language modeling"

Deciding whether *kg* should be expanded as singular or plural depends upon some analysis of context. We will have more to say about lexical disambiguation in Chapter 4. For now we will present a couple of methods — *rules* and *filters* — that have been used fairly extensively in the development of language-model components in our multilingual systems.

The first method is simply to write rules that describe which alternative to use given a local context. A general rule of thumb for abbreviations after numbers (usually measure terms like *kg*) is that they are singular after *1* and plural otherwise; see (17a) and (17b). The main exception is if the measure phrase modifies a following noun, in which case the expansion should be singular (in English) since plural modifiers tend to be avoided; see (17c).

(17) (a) *one kilogram*, **one kilograms*

(b) **two kilogram*, *two kilograms*

(c) *a two kilogram box*, **a two kilograms box*

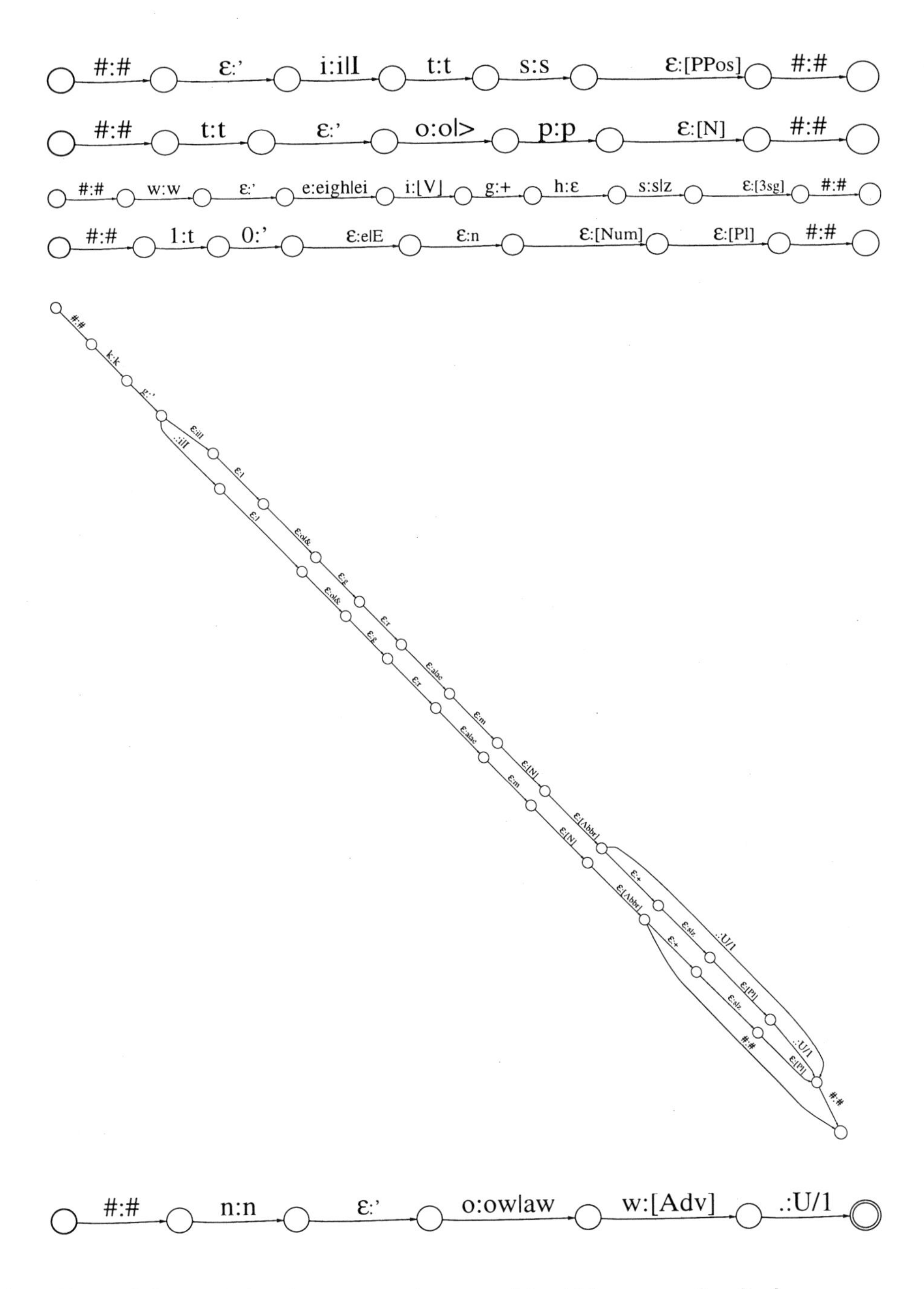

Figure 3.9: $I \circ L$ for the sentence *its top weighs 10 kg. now.* For display reasons the transducer is cut into sections, starting from top to bottom. The final arc labeled '#:#' in each section is the same arc as the first arc in the next section.

A set of rules that models these facts is given below:

(18)
$$\begin{array}{llll} \epsilon & \rightarrow & \boxed{\star} & / \ [\text{Abbr}] \ldots [\text{Pl}] \_\_\_ \# \ldots [\text{N}] \\ \epsilon & \rightarrow & \{\text{ok}\} & / \ [\text{Abbr}] \_\_\_ \# \ldots [\text{N}] \\ \epsilon & \rightarrow & \boxed{\star} & / \ [\text{Num}][\text{Pl}] \ \# \ldots [\text{Abbr}] \_\_\_ \ bd \\ \epsilon & \rightarrow & \boxed{\star} & / \ [\text{Num}][\text{Sg}] \ \# \ldots [\text{Abbr}] \ldots [\text{Pl}] \_\_\_ \ bd \end{array}$$

In these expressions, *bd* represents one of the boundaries '#' or 'U', and '...' represents any non-boundary elements, possibly including a morpheme boundary '+'. The rules make use of two tags '$\boxed{\star}$' and '{ok}'; the first of these marks constructions that are disallowed, and the second constructions that are allowed. The first two rules state that plural versions of abbreviations are disallowed before a following noun, and conversely that a singular expansion is allowed.[13] The next two rules mark as disallowed singular expansions of abbreviations after plural numbers and plural expansions of abbreviations after singular numbers. (Note that in this formulation we do not make use of an explicit singular tag for abbreviations: Singular abbreviations are just those which do not have any material intervening between the [Abbr] tag and the following word or phrase boundary.) In the construction of a language model WFST for gentex such rules are compiled into a transducer using the rule compilation algorithm described in (Mohri and Sproat, 1996).

Having tagged analyses, it is necessary to actually weed out the undesirable ones, and simply delete the tag *ok*. The first of these is accomplished by a *filter* that can be expressed as a simple regular expression (19a); the second is expressed as a rule (19b):

(19) (a) $\neg[\Sigma^* \boxed{\star} \Sigma^*]$

(b) $\{\text{ok}\} \rightarrow \epsilon$

We can then define the language model machine for this simple example as simply the composition of these three machines:

(20) $\Lambda = (18) \circ (19\text{a}) \circ (19\text{b})$

The composition $I \circ L \circ \Lambda$ will produce analyses where the singular version of *kg.* has been weeded out. However, it will not eliminate the ambiguity of whether to analyze the period after *kg.* as an end of sentence marker, or as a part of the abbreviation itself. This is where the cost assigned to the .-to-U transduction in (15) comes in. Under this scheme, it will always be possible to analyze a period as an end-of-sentence marker, but only at some cost; in contrast, the abbreviation model allows us to analyze (and delete) the period after *kg.* at no cost. We can therefore select the correct analysis, where the period is part of the abbreviation, by computing $BestPath[I \circ L \circ \Lambda]$.

The final operation that needs to be performed by the language model for this toy example is accenting. In the example, we have chosen a simple model

[13] Of course, a real statement of these rules would have to be more complex than this.

of accenting where all content words are accented; this is in any event a reasonable first-approximation model for accenting in a language like English. The following rule, where [Cont] denotes the set of content-word parts of speech, will accomplish this:

(21) $\epsilon \rightarrow$ {Acc|*} / *bd* ___ ... [Cont]

Λ, can then be redefined as follows:

(22) $\Lambda = (18) \circ (19a) \circ (19b) \circ (21)$

3.4.3 Phonology

Having computed a lexical analysis, one then needs to compute a phonological representation. The toy example we are considering avoids the complexities of English lexical phonology. All the phonological model needs to do for this example is:

- Delete grammatical labels and accent tags.[14]
- Convert lexical orthographic/phonological symbols into a phonemic representation.[15]
- Syllabify polysyllabic words such as *kilogram*.

The first two operations can be handled by rewrite rules:

(23)

[Feat]	$\rightarrow$	ϵ
[Accent]	$\rightarrow$	ϵ
{i|ɪ}	$\rightarrow$	ɪ
{o|ɔ}	$\rightarrow$	ɔ
{eigh|ei}	$\rightarrow$	ei
{e|ɛ}	$\rightarrow$	ɛ
{a|æ}	$\rightarrow$	æ
{ow|aʊ}	$\rightarrow$	aʊ
{s|z}	$\rightarrow$	z
{o|ə}	$\rightarrow$	ə

Syllabification can be handled by providing a declarative grammar of the possible locations of syllable boundaries within an utterance. A very simple model is as follows:

(24) $(\#(C^{*}\ '\ V\ C{<}1.0{>}^{*}\ \epsilon : \$)^{*}\ C^{*}\ '\ V\ C^{*})^{+}\ U$

[14] Gentex ultimately reads accentual information off the lexical transducer rather than the phonological transducer.

[15] More precisely, a *lexical* phonemic representation: subsequent phonological rules can modify this representation.

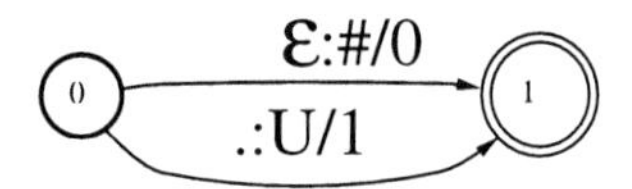

Figure 3.10: A modification of the transducer in 3.8 to handle the case of "pseudo-Chinese".

Here ˌ̓ represents an optional stress mark. What this expression says is that an utterance consists of one or more words, which consists of one of more syllables of the form C(onsonant)*V(owel)C*. In the case where there is more than one syllable in a word, all non-final syllables must be terminated by a syllable boundary (inserted by the expression $\epsilon : \$$). The cost ($< 1.0 >$) on the final C of each syllable favors placing the syllable boundary as early as possible, since each additional C in the coda of the syllable costs more; this is, of course, an implementation of the *maximal onset principle*. The best path computation will select the analysis where the syllable boundary is as far to the left as possible. Needless to say, this is a hopelessly simplistic model of syllabification: a more realistic model — but one that is nonetheless constructed along the same lines — will be presented in Section 3.6.1.

We can now define Φ as follows:

(25) $\Phi = (23) \circ (24)$

Having presented the construction of the lexical, language model and phonological transducers, our toy example is now complete.

3.4.4 If English were Chinese

What if English, like Chinese, Japanese or Thai, did not mark word boundaries? The input sentence for our toy example would then become: *itstopweighs10kg.now.* Only a couple of changes would be required to the model that we have just presented in order to handle the problem of word-segmentation in "pseudo-Chinese". The first change necessary is to the interword model, which would now have to be as depicted in Figure 3.10:[16] This will now allow us to produce a lexical analysis of *itstopweighs10kg.now.* identical to that in (2a). Of course, for a fuller English lexicon we will produce more than one lexical analysis: note that the current sentence could be analyzed in two ways:

[16] When we combine this model into the full lexical model as in (16), we actually compute the closure of *Punc*. Since '#' now transduces from nothing on the input side, an unbounded number of '#' can be introduced. This problem can easily be fixed by composing on the lexical side a filter that disallows sequences of two or more '#'.

(26) (a) Its top weighs 10 kg. now.
     (b) It stop weighs 10 kg. now.

There are various ways one might approach selecting among these alternatives:

- Part-of-speech disambiguation to determine, for example, if the sequence [PPos]-[N] is more likely than the sequence [P]-[V].
- An n-gram lexical model.
- "Local" grammatical constraints.
- Full syntactic parsing.

All of these approaches can be represented as WFST's and incorporated directly into the language model.[17] As long as part-of-speech disambiguation depends on local information, a part-of-speech model can be represented as a (weighted) finite-state automaton; n-gram models are similarly representable as weighted finite-state automata. Finite-state constraint grammars are reported in (Voutilainen, 1994) and finite-state models of local grammars are discussed in (Mohri, 1994). An approach to full-syntactic parsing is discussed in (inter alia) (Roche, 1997); or one might consider a finite-state approximation of a context-free grammar along the lines of (Pereira and Wright, 1997).

For concreteness in the present discussion let us consider adding to the language model a constraint that rules out the sequence *it* [V], where [V] is a bare verb, when this sequence occurs at the beginning of a sentence. (We do not want to rule out this sequence in just any position, as examples like *make it stop* show.) Such a constraint could be expressed as follows, where we use a special symbol $\boxed{\text{START}}$ to refer to the beginning of the sentence:[18]

(27) $\neg[\boxed{\text{START}}\ \#\ '\{\text{i}|\text{ɪ}\}\text{t}\ [\text{P}]\ \#\ \ldots[\text{V}]\ \#\ \Sigma^*]$

The language model transducer Λ thus becomes:

(28) $\Lambda = (18) \circ (19a) \circ (19b) \circ (27) \circ (21)$

We will discuss a real example of word segmentation in our Mandarin text-to-speech system in Section 3.6.6.

Note that while we have presented the constraint in (27) as a reasonable sort of solution to the problem of word segmentation, such constraints actually have nothing in particular to do with this specific problem: these sorts of constraints will also in general be necessary to handle ambiguities that, for example, affect pronunciation.

---

[17] Though the resulting transducers for any real application will be large and it would not be possible to precompose all of them beforehand with the rest of the language-model transducers. However, composition of a series of WFST's with a particular input at runtime is a perfectly feasible alternative to offline precomposition.

[18] For some discussion of the implementation of boundary conditions in a finite-state framework see (Kaplan and Kay, 1994, page 359).

3.5 A Refinement of the Model

For ordinary words — i.e., lexical elements that are not abbreviations, numbers, and so forth — the model just described has a certain amount of redundancy. Observe that the lexical transducers introduce a large number of lexical features, including part of speech labels, boundary symbols, and stress marks. But only a subset of these lexical features are relevant to pronunciation. For the toy example at hand (though certainly not for English in general), the only parts of the lexical string that are really necessary to pronounce the words correctly are the following:

(29) # '{i|ɪ}ts # t'{o|ɔ}p # w'{eigh|ei}{s|z} # t'{e|ɛ}n # k'{i|ɪ}l{o|ə}gr{a|æ}m {s|z} # n'{ow:aʊ} U

The lexical transducers introduce several lexical features, only to have them deleted again by the phonological transducers. The representation in (29) is nothing more or less than an *annotation* of the input orthographic string with the *minimal* set of features that allow one to decide how to pronounce it; this annotation is mediated by *morphological* information. We will henceforth refer to this level of representation as a *Minimal Morphologically-Motivated Annotation*, or MMA. Note that the possibly relevant features are not in principle restricted to stress marks and phonological annotations of the orthographic symbols: they could include morpheme-boundary information as well as morphosyntactic feature information.

Rather than continue with this toy English example, let us illustrate the point further with a real example, in this case from Russian, a language whose standard orthography is typically described as "morphological" (see, e.g., (Daniels and Bright, 1996)). What this means is that the orthography represents not a surface phonemic level of representation, but a more abstract level. While this description is accurate, it is worth noting that — with a well-circumscribed set of exceptions that includes loanwords — Russian orthography is almost completely phonemic in the sense that one can predict the pronunciation of most words in Russian based on the spelling of those words. The catch is that one needs to know the placement of word stress, since, among other things, several Russian vowels undergo reduction to varying degrees depending upon their position relative to stressed syllables.[19] Word-stress placement usually depends upon knowing lexical properties of the word, including morphological class information. Consider, for example, the word *kostra* (Cyrillic **костра**) (bonfire$_{[gen,sg]}$). This word belongs to a class of masculine nouns where the word stress is placed on the inflectional ending, where there is one. In this case the inflectional ending is *-a*, the genitive singular marker for this paradigm. Thus the stress pattern is *kostr'a*, and this serves as the MMA: the pronunciation /kʌstr'a/, with the first /o/ reduced to /ə/, is straightforwardly predictable from this. The morphological representation

[19] One other catch is that the distinction between е /je/ and ё /jo/ is not normally made in writing: see also Section 4.2.

— *kostËr{noun}{masc}{inan}+'a{sg}{gen}* — contains a lot more information, including grammatical information and morpheme boundary information. (The symbol 'Ë' represents a vowel that shows up in some forms — e.g., the nominative singular — but is deleted in others, including the genitive singular.) Usually, such information is irrelevant to pronunciation in Russian. Going from the MMA to the surface orthography is straightforward: in this example we just want to delete the stress mark. Call the transducer that accomplishes this *MMA2Surf*. Going from the lexical representation to the MMA is similarly straightforward: one needs merely to delete certain grammatical markers and, in appropriate contexts, elements like 'Ë'. Call this transducer *Lex2MMA*. Finally, phonological rules need to be constructed to go from the MMA to the phonological representation. Call this *MMA2Phon*. In terms of the model that we presented in the previous section, we would replace the *Lex2Surf* term in the definition for *Word* in (14) with $Lex2MMA \circ MMA2Surf$, so that the revised definition of *Word* becomes:

(30) $Word = [Lexicon \circ Lex2MMA \circ MMA2Surf]^{-1} \cup Abbr \cup Num$

The phonological transducer can be redefined as follows, where *MMA2Phon* contains the real phonological rules and declarations, with the deletion of irrelevant lexical material factored out:

(31) $\Phi = Lex2MMA \circ MMA2Phon$

Note that Φ applies to all lexical elements, including elements that are expanded from abbreviations and numerals: the *Lex2MMA* transducer therefore needs to be designed so as to produce appropriate MMA's for these elements also. Figure 3.11 summarizes the discussion in this section, showing the relationship between the surface, MMA, lexical and phonological levels of representation.[20]

3.6 Case Studies

In this section we present some real case studies of the application of the gentex model to various languages: Spanish, French, German, Russian, Mandarin and Japanese. We also discuss the problem of numeral expansion in Malagasy, and some possible solutions: it should be stressed that Malagasy is *not* a language for which we have (to date) pursued any text-to-speech work.

3.6.1 Spanish syllabification

In Section 3.4.3 we presented a toy example of how one might design a constraint-based model of syllabification within the finite-state framework developed here. In this section we present a more realistic example, this time

[20] Of course, since lexical and phonological transducers are built offline, there is strictly speaking nothing that *requires* one to abide by this model. The main advantage of the structure described here is that the lexical-to-MMA transduction can in general be *shared* between the surface-to-lexical and lexical to pronunciation transductions.

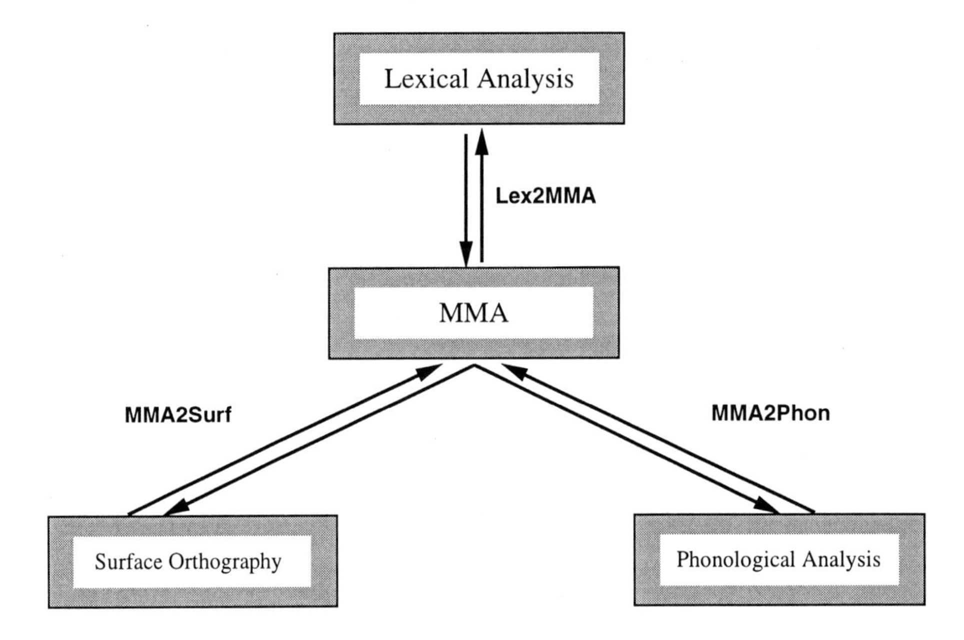

Figure 3.11: The gentex architecture for ordinary words.

from our Spanish TTS system. The description presented here is not complete (the real description is somewhat more complex) but it serves to illustrate well the general approach.

The statement of syllabification consists of a set of constraints, listed below:

(32) (a) $(C^* \ V \ C{<}\ 1.0 >^* \ \$ \ )^+$

(b) $\neg \ (\Sigma^* \ \$ \ (CC \cap \neg \ (CG \cup OL)) \ \Sigma^*)$

(c) $\neg \ (\Sigma^* \ \$ \ ([+cor] \cup /x/ \cup /\beta/) \ /l/ \ \Sigma^*)$

(d) $\neg \ (\Sigma^* \ \$ \ ([+cor,+strid] \cup /x/ \cup /\beta/) \ /r/ \ \Sigma^*)$

The first constraint (32a) states that syllables consist of zero or more consonants followed by a vowel, followed by zero or more consonants, followed by a syllable boundary marker $; and that a phrase consists of one or more syllables. (For the present discussion we ignore the possible presence of word or phrase boundary markers.) As in the toy example above, we assign a cost to each consonant assigned to the coda, assuring that the selected analysis will have as long an onset as possible. The constraints (32b)-(32d) are further constraints that are intersected with (32a). The first, (32b), disallows analyses where two consonants (CC) which *are not* either a sequence of a consonant and a glide (CG) or an obstruent and a liquid (OL), are assigned to the initial portion of the syllable.[21] Constraint (32c) states that coronal, /x/ or /β/, followed by /l/ is not a possible onset.[22] Finally, constraint (32d) states that strident coronals (/s/ and /č/), /x/ or /β/, followed by /r/ cannot form a legitimate onset.

Assuming a transducer that freely introduces syllable boundary symbols — call this *Intro*($) —, we can construct the syllabification of a particular word, say *estrella* /estreja/ 'star', as follows:

(33) $es\$tre\$ja =$
$\pi_2[BestPath(estreja \circ Intro(\$) \circ ((32a) \cap (32b) \cap (32c) \cap (32d)))]$

3.6.2 French liaison

Liaison in French has been defined as the linking of a word final orthographic consonant in a word w_1 with an immediately following word w_2, that begins with a vowel. When liaison occurs, a word-final consonant from w_1, which is not generally pronounced, is realized and resyllabification with w_2 takes place, thereby connecting the two words. Liaison is a complex phonological process (Schane, 1968; Tranel, 1981; Tranel, 1986; Tranel, 1990; Kenstowicz, 1994) resulting in surface phonetic alterations at the juncture of the two words. It can occur inside phrase boundaries as well as across major phrase boundaries. In addition to the complexity of strictly linguistic processes, other factors influence liaison, including socio-cultural background and dialect preferences.

Liaison can be:

1. **Obligatory**. For example the words *un* 'a' and *ami* 'friend', when pronounced in isolation, are respectively /œ̃/ and /ami/. Notice that the orthographic /n/ is not realized explicitly at the surface, and that the vowel /œ̃/ is nasalized. When the word *un* is a masculine singular article (and not a masculine singular pronoun), it will undergo liaison with the following word, yielding the pronunciation /œ̃ **n**ami/ with the liaison phoneme /n/ in boldface. Notice also the resyllabification at the juncture of w_1 and w_2 of the /n/, which now forms the phonetic onset to the syllable /na/.

2. **Optional**. The words *chez* 'at' and *Annie* 'Annie' in isolation are pronounced /še/ and /ani/. In succession, they can be pronounced either /še ani/ or /še **z**ani/ where the word-final orthographic /z/ is fully realized by the insertion of the phoneme /z/. Again, notice resyllabification of the /z/ which now forms the onset of the initial syllable of w_2, *Annie*. The optional case is largely dependent on socio-cultural background and dialect preferences.

[21] One cannot use the term *onset* here since consonant-glide sequences such as the /bw/ in *buey* 'ox' do not form an onset; see (Harris, 1983).

[22] Helen Karn (personal communication) has noted that the constraint against /tl/ onsets, subsumed by this restriction, may not correctly characterize Mexican Spanish.

3. **Forbidden**. In the phrase *j'en ai un aussi* 'I have one also' pronounced /zɑ̃ nɛ œ̃ osi/, liaison is blocked between *un* which is the masculine singular pronoun and the adverb *aussi*. It would be considered an error to pronounce these two words /œ̃ nosi/, although this type of error is typical of foreign speakers due to hypercorrection. Notice the apparent surface (i.e. phonemic) similarity between *un* and *aussi* and *un* and *ami*. In each case the form *un* is followed directly by a word beginning with a vowel. However, in cases where *un* is a pronoun, liaison does not occur with w_2, regardless of surface phonemic context. This clearly demonstrates the impact that part-of-speech has on liaison.

In addition to the surface realization of the final consonant of w_1, other phonological changes may occur, such as vowel opening or devoicing. For example, if w_1 ends in an oral (i.e. non-nasal) vowel, such as the half closed front /e/, a more open front vowel /ɛ/ will be used during liaison. Consider a w_1 such as *léger* 'light' or 'slight', which is pronounced /leže/ in isolation. However, in front of a w_2 starting with a vowel such as *ennui* 'boredom', it undergoes opening, and is then pronounced /ležɛ **r**ɑ̃nɥi/ 'slight boredom' with the additional realization of the orthographic /r/ and resyllabification at the juncture of w_1 and w_2. This illustrates an example of vowel change caused by liaison. Another phonological change that can be observed is devoicing. Specifically, when liaison occurs between two words, the final consonant of w_1 is linked with the initial vowel of w_2, and voiced consonants are then devoiced. The underlying final consonant of w_1 is thus realized by the voiceless counterpart. For instance, *grand* 'tall' is pronounced /grɑ̃/ and in liaison with *homme* 'man' /ɔm/, the orthographic *d* of the first word is realized as /t/:
/grɑ̃ **t**ɔm/. Such phonological complexities are a challenge for a text-to-speech system since several operations are required for correct realization.

Within the French text-to-speech system developed at Bell Laboratories (Tzoukermann, 1994), a multi-pass approach has been implemented to account for the rules associated with liaison. Morphological analysis is first performed; then a set of phonetic rules are applied at the morpheme level, syllable level, and phonemic level from left to right. Rules such as schwa deletion and liaison can be accurately applied at given stages in the processing. Since the principles governing the application of liaison depend largely on the part of speech of w_1 and w_2, a trace is kept of the underlying consonant and optionally the penultimate vowel throughout processing. In this way, the word final character of w_1 can be identified if it is to appear phonetically and undergo phonological change; similarly, the penultimate vowel is thereby available for phonological rule application. Every word undergoing possible liaison is marked with a symbol $[L_{(V)C}]$ where $[L]$ represents potential liaison, the subscript $_{(V)}$ represents the optional presence of a vowel for phonological changes, and the subscript $_C$ represents the value of the consonant to be realized at the juncture. For example, the word *mon* 'my' will first be rewritten as $mɔ[L_n]$, where $[L_n]$ is the trace of the orthographic <n>. Phonological rules apply at the juncture of w_1 and w_2 to the traces of w_1.

An example of how processing is achieved can be seen in the following words. Liaison is obligatory between plural or singular adjectives and nouns. For example the words *parfait accord* 'perfect agreement' (adjective-noun) in isolation are pronounced /parfɛ/ and /akɔr/. In the system, the representation of *parfait* will have the trace of the final orthographic *t* *parfɛ*$[L_t]$. Once the liaison rule applies, the two words are pronounced /parfɛ **t**akɔr/ with the realization of the orthographic *t*, and resyllabification is applied at the w_1 w_2 juncture. The same will apply between articles, possessive adjectives, demonstrative adjectives in w_1 position. For instance, if w_1 is *mon* 'my' *mɔ̃*$[L_n]$ and w_2 is *école* 'school' /ekɔl/, the surface phonetic realization will be /mɔ̃ **n**ekɔl/ after application of the liaison rules. Other rules also apply, as in the word *léger*, which is first rewritten as *lež*$[L_{\varepsilon r}]$ (where the liaison will act on the two final characters) and then realized as /ležɛr/ under liaison conditions.

In the following example, the disambiguation process is highlighted to show where the treatment of liaison occurs within the different steps of processing. In the sentence *Les enfants aimables et les adultes se promenaient dans la cour quand soudain un vent impétueux les emporta* 'The nice children and the adults were walking in the courtyard when suddenly a raging wind swept them away', morphological analysis is first performed. In the example below, all the possible output forms for each word are given:

#les[pron][3rd][masc][fem][pl]#
  #les[art][masc][fem][pl]#
#enfant[noun][masc]s[pl]#
#aimable[noun]s[pl]#
#et[conj]#
#les[pron][3rd][masc][fem][pl]#
  #les[art][masc][fem][pl]#
#adulte[adj]s[pl]#
  #adulte[noun][masc]s[pl]#
#se[pron][3rd][sg]#
  #se[pron][3rd][pl]#
#promen[V]aient[3rd][pl][imp][ind]#
#dans[prep#
#la[art][fem][sg]#
  #la[pron][3rd][sg]#
#cour[noun][fem]#
#quand[adv]#
  #quand[csub]#
#soudain[adv]#
#un[adj][num]#
  #un[art][masc][sg]#
  #un[pron][masc][sg]#
#vent[noun][masc]#
#impétueu[adj]x[masc][sg][pl#

#les[pron][3rd][masc][fem][pl]#
  #les[art][masc][fem][pl]#
#emport[V]a[3rd][sg][sp][ind]#

Next, pronunciation rules are applied:

#le[pron][3rd][masc][fem][pl]#
  #le[art][masc][fem][pl]#
#ɑ̃fɑ̃[L_s][noun][masc][pl]#
#ɛmabl[L_s][noun][pl]#
#e[conj]#
#le[L_s][pron][3rd][masc][fem][pl]#
  #le[L_s][art][masc][fem][pl]#
#adylt[L_s][adj][pl]#
  #adylt[L_s][noun][masc][pl]#
#sə[pron][3rd][sg]#
  #sə[pron][3rd][pl]#
#promənɛ[V][3rd][pl][imp][ind]#
#dɑ̃[prep#
#la[art][fem][sg]#
  #la[pron][3rd][sg]#
#kur[noun][fem]#
#kɑ̃[adv]#
  #kɑ̃[csub]#
#sudɛ̃[adv]#
#ɛ̃[adj][num]#
  #ɛ̃[art][masc][sg]#
  #ɛ̃[pron][masc][sg]#
#vɑ̃[noun][masc]#
#ɛ̃petyœ[adj][masc][sg][pl#
#le[L_s][pron][3rd][masc][fem][pl]#
  #le[L_s][art][masc][fem][pl]#
#ɑ̃pɔrta[V][3rd][sg][sp][ind]#

The third step consists of applying language model rules. These rules serve two distinct functions. The first function is to disambiguate homographic words having two different pronunciations, such as the word *est* 'east' or 'is', which can be either a noun /ɛst/ or a verb /ɛ/. The second function is to disambiguate parts of speech so that the liaison rules can be processed.

In the example, the words *les* and *la* can be either pronouns or articles. One rule states that if w_2 is a verb, w_1 cannot be an article. At the same time, if w_2 is a noun, w_1 cannot be a pronoun. The words *adultes*, *se*, *quand*, and *un* are also ambiguous and disambiguation rules will apply if necessary for the treatment of liaison. The disambiguated output appears as follows:

#le[L_s][art][masc][fem][pl]#
#ɑ̃fɑ̃[L_s][noun][masc][pl]#
#ɛmabl[L_s][noun][pl]#
#e[conj]#
#le[L_s][art][masc][fem][pl]#
#adylt[L_s][noun][masc][pl]#
#sə[pron][3rd][pl]#
#promənɛ[V][3rd][pl][imp][ind]#
#dɑ̃[prep#
#la[art][fem][sg]#
#kur[noun][fem]#
#kɑ̃[csub]#
#sudɛ̃[adv]#
#ɛ̃[art][masc][sg]#
#vɑ̃[noun][masc]#
#ɛ̃petyœ[adj][masc][sg][pl#
#le[L_s][pron][3rd][masc][fem][pl]#
#ɑ̃pɔrta[V][3rd][sg][sp][ind]#

The last step consists of applying liaison rules. Liaison applies if w_1 is an article and w_2 is a noun, as in the case of /le **z**ɑ̃fɑ̃/ and /le **z**adylt/. Liaison does not apply if w_1 is a noun and w_2 an adjective, as in /ɑ̃fɑ̃ ɛmabl/. Similarly, liaison does not apply if w_1 is a noun, w_2 an adjective, w_3 is a coordinating conjunction such as *et* 'and' or *ou* 'or', w_4 an article, and w_5 a noun. For example: /le **z**ɑ̃fɑ̃ ɛmabl e le **z**adylt/. Finally, liaison applies if w_1 is a pronoun and w_2 a verb, such as in /le **z**ɑ̃pɔrta/.

The whole disambiguated sentence appears as follows:

#le#zɑ̃fɑ̃#ɛmabl#e#le#zadylt
#sə#promənɛ#dɑ̃#la#kur#kɑ̃#sudɛ̃#ɛ̃
#vɑ̃#ɛ̃petyœ#le#zɑ̃pɔrta#

3.6.3 German numbers

Many aspects of the treatment of numerals in German are quite similar to English. In fact, the *Factorization* transducer, which represents the first of two phases of numeral expansion and converts strings of digits into factorized representations (see Section 3.4), is identical for the two languages. English and German also share the structure of the *NumberLexicon*, which constitutes the second phase and maps the factored representation into number names. The commonality extends to the level of providing simplex lexical items for small numbers up to and including 12 (*twelve, zwölf*); Romance languages have simplex words for numbers up to and including 15 (Spanish) or even 16 (French). Obviously, the lexical entries themselves are language-specific.

However, there are a few peculiarities specific to numeral expansion in German (and several other Germanic languages). Let us first consider the phenomenon earlier referred to as "decade flop" (Section 3.4.1), i.e., the reversal of the order of decades and units in numbers between 13 and 99[23]. To give an example, the correct expansion of the number 21 is *einundzwanzig*. In our German TTS system, the pertinent factorization yields

(34) $[2 \times 10^1 + 1]$.

The output of the *Factorization* transducer is modified by a language-specific filter which performs the reversal of decades and units:

(35) $2 \times 10^1 + 1 \rightarrow 1 + 2 \times 10^1$

The result of the filtered transducer output is *einszwanzig*.

A second process concomitant with the decade flop phenomenon is the obligatory insertion of *und* between the unit and the decade. In the German TTS system, this process is implemented in the form of a rewrite rule:

(36) $\epsilon \rightarrow$ und [conj][clit] / (s|(ei)|(ier)|(nf)|n|(cht)) [num][Grammatical]* ___ (Sigma & ! #)* ig [num]

The rule states that the cliticized conjunction *und* is inserted after certain orthographic substrings, namely the ones that represent unit digits and are tagged in the lexicon with the label [num] and possibly other grammatical labels ([Grammatical]), and before any sequence of symbols, with the exception of a word boundary (#), that ends with the substring *-ig* (indicating decades) and again is tagged with the label [num]. After application of this rule the expansion of our example 21 now reads *einsundzwanzig*.

Additional rules are required that model certain allomorphic processes in German numeral expansion. For instance, *eins* has to be rewritten as *ein* if it is followed by *hundert*, *tausend* or *und*. This rule generates the correct expansion of 21 into *einundzwanzig*. Furthermore, the number words for 10^6 (*Million*), 10^9 (*Milliarde*), 10^{12} (*Billion*), 10^{15} (*Billiarde*) require a preceding *eins* to agree in gender (*eine*); simultaneously, these number words also have to agree in grammatical number with the preceding unit: 2×10^6 *zwei Millionen*, but 1×10^6 *eine Million*. All these number filter rules are compiled into a WFST *NumFilters*. The *Num* transducer, which is the composition of the *Factorization*, the decade flop, and the *NumberLexicon* transducers, is then composed with *NumFilters*.

A rather minor difference between the writing conventions for numbers in English and German is the opposite use of the decimal point and number comma

[23] Interestingly, in its section on the formation of cardinal numbers, the Duden Grammatik (Duden, 1984) does not mention the "decade flop" process at all; it may not be too far-fetched to interpret this oversight or omission as an indication that the phenomenon is prone to be taken for granted by native speakers of German, and even by German linguists. Coincidentally, it is often ignored that English also displays the decade flop property, if only for the numbers between 13 and 19, at least from the point of view of etymology.

symbols. The decimal point symbol in German is in fact a comma, and it is customary to enhance the legibility of long numbers by separating triples of digits by means of a period.

Numeral expressions in German can consist of combinations of digits, periods, commas, slashes, percent symbols, and letters. A complex numeral expression like

(37) 4.135.678,749%igem

would be analyzed by the system as follows:

(38) # v'ier # milli'onen [pl][fem] # 'ein [masc][sg] # h'undert # f'ünf # und [conj][clit] # dr'ei + ßig # t'ausend # s'echs # h'undert # 'acht # und [conj][clit] # s'ieb + zig # Komma # s'ieben # v'ier # n'eun [num] # pro + z'ent [N] igem [adj][masc][sg][dat] #

and expanded as *vier Millionen einhundertfünfunddreißigtausend sechshundertachtundsiebzig Komma sieben vier neun prozentigem.*

Since German is a highly inflected language, it is important to select the correct word form for expanded cardinal and ordinal numbers — correct in terms of agreement with the syntactic context. For example, the phrases *mit 1 Katze* and *mit 1 Hund* have to be expanded such that the word form of the number *eins* agrees with the grammatical gender of the noun (feminine in the case of *Katze*, masculine for *Hund*). The correct word form is determined by applying language model filters. The problems one encounters in this area of German text analysis are similar to the ones found in Russian, if not quite as complex; see Section 3.6.5.

3.6.4 German unknown words

The correct pronunciation of unknown or novel words is one of the biggest challenges for TTS systems. Generally speaking, correct analysis and pronunciation can only be guaranteed in the case of a direct match between an input string and a corresponding entry in a pronunciation dictionary. However, any well-formed text input to a general-purpose TTS system in any language is extremely likely to contain words that are not explicitly listed in the lexicon.

There are two major reasons why this statement holds. First, the inventory of lexicon entries of every language is infinite; every natural language has productive word formation processes, and the community of speakers of a particular language can, via convention or decree, create new words as need arises. Second, the set of (personal, place, brand, etc.) names is, from a practical point of view, too large to be listed and transcribed in a pronunciation dictionary; even more importantly, names are subjected to the same productive and innovative processes as regular words are. The issue is not one of storage for such a large list. Rather, a priori construction of such a list is simply impossible.

The German language lends itself to a case study because it is notorious for its extensive use of compounds. What makes this a challenge for linguistic analysis is the fact that compounding is extraordinarily productive and speakers of

German can, and in fact do, 'invent' new compounds on the fly. The famous *Donaudampfschiffahrtsgesellschaftskapitän* 'Danube Steamboat Company Captain' is actually less typical than a spontaneously created word like *Unerfindlichkeitsunterstellung* ('allegation of incomprehensibility'). Therefore, linguistic analysis has to provide a mechanism to appropriately decompose compounds and, more generally, to handle unknown words. We include here the analysis and pronunciation of names as a special category of unknown words, but we exclude the problem of *detecting* names, i.e., of recognizing such words as names, in arbitrary text.

The unknown word analysis component of the German TTS system is based on a model of the morphological structure of words. We also performed a study of the productivity of word forming affixes, applying the productivity measure suggested by Baayen (1989).

Compositional model of German words

The compositional model is based on the morphological structure of words and the phonological structure of syllables. The core of the module is a list of approximately 5000 nominal, verbal, and adjectival stems that were extracted from morphologically annotated lexicon files. To this collection we added about 250 prefixes and 220 suffixes that were found to be productive or marginally productive. Finally, we included 8 infixes (*Fugen*) that German word formation grammar requires as insertions between components within a compounded word in certain cases, such as *Arbeit*+**s**+*amt* 'job center' or *Sonne*+**n**+*schein* 'sunshine'.

Probably the two most important aspects of this linguistic description are the decision which states can be reached from any given current state, and which ones among the legal paths through the graph should be preferred. The first aspect can be regarded as an instantiation of a declarative grammar of the morphological structure of German words; the second aspect reflects the degrees of productivity of word formation, represented by costs on the transitions between states.

Figure 3.12 shows parts of the *arclist* source file for unknown word decomposition. In each line, the first and second column are labels of two states, the state of origin and the state of destination, respectively. The third column contains a string of symbols on the arc (or transition) between the two states. These strings consist of regular orthography, which is annotated with lexical and morphological labels including morpheme boundaries ('+'), symbols for primary (') and secondary (") lexical stress, and optional costs for the transition.

The transition from the initial state START to the state PREFIX is defined by a family of arcs which represent productive prefixes. One of the prefix arcs is empty, allowing for words or components of complex words that do not have a prefix. Multiple consecutive prefixes can be modeled by returning from PREFIX to START; by assigning a relatively high cost to this path, analyses that require

```
START    PREFIX   ε
START    PREFIX   "a [pref]+<0.3>
START    PREFIX   'ab [pref]+<0.3>
START    PREFIX   "aber [pref]+<0.3>
START    PREFIX   er [pref]+<0.3>
START    PREFIX   "un [pref]+<0.3>
START    PREFIX   "unter [pref]+<0.3>
START    PREFIX   zw"ei [pref]+<0.3>
START    PREFIX   zw"ischen [pref]+<0.3>
         ...
PREFIX   START    ε<1.0>
PREFIX   ROOT     SyllableModel [root]<10.0>
PREFIX   ROOT     'aal [root]
PREFIX   ROOT     'aar [root]
PREFIX   ROOT     f'ind [root]
PREFIX   ROOT     st'ell [root]
PREFIX   ROOT     zw'eig [root]
PREFIX   ROOT     z'yste [root]
         ...
ROOT     PREFIX   [comp]+<0.1>
ROOT     SUFFIX   ε<0.2>
ROOT     SUFFIX   +"abel [suff]<0.2>
ROOT     SUFFIX   +ab"il [suff]<0.2>
ROOT     SUFFIX   +lich [suff]<0.2>
ROOT     SUFFIX   +keit [suff]<0.2>
ROOT     SUFFIX   +ung [suff]<0.2>
ROOT     SUFFIX   +zig [suff]<0.2>
ROOT     END      [N]<1.0>
         ...
SUFFIX   ROOT     ε<0.2>
SUFFIX   END      [N]<1.0>
SUFFIX   FUGE     n<0.2>
SUFFIX   FUGE     +s<0.2>
SUFFIX   FUGE     en<0.2>
         ...
FUGE     START    [comp]+<0.5>
END
```

Figure 3.12: Parts of a declarative arclist grammar for unknown word decomposition in German. Column 1: state of origin; column 2: state of destination; column 3: string on arc between states.

only one prefix are favored. The transition back to START is labeled with ϵ.

A large family of arcs from PREFIX to ROOT represents nominal, verbal, and adjectival stems. Sequences of stems are modeled by returning to PREFIX from ROOT without intervening affixes; in this case, the tag [comp] indicates the end of a complete subcomponent in a complex word. If a word terminates in a stem, the appropriate path through the graph is from ROOT to END.

The transition from PREFIX to ROOT which is labeled *SyllableModel* is a place holder for a phonetic syllable model which reflects the phonotactics and the segmental structure of syllables in German, or rather their correlates on the orthographic surface. This allows the module to analyze substrings of words that are unaccounted for by the explicitly listed stems and affixes in arbitrary locations in a morphologically complex word. Applying the syllable model is expensive because we want to cover the orthographic string with as many known components as possible. The costs actually vary depending upon the number of syllables in the residual string and the number of graphemes in each syllable. For the sake of simplicity we assign a flat cost of 10.0 in our example.

Productive suffixes are represented by a family of arcs between the states ROOT and SUFFIX. Analogous to the case of prefixes, one such arc carries the empty string to skip suffixes, and consecutive suffixes can be covered by returning to ROOT from SUFFIX. The iteration over suffixes is less expensive than the one over prefixes because sequences of suffixes are significantly more common than multiple prefixes. If the word ends in a suffix, the path through the graph continues from SUFFIX to END.

The last family of arcs represents the *Fugen* infixes as transitions from SUFFIX to FUGE. There is only one legal continuation through the graph from FUGE, i.e., back to the beginning of the graph. This design reflects the fact that *Fugen*, by definition, can only occur between major subcomponents of a complex word, each of which have their own stem; thus also the indication of the completion of the subcomponent by means of the tag [comp].

On termination the machine labels the word as a noun by default. In a more sophisticated word model it should be possible in principle to assign the part-of-speech category of the unknown word on the basis of the types of stems and affixes involved, by distinguishing between noun, verb, and adjective forming affixes. In any event, it is sufficient for the prosodic components of the TTS system to know that the word is a content word, a relatively safe assumption for novel words.

Most arc labels are weighted by a cost. Weights in the unknown word analysis module of our German TTS system are currently based on linguistic intuition; they are assigned such that direct matches of input strings to entries in the lexicon will be less expensive than unknown word analysis. There is no legal path through the unknown word graph that comes with zero cost; the minimal cost of 1.0 would be for a simplex stem that is explicitly listed and does not have any affixes.

The finite-state transducer implementing this declarative grammar is far too complex to be usefully diagrammed here. For the sake of example, let

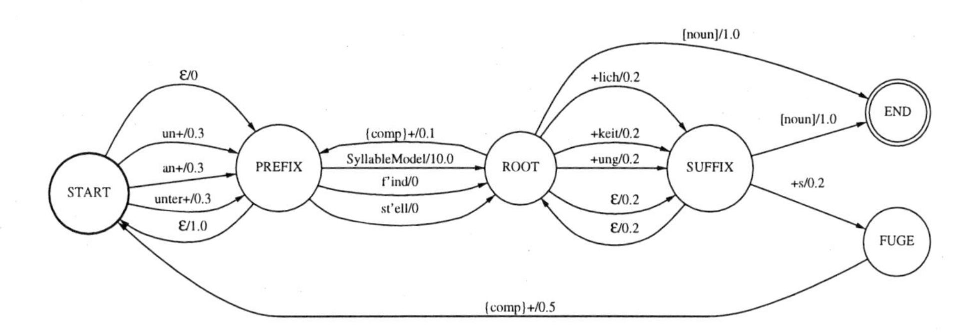

Figure 3.13: The transducer compiled from the sub-grammar that performs the decomposition of the morphologically complex word *Unerfindlichkeitsunterstellung* 'allegation of incomprehensibility'.

us instead consider the previously mentioned complex, and probably novel, word *Unerfindlichkeitsunterstellung* 'allegation of incomprehensibility'. Figure 3.13 shows the transducer corresponding to the sub-grammar that performs the decomposition of this word.

Application to names

The approach to unknown word decomposition described above is also applied to the analysis of names in our German TTS system. Arguably, names are not equally amenable to morphological processes, such as word formation and derivation or to morphological decomposition, as regular words are. That does not render such an approach infeasible, though, as was shown in a recent evaluation of the system's performance on street names. Street names are an interesting category because they encompass aspects of geographical and personal names. In their study, Jannedy and Möbius (1997) report a pronunciation error rate *by word* of 11–13% for unknown names. In other words, roughly one out of eight names will be pronounced incorrectly. This performance compares rather favorably with results reported in the literature, for instance for the German branch of the European Onomastica project (Onomastica, 1995). Onomastica was funded by the European Community from 1993 to 1995 and aimed to produce pronunciation dictionaries of proper names and place names in eleven languages. The final report describes the performance of grapheme-to-phoneme rule sets developed for each language. For German, the accuracy rate by word for quality band III — names which were transcribed by rule only — was 71%, yielding an error rate of 29%. The grapheme-to-phoneme conversion rules in Onomastica were written by experts, based on tens of thousands of the most frequent names that were manually transcribed by an expert phonetician. In our TTS system, the phonological or pronunciation rules capitalize on the extensive morphological information provided by annotated lexica *and* the unknown word analysis component.

One obvious area for improvement is to add a name-specific set of pronunciation rules to the general-purpose one. Using this approach, Belhoula (1993) reports error rates of 4.3% (by word) for German place names and 10% for family names. These results are obtained in recall tests on a manually transcribed training corpus. The addition of name-specific rules presupposes that the system knows which orthographic strings are names and which are regular words. The problem of name detection in arbitrary text (see (Thielen, 1995) for an approach to German name tagging) has not been addressed so far; instead, it is by-passed for the time being by integrating the name component into the general text analysis system and by adjusting the weights appropriately.

3.6.5 Russian percentage terms

We return here to the problem of the interpretation of the Russian percentage symbol, introduced in Section 3.3. Consider the example **с 5% скидкой** — *s 5% skidkoj* — (with 5% discount) 'with a five-percent discount'.[24] This is first composed with the lexical analysis WFST to produce a set of possible lexical forms; see Figure 3.14. By default the lexical analyzer marks the adjectival readings of '%' with '$\boxed{\star}$', meaning that they will be filtered out by the language-model WFST's, if contextual information does not save them. Costs on analyses (here represented as subscripted floating-point numbers) mark constructions — usually oblique case forms — that are not in principle ill-formed but are disfavored except in certain well-defined contexts. The correct analysis (boxed in Figure 3.14), for example, has a cost of 2.0 which is an arbitrary cost assigned to the oblique instrumental adjectival case form: the preferred form of the adjectival rendition of '%' is masculine, nominative, singular if no constraints apply to rule it out.

Next, the language model WFST's Λ are composed with the lexical analysis lattice. The WFST's in Λ include transducers compiled from rewrite rules that ensure that the adjectival rendition of '%' is selected whenever there is a noun following the percent expression, and rules that ensure the correct case, number and gender of the adjectival form given the form of the following noun. In addition, a filter expressible as $\neg(\Sigma^* \boxed{\star} \Sigma^*)$ removes any analyses containing the tag '$\boxed{\star}$' (cf. the discussion of the disambiguation of *kg.* for the toy English analysis in Section 3.4.2); see Figure 3.15. The best-cost analysis among the remaining analyses is then selected. Finally, the lexical analysis is composed with *Lex2MMA* $\circ$ *MMA2Phon* to produce the phonetic transcription; see Figure 3.16.

[24] A fuller discussion of this topic is given in (Sproat, 1997b).

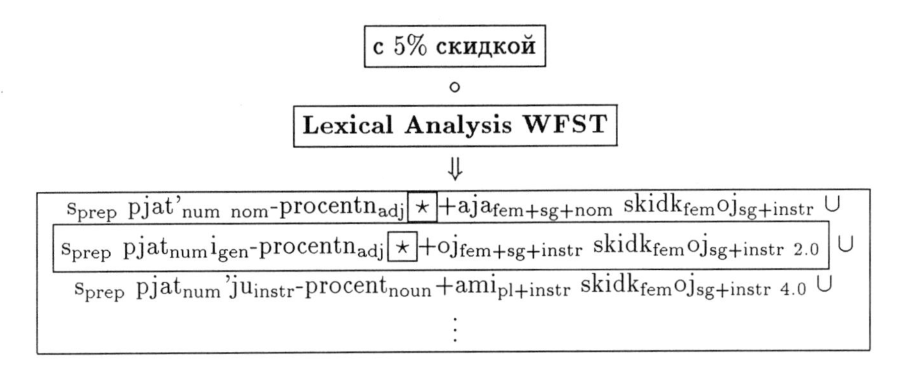

Figure 3.14: Composition of Russian **с** 5% **скидкой** — *s 5% skidkoj* — 'with a 5% discount' with the lexical analysis WFST. By default the adjectival readings of '%' are marked with '⋆', which means that they will be filtered out by the language-model WFST's; see Figure 3.15. The boxed analysis is the correct one.

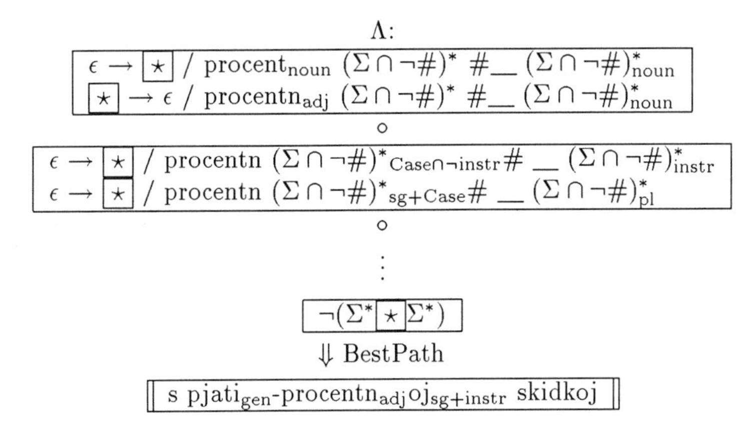

Figure 3.15: A subset of the language model WFST's related to the rendition of Russian percentages. The first block of rules ensures that adjectival forms are used before nouns, by switching the tag '⋆' on the adjectival and nominal forms. The second block of rules deals with adjectival agreement with the adjectival forms. The final block is a filter ruling out the forms tagged with '⋆'. The (correct) output of this sequence of transductions is shown at the bottom.

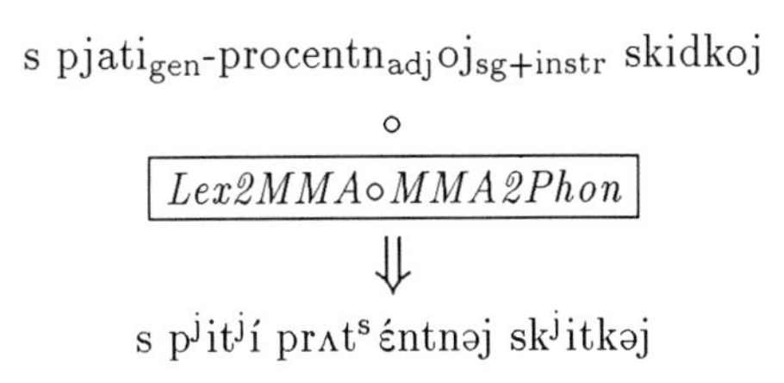

Figure 3.16: Mapping the selected Russian lexical analysis to a phonetic rendition.

3.6.6 Mandarin Chinese word segmentation

In this section we give only a brief overview of Mandarin segmentation: a much more extensive discussion can be found in (Sproat et al., 1996), and a somewhat more extensive discussion with more particulars related to gentex is given in (Shih and Sproat, 1996).

Consider the Mandarin sentence 日文章魚怎麼說 (*rìwén zhāngyú zěnmo shuō*) "How do you say octopus in Japanese?". This consists of four words, namely 日文 *rì-wén* 'Japanese', 章魚 *zhāng-yú* 'octopus', 怎麼 *zěn-mo* 'how', and 說 *shuō* 'say'. The problem with this sentence is that 日 *rì* is also a word (e.g. a common abbreviation for Japan) as are 文章 *wén-zhāng* 'essay', and 魚 *yú* 'fish', so there is not a unique segmentation (cf. the toy example of "pseudo-Chinese" in Section 3.4.4). In the current version of Mandarin text analysis, we model the lexicon as a WFST, where costs associated with each word correspond to the negative log probabilities of those words as estimated from a training corpus. For instance, the word 章魚 'octopus' would be represented as the sequence of transductions 章:章 魚:魚 ϵ:noun< 13.18 >, where 13.18 is the estimated cost of the word. Entries in the lexicon are annotated with information on word-pronunciation, as described in Section 3.4. Thus the character 乾 is pronounced as *gān* in 餅乾 *bǐnggān* 'cookie', but as *qián* in 乾坤 *qiánkūn* 'universe'. The lexical entries for these words are, respectively, 餅乾$_1$ [N] 'cookie' and 乾$_2$坤 [N] 'universe'. In the *MMA2Surf* map (see Section 3.5) both 乾$_1$ and 乾$_2$ map to 乾; in the *MMA2Phon* map the former maps to *gān* and the latter to *qián*.

This lexicon is combined with other lexical models (e.g. for numbers) and this is further combined with the "punctuation" model as in the toy example discussed previously. In the current implementation, segmentations are disambiguated simply by selecting the lowest-cost string in $S \circ L$, which gives the correct result in the example at hand. As argued in (Sproat et al., 1996), this approach agrees with human judges almost as well as the human judges agree among themselves, but there is clearly room for more sophisticated models.

Of course, as is the case with English, no Chinese dictionary covers all of the words that one will encounter in Chinese text. For example, many words that are derived via productive morphological processes are not generally to be found in the dictionary. One such case in Chinese involves words derived via the nominal plural affix 們 *-men*. While some words in 們 will be found in the dictionary (e.g., 他們 *tā-men* 'they'; 人們 *rén-men* 'people'), many attested instances will not: for example, 小將們 *xiǎo-jiāng-men* 'little (military) generals', 青蛙們 *qīng-wā-men* 'frogs'. To handle such cases, the dictionary is extended using standard finite state morphology techniques. For instance, we can represent the fact that 們 attaches to nouns by allowing ϵ-transitions from the final states of noun entries, to the initial state of a sub-transducer containing 們. However, for our purposes it is not sufficient merely to represent the morphological decomposition of (say) plural nouns, since one also wants to estimate the cost of the resulting words. For derived words that occur in the training corpus we can estimate these costs as we would the costs for an underived dictionary entry. For non-occurring but possible plural forms we use a value based on the Good-Turing estimate (e.g. (Baayen, 1989; Church and Gale, 1991)), whereby the aggregate probability of previously unseen members of a construction is estimated as N_1/N, where N is the total number of observed tokens and N_1 is the number of types observed only once; again, we arrange the automaton so that noun entries may transition to 們, and the cost of the whole (previously unseen) construction sums to the value derived from the Good-Turing estimate. Similar techniques are used to compute analyses and cost estimates for other classes of lexical construction, including personal names, and foreign names in transliteration.

3.6.7 Accentual phrasing in Japanese compounds

Japanese words can be classified as *accented* or *unaccented* (Pierrehumbert and Beckman, 1988). Accented words have a sharp fall in pitch while unaccented words do not. To mark this information in the lexicon, we use the tag {a0} for unaccented entries and the tags {a1}, {a2}, {a3}, and so on, for accented entries. (The numbers indicate which mora an accent falls on.) These tags are later transduced to the actual accent symbol ' by the *MMA2Phon* transducer. For example, if we ignore the costs of the words for the moment, the initially-accented word 社会 *sya'kai* 'society' and the unaccented word 大学 *daigaku* 'university' would be listed as follows in the lexicon.[25]

(39) 社$_1$会$_1${a1}[N]
大$_1$学$_1${a0}[N]

When a word is pronounced in isolation, it forms one *accentual phrase*. (For the sake of simplicity, we will not deal with prosodic phrasing at higher levels here.) Each accentual phrase has a rise in pitch at the beginning, and contains at most one accent.

When two nouns form a compound noun, in general, it becomes *prosodically unified*, forming one accentual phrase. The following is a somewhat simplified version of the general principles for the reassignment of the accent — the *Compound Accent Rules* — within the unified accentual phrase (cf. (Kubozono, 1988)).

1. If the second element is short (either one-mora or two-morae long):
   (a) If the second element belongs to a group of de-accenting morphemes, such as 色 *iro'* 'color', the whole compound becomes unaccented.
   (b) Otherwise, the new accent is on the last mora of the first element regardless of the original accent type of either element.

[25] As with the Chinese system, each Chinese character is annotated according to its pronunciation. For the remainder of this section, we will assume that the annotated characters are already transduced to a phonemic representation.

2. If the second element is long (three-morae long or longer):
   (a) If the second element is unaccented or accented on its final syllable, the resulting accent is on the first mora of the second element.
   (b) Otherwise, the original accent of the second element remains while the first element is de-accented.

These rules can be implemented with WFST's. For instance, (2a) can be partially expressed as the composition of the WFST's compiled from the rewrite rules in (40). {Acc} here subsumes all accent tags, and μ represents a mora.

(40) $\{\mathrm{Acc}\} \rightarrow \epsilon \;/\; \mu^{+} \;\_\; [\mathrm{N}] \;\#\; \mu\mu\mu^{+} \;\{\mathrm{a0}\}\; [\mathrm{N}]$
$\{\mathrm{a0}\} \rightarrow \{\mathrm{a1}\} \;/\; \mu^{+} \;[\mathrm{N}] \;\#\; \mu\mu\mu^{+} \;\_\; [\mathrm{N}]$
$[\mathrm{N}] \rightarrow \epsilon \;/\; \mu^{+} \;\_\; \#\; \mu\mu\mu^{+} \;\{\mathrm{a1}\}\; [\mathrm{N}]$

Using such WFSTs, the initially-accented word 社会 *sya'kai* 'society' followed by the unaccented word 問題 *mondai* 'problem' becomes the compound 社会問題 *syakaimo'ndai* 'social problem' as in (41).

(41) syakai {a1} [N]# mondai {a0} [N] →
syakai # mondai {a1} [N]

There are also cases where a compound is not prosodically unified at this stage of analysis. For example, certain prefixes do not undergo such prosodic unification. In (42), 元 *mo'to* 'former' modifies 大統領 *daito'oryoo* 'president', but there is an accentual phrase boundary (which we represent here with the symbol '$]_a$') between them.

(42) 元 大統領
mo'to daito'oryoo → mo'to $]_a$ daito'oryoo
'former president'

We can list all such prefixes in the lexicon, and model the accentual phrase boundary placement by inserting the symbol $]_a$ between the prefix and the noun.

If a compound is made of three or more elements, the internal structure of the compound also affects whether or not it is prosodically unified to form one accentual phrase. A three-element compound noun whose structure is left-branching in general becomes one accentual phrase as in (43). On the other hand, a three-element compound noun whose structure is right-branching in general has an accentual phrase boundary, at this stage of analysis, before the (heavy) right branch, as in (44).[26]

(43) [[安全 保証] 条約]
anzen hosyoo zyooyaku → anzenhosyoozyo'oyaku
safety guarantee treaty
'security treaty'

[26] The examples are taken from Kubozono (1988).

(44) [名古屋 [工業 大学]]
na'goya ko'ogyoo da'igaku → na'goya $]_a$ koogyooda'igaku
nagoya industry university
'Nagoya Institute of Technology'

No modifications in the WFSTs are necessary for the left-branching structure if the rules are compiled to work left-to-right. The following rendition illustrates how the rules in (40) work for the compound in (43):

(45) anzen {a0} [N] # hosyoo {a0} [N] # zyooyaku {a0} [N] →
anzen [N] # hosyoo [N] # zyooyaku {a0} [N] →
anzen [N] # hosyoo [N] # zyooyaku {a1} [N] →
anzen # hosyoo # zyooyaku {a1} [N]

On the other hand, the right-branching structure in (44) requires more care. We would like to represent somehow that the second and third elements are more closely bound than the first and second elements are. We might simply list the compound 工業大学 in the lexicon, or we might treat 大学 as if it were a suffix to a certain class of nouns. Either way, we use an additional tag to introduce the accentual phrase boundary and block the Compound Accent Rules if there is a preceding noun. Between the two analyses in (46) and (47), if the estimated cost of 工業大学 *koogyoodaigaku* is lower than the sum of the costs of 工業 *koogyoo* and 大学 *daigaku* individually, the incorrect analysis (47) is filtered out when the best path is chosen in the language-model analysis.

(46) nagoya {a1} # $]_a$ koogyoodaigaku {a5} [N]

(47) nagoya # koogyoo # daigaku {a1} [N]

3.6.8 Malagasy numbers

Malagasy[27] has a decimal number system with simplex number names for all powers of ten up to 10^6. The interesting property of Malagasy number names is that they are read "backwards", starting with the units, and proceeding to the higher powers of ten (Rajemisa-Raolison, 1971):

		9.543.815	
		dimy	five
		ambin' ny folo	and ten
		amby valon-jato	and eight-hundred
(48)	(a)	*sy telo arivo*	and three thousand
		sy efatra alina	and four ten-thousand
		sy dimy hetsy	and five hundred-thousand
		sy sivy tapiritsa	and nine million

[27] Austronesian, Madagascar.

	44.725.086
enina	six
amby valopolo	and eighty
sy dimy arivo	and five thousand
sy roa alina	and two ten-thousand
sy fito hetsy	and seven hundred-thousand
sy efatra	and four
amby efapolo tapiritsa	and forty million

(b)

The word order of the second case, for example — *enina amby valopolo sy dimy arivo sy roa alina sy fito hetsy sy efatra amby efapolo tapiritsa* — implies the following factorization; this is, mutatis mutandis, the exact mirror image of what one would find in English and the opposite to the order in which the digits appear on the printed page:

(49) $6 + 8 \times 10^1 + 5 \times 10^3 + 2 \times 10^4 + 7 \times 10^5 + [4 + 4 \times 10^1] \times 10^6$

What seems necessary for Malagasy is a generalization of the "decade flop" filter that is used in German (see Section 3.6.3) to reverse decades and units: for Malagasy, the equivalent filter will need to reverse *all digit* strings. Up to a finite length, this is of course possible to do with a finite-state transducer — but at a substantial price. Mapping an arbitrary string w to its reversal is *not* a regular relation, and the only way to implement the reversal is to build a machine that in effect stores every digit string and its reversal, up to the predetermined length.[28] Figure 3.17 illustrates the problem, demonstrating that the size of a "number-flipping" machine grows linearly with the largest power of ten which is handled by the machine.[29] Clearly the finite-state approach breaks down if we require it to handle this problem.

The only viable solution to this problem would be to filter the text with a special-purpose preprocessor that reverses digit sequences; with such a device, Malagasy digit-sequence-to-number-name conversion becomes straightforward within the framework we have presented. As ad hoc as this may seem, it is not actually such an unreasonable solution. Note that in Semitic languages like Hebrew and Arabic, although ordinary writing proceeds from right to left, numbers (since the numeral system of Arabic was borrowed from India) are written from left to right. Semitic languages construct their number names "from left to right", as do European languages, so the situation in Semitic languages is an exact analog of the situation in Malagasy. Interestingly, though the rendering order for digit strings is the reverse of the normal orthographic order, it is standard practice to represent the text electronically in logical order (see, e.g., (The Unicode Consortium, 1996)), and leave it up to the rendering software to make sure that Semitic alphabetic symbols appear in right-to-left

[28] Note that this is distinct from the fact that regular languages and relations are closed under *reversal*; see Sections 2.5.1 and 2.5.2.

[29] The slight dips at 1000 and 10000 are not artifacts: the machines are slightly smaller than expected in these cases because the largest subset of numbers in these sets have an odd number of digits, and mapping the middle digit to itself requires only ten arcs.

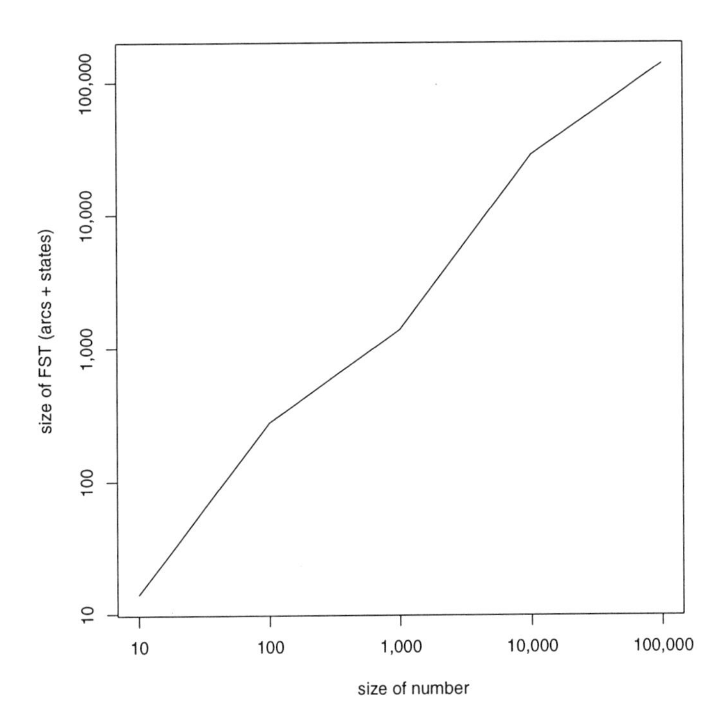

Figure 3.17: The size (number of states + number of arcs) of a transducer that reverse digit strings, plotted as a function of the largest power of 10 that is handled by the machine.

	States	Arcs
Spanish	31,601	71,260
Mandarin	45,806	277,432
Japanese	93,720	415,381
German	111,190	623,643
Italian	112,659	269,487
Russian	139,592	495,847

Table 3.2: Sizes of lexical analysis WFST's for selected languages

order, but the digit sequences in left-to-right order. The suggested solution to the problem in Malagasy, then, is basically just a simple version of what rendering software for Semitic languages routinely does.

3.6.9 Some sizes of lexical transducers

Table 3.2 gives the sizes of the lexical analysis WFST's for the languages Spanish, Mandarin, Japanese, German, Italian, and Russian. While the differences in size to a large extent reflect the amount of work that has been done on each language, it is also true that the sizes accord somewhat with our intuitions of the difficulties of lexical processing in the various languages. So the number of states in Russian is very large, correlating with the complexity of the morphology in that language. German and Italian have somewhat fewer states (though German has more arcs than Russian). Mandarin has a small number of states, correlating with the fact that Mandarin words tend to be simple in terms of morphemic structure; but there are a relatively large number of arcs, due to the large character set involved. Sizes for the Spanish transducer are misleading since the current Spanish system includes only partial morphological analysis: note, though, that morphological analysis is mostly unnecessary in Spanish for correct word pronunciation (though it is certainly important for other reasons).

While the transducers can be large, the performance on a (by now standard) 100Mhz machine is acceptably fast for a TTS application. Slower performance is observed, however, when the system is required to explore certain areas of the network, as for example in the case of expanding and disambiguating Russian number expressions.[30]

[30] To date we have performed no evaluations on the text analysis modules of our non-American-English synthesizers, though some components of these systems have been evaluated: see, e.g., (Sproat et al., 1996) for an evaluation of the Chinese word-segmentation method discussed in Section 3.6.6; and (Jannedy and Möbius, 1997) and Section 3.6.4 for an evaluation of name pronunciation in the German system. The reason for this is that work is proceeding on all of our systems, and formal evaluations are, we feel, premature. We discuss

3.7 On Automatic Learning Methods

Most of the discussion in this chapter has focused implicitly on hand-constructed grammars. This may seem like an oversight in the light of the large body of recent work on trainable methods. Such work has been particularly concentrated on the problem of word pronunciation, especially for languages like English where rule-based or dictionary-based methods require a substantial investment of time in their development; see (Luk and Damper, 1993; Andersen et al., 1996; Luk and Damper, 1996; Daelemans and van den Bosch, 1997) for some recent examples.

Putting to one side the perfectly valid interest in automatic inference methods, the major premises behind much of this work appear to be as follows:

- Rule-based methods are costly to develop. Furthermore they are prone to bugs since in any large set of rules, the interactions among rules can be complex and unpredictable.
- Automatic methods are much cheaper to develop. Furthermore, their behavior is (by implied contrast) much more robust and predictable.

We agree with the first premise: rule systems are indeed costly to develop, especially if one includes under this rubric dictionary-based methods where considerable time must be spent in editing and correcting machine-readable dictionaries. Furthermore, one has to be very careful in their development to avoid unexpected interactions between rules. But we disagree with most of the second premise: learning an accurate model may ultimately not be much cheaper than developing a model by hand, and there is certainly no reason to believe that the results of training a decision tree, a decision list, an inferable stochastic transducer or a neural net will necessarily result in a system that is more robust or predictable on unseen data than a carefully developed set of rules.

Consider, as a concrete example, the recent work on *stochastic phonographic transduction* reported in (Luk and Damper, 1996). Luk and Damper's method was trained and tested on English words paired with their pronunciations. The system infers from a set of training pairs a (formally regular) stochastic grammar by a three-phase algorithm (pages 139–145):

1. Count all possible transitions between all possible single letter-phoneme correspondences in the training data. These counts are used to estimate probabilities for the initial grammar.
2. Using the initial grammar, stochastically align all letter-phoneme combinations in the training data; from these alignments, compute multi-letter to multi-phoneme correspondences, given some method for deciding how to "break" the initial alignment into segments.

---

work on evaluation of our American English system in Chapter 8, along with some general methodological issues on how we believe TTS evaluation should be carried out.

3. Estimate transition probabilities for the newly inferred multi-letter/multi-phoneme correspondences.

The final grammar can then be used to predict pronunciations for new words, by assigning a weight to each *sequence* of all possible letter-sequence to phoneme-sequence correspondences which span the word. (As far as we can tell, the inferred grammar would be straightforwardly implementable as a WFST.)

The system has a performance of just under 80% correct pronunciation *by word* (about 95% correct by symbol) for the best test set, a *previously unseen* portion of the same dictionary of ordinary words as was used for training. Luk and Damper claim that this is comparable to the best reported scores for systems based purely on letter-to-sound rules, and this assessment may well be correct. It is not, however, comparable to the best methods that depend upon a judicious combination of rule-based and dictionary-based methods, such as the Bell Labs American English word-pronunciation system discussed in (Coker, Church, and Liberman, 1990) (and see also Section 3.8.2), which reports 95% "good" or "ok" judgments *by word* for the much harder task of pronouncing personal names. It remains to be seen whether a method such as Luk and Damper's can approach the performance of a carefully constructed rule- and dictionary-based method.

But there is a much more fundamental problem with Luk and Damper's approach in that it crucially assumes that letter-to-phoneme correspondences can in general be determined on the basis of information local to a particular portion of the letter string. While this is clearly true in some languages (e.g. Spanish), it is simply false for others. A good example of this is Russian where, as we have noted in Section 3.5, placement of lexical stress crucially depends upon which morphological category the word belongs to. There is nothing in the letter string <**карандашом**> *karandašom* 'with a pencil' that tells one that stress should be on the final syllable (/kərəndʌšóm/); or in the letter string <**городом**> *gorodom* 'with a city' (/górədəm/), that tells one that it should fall on the first. Indeed, homographic pairs such as /górədʌ/ 'of a city' and /gərʌdá/ 'cities', both written <**города**> (*goroda*) abound in Russian, so clearly the *letter* string cannot be sufficient in and of itself to predict placement of stress.[31] And without correct stress placement, correct prediction of vowel pronunciation is impossible: stress strongly affects the rendition of the vowels, as attested by the difference in the rendition of the ending *-om* (/-óm/ vs. /-əm/) in the examples *gorodom* and *karandašom*. So, to pronounce words correctly in Russian we need more than just the letter string. We need access to the morphological analysis, including information on which morphological class a given word belongs to. For *gorod* 'city', we need to know, for instance, that it belongs to the class where singular case forms have lexical stress on the stem rather than the ending, whereas for *karandaš* 'pencil' stress is assigned to the ending. Luk and Damper's approach, and other similar approaches are thus

[31] In a corpus of only two million words, we have collected close to six hundred such homographs, each of which occurs at least fifty times.

unlikely to succeed on this problem.

Of course, there is one way in which such stochastic induction methods could succeed, and that is if we were to add to the letter string relevant lexical information, in the form of lexical tags; see also Section 3.9.3. Then the system could in principle learn the correspondence between these lexical tags, and the various possible renditions of sequences of letters. Unfortunately, this would hardly result in much savings of labor, since once the lexical information is available, computing the pronunciation by a set of hand-constructed rules is not terribly onerous. In our experience, the vast majority of the work on Russian lexical analysis went into constructing the morphological analyzer given the at times very incomplete (and at times completely wrong) information in our electronic dictionary. This involved many stages, including constructing the morphological paradigms, building rules to handle a large number of sets of highly restricted morphophonemic alternations, and marking a handful of common but sporadically distributed lexical exception features.[32] Compared to the amount of work that went into building the lexical analyzer, the required set of pronunciation rules themselves was easy to construct.

We want to make it clear that we have nothing in principle against automatic or statistical methods. Indeed, as discussed in Section 3.6.6 and as we will further discuss in Chapter 4, we have done substantial work applying such methods to problems in TTS. However, we do not believe such approaches can be done in the absence of significant human intervention in the sense that it is, in many cases, a human who needs to decide which features are likely to be important in making a decision. It is unreasonable to expect that good results will be obtained from a system trained with no guidance of this kind, or (as in the case of Luk and Damper's work) with data that is simply insufficient to the task.

3.8 Relation to Previous Bell Labs English TTS

The focus of this chapter has been on the model of text-analysis incorporated in gentex. At the time of writing, the one language for which we have extensive experience in building text-to-speech systems, yet for which we do *not*, paradoxically, have a gentex-based model of text analysis is English. The purpose of this section is to give a brief overview of the text-analysis component of our American English system.

As we have argued, the gentex model is rather novel in that all aspects of text-analysis are handled in one component, including those aspects commonly relegated to a "preprocessing" phase. In many ways the model of text analysis in the Bell Labs American English system is a much more traditional one. What

[32]For example, quite a few Russian nouns have a so-called "second prepositional" form, which is spelled the same way as the regular *dative* (with the suffix -*u* or a variant), but which has consistent final stress. But the distribution of this form is fairly unpredictable, and one simply has to list the words with which it occurs.

we have been calling text analysis is spread over five modules:[33]

Front end, the preprocessing component, which includes the following:

1. Sentence-boundary detection
2. Abbreviation expansion
3. Expansion of numbers
4. Grammatical part-of-speech assignment.

Lemmatization: deriving uninflected (base) forms from inflected words.

Accentuation.

Prosodic phrasing.

Word pronunciation, including:

1. Detection and analysis of homographs
2. Pronunciation of ordinary words and names.

Detailed descriptions of the accentuation, prosodic phrasing and homograph disambiguation components are given in Chapter 4. In this section we give a brief outline of the front end and the pronunciation of ordinary words and names.

3.8.1 Front end

Let us illustrate the operation of the front end by example. Consider the sentence *Dr. Smith lives at 123 Lake Dr.* The first operation that is performed is to tokenize the input into words. This is done on the basis of white space, and in addition certain classes of symbols, including marks of punctuation, are represented as separate words. For the case at hand, the input will be tokenized as follows, where we will use brackets to delimit the individual "words":

(50) [Dr][.][Smith][lives][at][123][Lake][Dr][.]

After tokenization, abbreviations are disambiguated and expanded and sentence boundaries are detected. In the case at hand, *Dr.* can be disambiguated by considering the capitalization of the word immediately to the right or left: if the word to the right is capitalized then it will be expanded as *doctor*, otherwise if the word to the left is capitalized it will be expanded as *drive*. Periods that correspond to sentence boundaries are retained, but periods after abbreviations that occur in other positions are deleted. This will leave us with:

(51) [doctor][Smith][lives][at][123][Lake][drive][.]

[33] The interface between these modules is identical to that described in Section 2.6.

The next phase of analysis involves detection and treatment of various predefined sets of *text classes*. These include things like addresses, dates, fractions and telephone numbers. A text class is detected by pattern matching using a finite-state grammar. The precise set of patterns that will be detected for a given class depends upon a user-specifiable degree of "risk" desired for that class. For example, if one sets address expansion to "risky" mode, the front end will classify the sequence *123 Lake Drive* as a portion of an address, and will split the sequence *123* into *1* and *23*:

(52) [doctor][Smith][lives][at][1][23][Lake][drive][.]

When numerals are expanded later in the front end processing, the number will be expanded as *one twenty three*, rather than *one hundred twenty three*, which would be less normal in reading addresses. In the (default) "conservative" mode, however, only full addresses (those containing at least a street address, a city and a state code) are detected[34] In this case the number would be left alone and later expanded as *one hundred twenty three*.

The next stage of preprocessing involves the expansion of numbers, and the treatment of other punctuation symbols. Contiguous strings of digits — i.e., those that have not been split by one or another text class analysis component — are expanded according to various built-in rules. For example, numbers with leading zeroes (*0123*) will be translated serially (*zero one two three*), as will numbers after a decimal point; other numbers will generally be translated as ordinary number names (*one hundred twenty three*), with the exception of the special class of four digit numbers between *1100* and *9999*, which are (with the exception of the thousands) translated in "hundreds" mode (e.g., *nineteen hundred and ninety seven*);

One of the final processes that the front end performs is the assignment of grammatical parts of speech. The algorithm, due to Church (Church, 1988), is a simple probabilistic model that attempts to maximize the product of

- the a priori (lexical) probability of the j-th part of speech being assigned to the i-th word, for all i, and
- the probability of the j-th part of speech of the i-th word given the k-th part of speech of the $i-1$-st word and the l-th part of speech of the $i-2$-nd word.

In other words:

$$argmax \prod_{path} prob(cat_j|word_i)prob(cat_{j_i}|cat_{k_{i-1}}cat_{l_{i-2}}) \qquad (3.1)$$

[34]Note that it is risky to assume that a string of the form *number capitalized-word possible-street-term* is necessarily an address. Thus the conservative mode is a reasonable default for unrestricted text. However, if one has more knowledge about the input text — e.g., if the application is one that involves reading names and addresses — then the riskier setting is reasonable.

(Note that a model of this kind is straightforwardly representable in terms of WFST's.)

The front end of the American English system thus performs several of the operations that are, in gentex, integrated with the rest of text-analysis. As one might expect the operations performed by the English front end are, to date, more complete than the comparable operations for our other languages. The main drawback of the system is that a great deal of language-specific information particular to English is hard-coded in the system, in ways that would make it relatively difficult to adapt to other languages. This, needless to say was a major motivation for the development of the gentex model in the first place.

3.8.2 Word pronunciation

The first stage of word pronunciation involves checking each word against a small customizable list (called the *cache*) to see if a pronunciation is listed for that word. Next, the word is tested to see if it belongs to a list of known homographs, and if it is, is processed by the homograph disambiguation component, which will be described more fully in Section 4.2. Most words, of course, will fall through these initial two stages to the main word pronunciation routines.

For the vast majority of words and names the word pronunciation module (described in (Coker, Church, and Liberman, 1990)) computes the pronunciation by dictionary-based lexical analysis. At the core of this analysis is a base lexicon of approximately sixty thousand words and fifty thousand proper names. Two methods are used to extend this dictionary to novel words. The first is morphological derivation. Several dozens of English affixes, including affixes that occur in names (e.g. *-son*, *Mc-*) are listed, along with their properties, in particular (in the case of suffixes) their effect on the stress of the preceding stem. For example, the noun-forming suffix *-ity* and the name-forming suffix *-ovich* are listed as having the property of attracting stress to the preceding syllable:

(53)
grammátical *grammaticálity*
Ádam *Adámovich*

The second technique, applied to names, is analogical extension. Given a name that is in the dictionary — *Trotsky* — one can infer the pronunciation of a similar, but unlisted name — *Plotsky* — by analogizing the pronunciation of the former to the latter. In this case the words share the same final letter sequence, and we can make the reasonable assumption that they rhyme. So the *otsky* portion of *Plotsky* will be pronounced by simply retrieving the pronunciation of the relevant portion of *Trotsky*; and the *Pl* portion will be pronounced by rule. Other techniques for analogy are also implemented, and described in (Coker, Church, and Liberman, 1990). Words that fail to get analyzed by either of these methods fall through to a default set of letter-to-sound rules.[35]

[35] Due to Doug McIlroy. Another method by which personal names may be pronounced is

3.9 Previous and Related Work

The topics that we have considered under the rubric of text analysis cover a wide set of areas, ranging from text "normalization", to morphological analysis, to "letter-to-sound" rules, to prosodic phrasing. In the broader literature on TTS, each of these areas tends to be viewed as a domain in its own right, and each area has its own community of practitioners. Since so many areas are involved it is difficult to give a review of the literature at large that is not only detailed and comprehensive, but concise. Our purpose here is to situate the work on text analysis in the Bell Labs multilingual TTS system among work in relevant areas done elsewhere. From that point of view comprehensiveness (breadth of coverage) is of paramount importance; reasons of space place concision a close second.

Thus, in this section, rather than focus on the details of particular pieces of work, we will instead outline concisely what we believe to be the dominant trends in most current work on text analysis for TTS, illustrating each trend with particular instances from the literature. We should stress at the outset that this section is not intended as a tutorial introduction to work on text analysis in TTS: for such an introduction, see (Dutoit, 1997).

3.9.1 Text "normalization"

As we have observed elsewhere, it is the norm in TTS systems to treat such operations as abbreviation expansion, numeral expansion or tokenization of the input into words, in a text normalization phase that is both separate from, and prior to real linguistic analysis. Such an architecture characterizes not only early systems such as MITalk (Allen, Hunnicutt, and Klatt, 1987) but also more recent systems, including (Gaved, 1993) (British English), (van Holsteijn, 1993) (Dutch), (Balestri et al., 1993) (Italian), (Lee and Oh, 1996) (Korean), and (Chou et al., 1997) (Mandarin).[36]

The typical justification given for this model is that it is necessary to convert special symbols, numbers, abbreviations, and the like into ordinary words, so that the linguistic analysis module (which only knows how to deal with ordinary words), can handle them.[37] Two quotations serve to illustrate this argument:

---

by guessing the national origin of the name, and then applying a set of "language"-specific pronunciation rules. Our own implementation of this approach is due to Ken Church: a published description of a related though independently developed system can be found in (Vitale, 1991). This mode of analysis is, however, rarely used: nearly all words and names are handled by the morphology and rhyming techniques described above.

[36] As we have discussed in Section 3.8, our own American English system also follows this model.

[37] It is interesting to speculate on why this bifurcated model has been the overwhelmingly preferred one. One reason may have to do with the fact that linguists would generally think of such phenomena as digit sequences or abbreviations as being peripheral to the interests of linguistics: the domain of linguistic analysis is ordinary words of a language, and constructions that can be built out of them. Since even early TTS systems, like MITalk, were heavily influenced by linguistic methodology and theory, it is a natural extension of this idea to

> In order to convert text to speech, it is necessary to find an appropriate expression in words for such symbols as "3", "%", and "&", for abbreviations such as "Mr.", "num.", "Nov.", "M.I.T.", and conventions such as indentations for paragraphs. This text processing must be done before any further analysis to prevent an abbreviation from being treated as a word followed by an "end-of-sentence" marker, and to allow symbols with word equivalents to be replaced by strings analyzable by the lexical analysis modules. (Allen, Hunnicutt, and Klatt, 1987, page 16)

> The text preprocessing submodule handles problem[s] of non-Chinese characters. Non-Chinese characters, such as number[s] (1,234, 363-5251, ...), symbols ($1.00, 95%, ...) and foreign languages (PM, TEL, ...) must be transformed to suitable Chinese character sequence[s]. The word identification and tagging submodule then generates the tagged word sequence ...(Chou et al., 1997, page 926)

The few exceptions to this general trend seem to be Traber's work in the SVOX system for German (1993; 1995); and the recent work on the INFOVOX system reported in (Ljungqvist, Lindström, and Gustafson, 1994). In Traber's system, for example, the problem of numeral expansion is treated as an instance of morphological analysis, and the same formal mechanisms are used to render a digit sequence as to handle, say, a German noun compound. Similarly in the INFOVOX work, expansion of numbers and abbreviations is (broadly speaking), integrated with the rest of linguistic analysis.

As we have argued, the evidence (especially from languages like Russian) favors these more radical approaches over the more commonly assumed models. Assuming one wants a model of text analysis that is readily extendable to a wide variety of languages, then one needs a model that delays the decision on what to do with special symbols, abbreviations, numbers, and the like, until a serious linguistic analysis of the context has been performed.

3.9.2 Morphological analysis

Morphological analysis in TTS systems is performed in the service of other analyses, typically either word pronunciation or subsequent syntactic or prosodic analysis. Most TTS systems (for whatever language) do some amount of morphological analysis, and in some systems the morphological analysis performed

---

assume that true linguistic analysis within a TTS system deals with canonically spelled words, and anything that is not canonically spelled must be first converted into a canonical spelling before it can be subjected to (true) linguistic analysis.

Perhaps consistent with this view of "text normalization" as a second-class citizen, it is hard to find extensive discussions of this topic in the literature, presumably because it is not viewed as being of much interest. The chapter on "text normalization" in (Allen, Hunnicutt, and Klatt, 1987), for example is very short. The paper by van Holsteijn (1993) seems to be among the longest discussions specifically devoted to this topic.

is quite significant. For example, recent work that includes significant amounts of morphological analysis (primarily) in the service of word pronunciation includes (Heemskerk and van Heuven, 1993; Nunn and van Heuven, 1993; Konst and Boves, 1994) (Dutch), (Traber, 1993; Meyer et al., 1993; Belhoula, 1993; Traber, 1995) (German), (Lee and Oh, 1996) (Korean) and (Ferri, Pierucci, and Sanzone, 1997) (Italian). Most morphological analysis systems are explicitly finite-state or at least, in most cases where context-free rewrite formalism is used, equivalent to finite-state. An interesting exception is the work of Heemskerk (1993), which uses probabilistic context-free grammars. As in our own work on probabilistic morphology in the context of Mandarin word segmentation (Section 3.6.6), probabilities are used in order to assign relative likelihoods to different morphological analyses.

3.9.3 Word pronunciation

The traditional approach to the problem of word pronunciation is to use a linguistically-informed hand-crafted analysis algorithm. Depending upon the language, the algorithm could be as simple as a set of "letter-to-sound" rules; or it could involve a morphological analyzer based on a dictionary that lists pronunciations for morphological stems. Some systems, like MITalk, contain some combination of both of these approaches, and in general, "rule-based" systems of this kind have remained a popular method of computing word pronunciations. Recent work on rule-based approaches includes (Williams, 1987) (English), (Belhoula, 1993) (German), (Konst and Boves, 1994) (Dutch), (Lee and Oh, 1996) (Korean), (Ferri, Pierucci, and Sanzone, 1997) (Italian), and (Williams, 1994) (Welsh). The reason for the popularity of such approaches must be, at least in part, because they work. That is, once one has gone through the labor of constructing and debugging the databases, and once one has verified that one's linguistic assumptions are generalizable, such systems typically perform quite well on the problem of pronouncing words in unrestricted text.

Nevertheless, the considerable labor involved in building rule-based systems for word pronunciation — especially for languages like English, which have a complex spelling-to-phonology correspondence — has prompted many to propose various self-organizing approaches. The most famous of these is undoubtedly NETtalk (Sejnowski and Rosenberg, 1987), and several researchers have followed in this vein; see, e.g., (McCulloch, Bedworth, and Bridle, 1987; Adamson and Damper, 1996) for just a couple of examples. The disturbing property of most of these systems is that they typically report correct scores in the mid to high 80% range *by phoneme*, which of course translates into a much poorer score by word. Compared to the best rule-based systems (which typically perform in the mid 90% range, *by word* — (Coker, Church, and Liberman, 1990)) this performance is not particularly impressive. Other self-organizing methods include the *stochastic phonographic transduction* work (Luk and Damper, 1993; Luk and Damper, 1996) (English) that we critiqued in Section 3.7; analogical methods such as (Dedina and Nusbaum, 1996; Golding, 1991) (English); and tree-based methods such as (Andersen et al., 1996; Daelemans and van den Bosch, 1997) (Dutch). The Daelemans and van den Bosch system is particularly interesting since it reports a performance (for Dutch word pronunciation) that is not only better than a connectionist approach, but one that is also better than the rule-based approach discussed in (Heemskerk and van Heuven, 1993; Nunn and van Heuven, 1993) in the context of the Spraakmaker project (van Heuven and Pols, 1993).[38]

As also discussed in Section 3.7, self-organizing methods for word pronunciation start with the assumption that the information necessary to pronounce a word can be found entirely in the letter string composing that word. Put in another way, such methods view standard orthography as a kind of glorified phonetic transcription. The problem is, that as we have seen, in many languages correct pronunciation of words may depend upon lexical features that have no direct manifestation in the spelling of the word. Now, as Daelemans and van den Bosch note in their discussion, it would be perfectly possible to include such lexical features (presumably derived via morphological analysis) in the training data for a self-organizing method, and the learning algorithm would hopefully deduce the appropriate association between these features and their phonetic effects.[39] The problem, as we have argued elsewhere, is that deriving these lexical features (e.g. building a morphological analyzer for Russian), can be a major effort, whereas the prediction of pronunciations once we know those features (again, consider the case of Russian), can be straightforwardly handled with a small set of hand-constructed rules. Under that scenario, a self-organizing technique would not afford much of a savings in effort. Add to this the fact that such techniques depend upon having a large database that gives the correspondence between the spelling and the pronunciation of words, and that such databases are often lacking for languages other than a few of the more intensively studied Western European ones. For example, we know of no database that gives orthography/phonology mappings for Russian *inflected* words: there are pronunciation dictionaries, but these almost exclusively list base forms. Yet such a database for inflected words would be critical for training a self-organizing method for the pronunciation of unrestricted Russian text.

We do not mean to be overly negative here: self-organizing techniques are a useful line of inquiry, and there is no question that statistical inferable models have an important function in linguistic analysis of all kinds, including word pronunciation models. But the apparent limited success of methods applied to English, and a few other languages where extensive training databases exist, should not mislead one into believing that future TTS work on new languages can eschew handwork in favor of totally self-organizing methods.

---

[38] Needless to say, it is not entirely clear how to interpret that result: could it be the case that the rule-based system in Spraakmaker simply needs further work?

[39] Interestingly, as Luk and Damper report (Luk and Damper, 1993), their stochastic phonographic transduction method can actually benefit from knowing something about the morphological structure of the words.

3.9.4 Syntactic analysis and prosodic phrasing

Syntactic analysis in TTS systems is typically done in the service of prosodic phrasing: even though it is widely understood that prosodic phrases are not in general isomorphic to syntactic phrases, there is nonetheless clearly a relationship.

The syntactic analysis can range from simple function-word-based 'parsing' of the form proposed, for example, by O'Shaughnessy (1989); to full-blown parsing of the type used in (Bachenko and Fitzpatrick, 1990). Yet another approach more similar to the O'Shaughnessy approach than to the Bachenko-Fitzpatrick approach is the use of local contextual windows, filled with part of speech or other lexical information, to determine the likelihood of a break being placed at a certain point (Wang and Hirschberg, 1992). O'Shaughnessy-like approaches include (Quené and Kager, 1992; Quené and Kager, 1993) (Dutch), (Horne and Filipsson, 1994) (Swedish), (Ferri, Pierucci, and Sanzone, 1997) (Italian), and (Magata, Hamagami, and Komura, 1996) (Japanese). The Magata et al. paper argues that good results can be achieved by using the final (function) word in each *bunsetsu* (roughly "clitic group") as a predictor of the strength of the boundary following that bunsetsu.

Systems that use more or less full syntactic parsing include the work of Meyer et al. (1993) and Traber (1993; 1995) on German, and the work for Japanese reported in (Fujio, Sagisaka, and Higuchi, 1995). Fujio et al.'s work uses a mixed strategy in that the syntactic parsing is done using a stochastic context-free grammar (SCFG) the parameters of which are estimated from a parsed corpus. The SCFG parameters are then fed into a feed-forward neural network, which performs the actual prediction of phrase boundary location.[40]

Approaches similar in spirit to Wang and Hirschberg's method include (Sanders and Taylor, 1995) (English) and (Karn, 1996) (Spanish). Sanders and Taylor's method, for example, compares five different statistical methods for relating the part-of-speech window around the potential boundary to the likelihood that there is actually a boundary at that location.

3.9.5 Accentuation

Compared with the large amount of work on prosodic phrasing, there are relatively few published descriptions on modeling accentuation. Most work in this area tends to be at a fairly high level, assigning accents to words on the basis of their lexical class: apart from the work by Hirschberg (Hirschberg, 1993) for English, work that uses lexical class-based accentuation includes (Quené and Kager, 1992) (for Dutch), and (Meyer et al., 1993; Traber, 1993; Traber, 1995) (for German). In addition to lexical class information, pragmatic or semantic information has also been used to model discourse effects on accentuation, such

[40]Another recent piece of work on syntactic parsing for TTS is reported by Heggtveit (1996) (Norwegian), who introduces an algorithm for probabilistic generalized LR parsing. The initial application is to part-of-speech tagging, but is is clear that it can also be applied to phrasing prediction.

as the effect of information status of particular words (Hirschberg, 1993; Meyer et al., 1993).

In many cases lexical class information, or some combination of lexical class information and higher-level discourse factors, seem to be sufficient for producing decent models of lexical prominence. But there are some thorny areas that are not so amenable to such broad models, the most notable of these being the case of premodified nominals in English (see Section 4.1.2). There has been significantly less published work in this area. Apart from the Bell Labs work (Sproat, 1994), the other main strand of work is that of Monaghan (1990). It is curious that less attention has been devoted to this problem, given the frequency of premodified nominal constructions in English, and the difficulty of predicting appropriate accentuation.[41] Of course, the factors determining accentuation in English premodified nominal constructions are very complex, and it is certainly the case that one must invest a great deal of effort to show even a minor improvement in performance; it may be in part for this reason that this has not been an overly popular area of research.

3.9.6 Further issues

There are of course other pieces of work that relate to text analysis that do not easily fit into the broad categories described here. One such piece of work which is somewhat unusual, but which is nonetheless of both practical and scientific interest is the system for the description of kanji reported in (Ooyama, Asano, and Matsuoka, 1996). In a language like English, when one is trying to describe an unknown word to a listener — for example, a personal name — one can resort to spelling that word letter by letter. In languages like Chinese or Japanese, which do not use an alphabetic writing system, the direct equivalent of spelling, namely reading the text character by character, is not very useful: in both languages, there are many sets of homophones or near-homophones: for example the Chinese characters 市 'city, market', 是 'be' and 事 'matter, affair', are all pronounced the same in Mandarin (*shì*), so simply reading them in their canonical form would not be informative. What Chinese or Japanese speakers do in such cases is to *describe* the character, either in terms of its shape, or in terms of a common and unambiguous word in which it occurs. Thus, one might describe 市 as being the 市 in 城市 *chéngshì* 'city'. Or one might describe the character 琳 *lín* 'fine jade', as being constructed out of the 林 *lín* 'woods' that occurs in the word 森林 *sēnlín* 'forest', along with the 'jade' radical 王 *yù* on the side. Modeling this in a TTS system is interesting from the point of view of modeling what humans do in such instances; and it is also

[41] Lest one assume that nominal accent assignment is perceptually of low importance for casual users of TTS, we offer the following anecdotal evidence that this is not the case. During an interview with *PC Week* at the April 1997 *Java One* trade show in San Francisco, John Holmgren of Lucent Technologies allowed the reporters to type their own sentences at the Bell Labs American English TTS system. The *only* error that the reporters explicitly pointed out was the incorrect accentuation of the show's name *Java One*. At the time of writing, the interview could be found at `http://www.pcweek.com/radio/0407/nmlucent.ram`.

of practical importance in TTS since TTS systems frequently include a "spell mode" that can be used in certain applications to spell out potentially difficult-to-understand portions of the text. Ooyama and his colleagues investigate the analogous problem for Japanese kanji. Their system uses information about the form and lexical usage of kanji to produce verbal descriptions of them.

Chapter 4

Further Issues in Text Analysis

Richard Sproat

In this chapter we describe some other issues in text analysis which have not been treated extensively, or at all, in our work on the synthesis of non-English languages. These topics are (lexical) accentuation, homograph disambiguation, and prosodic phrase boundary prediction. In each case our previous work on English is extensive, has been previously reported widely in the literature, and will be only sketched here. And in each case we describe how the different kinds of models that we have applied in the case of our American English synthesis system can be fit into the gentex model described in the previous chapter. In some cases the proposed method has been implemented, and is merely awaiting more extensive development: in other cases we merely present a proposal for how the implementation should be done, and leave it for future work to develop the ideas more extensively.

4.1 Accentuation

Pitch accenting decisions in many languages depend upon information about a word's lexical properties, as well as its usage in a particular (syntactic, semantic or pragmatic) context: the problem is therefore much akin to the problem of sense disambiguation, which we will treat in some depth in Section 4.2. To date we have done little serious work on accenting in languages other than English, and so this section will mostly review our work on pitch accent prediction for English; further and more detailed discussion of the topics treated herein can be found in (Sproat, Hirschberg, and Yarowsky, 1992; Hirschberg, 1993; Sproat,

Class	Default Accent	Examples
open	accented	most nouns, verbs, adjectives ...
closed accented	accented	*every*, *above*, *beneath* ...
closed deaccented	deaccented	*ago*, *there*, *so* ...
closed cliticized	cliticized	*the*, *a*, *of* ...

Table 4.1: Description of the four lexical accent classes for the American English TTS system.

1994). As with other aspects of our American English system (Section 3.8), the computational implementation of accenting models is rather different than the approach necessitated by gentex. We will therefore conclude this section by describing how some of these models might fit into the gentex framework.

4.1.1 General models for accentuation

The models of accentuation used in the Bell Labs American English TTS system are all corpus-based in that they are derived either fully automatically by decision trees trained on labeled and tagged data; or else they are based on human observations of corpora of read speech. The distinction between automatically-derived and hand-derived models is somewhat orthogonal to the discussion here, and so we will not dwell upon this: the interested reader is referred to (Hirschberg, 1993) for details. The main distinction of interest is between context-independent lexical-feature-based accentuation; and context-dependent accentuation that depends either upon the position of the word in question in the sentence, or else upon previous discourse context.

Context-independent accentuation depends upon a four-way classification into the categories, *closed-cliticized*, *closed-deaccented*, *closed-accented*, and *open*. Examples are given in Table 4.1. These lexical properties are computed by table lookup. Also handled by table lookup are *verb-particle* constructions such as *give in*, *look up*, *give out*. The particles in such constructions can usually also function as prepositions, and as such they would typically be deaccented or cliticized. When they function as particles, however, they would normally be fully accented.

Contextual models include information on coarse position within a sentence, whether or not the open-class word in question occurs within a premodified nominal, and the occurrence or non-occurrence of related words in the previous discourse context. Putting to one side the treatment of premodified nominals, which is discussed in Section 4.1.2, the specific context features computed are as follows:

- *Preposed adverbials* and *fronted PP's*, which often have a contrastive function, are identified by their previously computed grammatical part-of-speech sequences and by their position in the sentence.

- Discourse markers such as *well* or *now*, which provide cues to the structure of the discourse, are detected from their surface position: such *cue phrases* typically occur sentence initially, and are often implemented with a prominent pitch accented; see (Litman and Hirschberg, 1990).
- Lemmata (uninflected forms) of content words are stored in the current frame of a discourse stack, and such stored items are marked as *given*. Subsequent occurrence of given items are potentially deaccentable. *New* (non-given) items, in contrast, may have a contrastive function and are thus typically accented, even if that overrides the citation-form accenting predicted by the premodified nominal accent model. Cue phrases are associated with either a push or pop of the current focus frame (see (Grosz and Sidner, 1986)). Depending upon how the TTS system is invoked, either paragraph or sentence boundaries will cause the entire stack to be cleared.

The above-described contextual and lexical features are organized into a simple decision tree, whose behavior can be described as follows:[1]

1. Closed-cliticized and closed-deaccented items are accented according to their class.
2. Contrastive items — cue phrases, preposed adverbials, fronted PP's, and *new* content words within premodified nominals — are assigned emphatic accent.
3. Closed-class accented items are accented.
4. *Given* items are deaccented.
5. Elements within premodified nominals that are not contrastive or given are assigned their citation-form accent.
6. All other items are accented.

Note that if premodified-nominal internal citation-form accents are overridden, this will in turn trigger a reassessment of other accenting decisions within the

[1]The model described here is the one that is currently implemented within the Bell Labs American English TTS system. Now, models of accentuation are crucially intertwined with the model of intonation that they support. Thus the particular model of accentuation here is intended to support a Pierrehumbert-Liberman-style (Liberman and Pierrehumbert, 1984) model of intonation which depends upon a rather dense distribution of accents. In contrast, the new model of intonation which we describe in Chapter 6 depends upon rather sparser accent assignments. Thus the current accenting algorithm is not entirely compatible with this new intonation model, since the resulting synthesized speech tends to sound "over-accented". An area of future work will be the development of a model of lexical pitch accent assignment that is more compatible with this newer intonation model.

nominal. Thus in a nominal where the accent is normally on the righthand member, if the lefthand member is computed to be contrastive, then not only will this lefthand member be emphatically accented, but the righthand member will also be deaccented. Thus a contrastive use of the word *cherry* in *cherry píe*, will result in *chérry pie* with *cherry* accented, and *pie* deaccented.

Performance measures for the accenting system vary depending upon the style of speech being modeled. For citation-format utterances (e.g. sentences in isolation), correct prediction rates as high as 98% are possible; for more heavily stylistic speech, such as radio newsreader speech, rates are around 85%. See (Hirschberg, 1993) for further details.

4.1.2 Accentuation in premodified nominals

We use the term *premodified nominal* to refer to a sequence of one or more modifiers (excluding articles, possessive pronouns, and other specifiers) followed by a head noun, where *noun* includes proper names, and *modifier* may also include names: thus *dog house*, *city hall*, *green ball*, *Washington townhouse* and *Bill Clinton* would all be considered premodified nominals.[2] Premodified nominals in English are tricky to handle in TTS systems. Although there are general defaults — noun-noun sequences are generally accented on the left, whereas adjective-noun sequences are usually accented on the right — there are many classes of exceptions, and lexical accent can in principle be assigned to any member.[3] Nonetheless, speakers usually have fairly strong intuitions about which member should be stressed in a "discourse neutral" context (though there is widespread dialectal and even idiolectal variation). So, in the absence of pragmatic reasons to do otherwise, English speakers would generally accent the first element (and deaccent the second) in *cótton factory*; whereas the second element would generally be accented in *cotton clóthing*. Such accenting decisions are based partly on the semantic relationship between the words of the nominal, and partly on apparently arbitrary lexical factors. (Fudge, 1984; Liberman and Sproat, 1992). The nominal accenting program — NP — which is described fully in (Sproat, 1994) uses a mixture of lexical listing (e.g. *littleneck clám*, and *Whíte House* are both listed with their citation-form accents), and a database of quasi-semantic rules to predict accent placement. Thus, a combination of *ROOM* + *HOUSEWARE* is predicted to have righthand accent, as in the case of *kitchen phóne*;[4] this presumes a lexicon of semantic categories

[2]Unfortunately there is no really appropriate conventional linguistic term denoting this class of entities. In previous work we have used the terms *noun compound*, and *complex nominal*, but neither of these is really appropriate.

[3]The situation in English with accenting of premodified nominals appears to be exceptionally complicated: most languages for which the issue of phrasal accenting has been studied in detail do not exhibit this complexity. Nonetheless, English is not entirely unique: the prediction of accentual phrasing in Japanese within long nominals has similar properties to English nominal accenting (see Section 3.6.7).

[4]We are focusing here on nominals of length two, that is, nominals with one modifier and one head noun; these comprise about 90% of the nominals that one encounters in English text. Nevertheless, the nominal analyzer in the American English TTS system deals

associated with each word, and a partial lexicon of this sort is indeed part of the system. Still, while this approach fares well enough with nominals that it can analyze, there are a large residue of cases which it cannot analyze, and these revert to the default accenting pattern, which for noun-noun sequences is only correct about 75% of the time. The rule-based method was therefore supplemented with a simple statistical model trained on a hand-labeled corpus of 7831 noun compound types picked randomly from the 1990 Associated Press newswire, and tagged for lefthand or righthand accent by one of the authors. Of these, 88% (6891) compounds were used as a training set and the rest were set aside as the test set. For each compound in the training set and for each of the two lemmata in the compound we collect the set of broad topical categories associated with the lemma in *Roget's Thesaurus* (Roget, 1977). Then, for each element of the cross-product of these Roget categories we tally the accent (left or right) tagged for the compound. Intuitively, the cross-product of the categories gives a crude encoding of the set of possible semantic relationships between the two words. We also create entries for the first lemma (modifier) *qua* modifier, and the second lemma (head) *qua* head, and tally the accent for those entries. For a compound in the test set, we sum the accent pattern evidence accumulated for each of the elements in the cross product of the Roget categories for the compound, and the head/modifier entries for the two lemmata, selecting the accenting that wins. As an example, consider *cloth díapers*, which occurs in the test set and not in the training set. In the training set, the lemma *diaper* never occurs in the righthand position; the lemma *cloth* occurs twice in the lefthand position, once in a compound with lefthand accent (*clóth merchant*) and once in a compound with righthand accent (*cloth bánners*). This amounts to no evidence, so one would by default assign lefthand accent to *cloth diapers*; however this compound also matches the category sequence *MATERIALS+CLOTHING*, for which there is a score of 13 favoring righthand accent.[5]

For the held out data we compared the predictions of the model with those of six native speakers of English. The inter-judge similarity measure (defined as the mean of the *precision* and *recall* between any two judges) was 0.91; the agreement of judges with a baseline system that uniformly assigns lefthand accent was 0.76; the agreement with the statistical algorithm is 0.85, meaning that the algorithm covers about 67% of the difference in performance between humans and the baseline.[6] Again, the reader is referred to (Sproat, 1994) for

with longer nominals, attempting to parse them into appropriate binary trees and applying prosodically motivated modifications such as the rhythm rule (Liberman and Prince, 1977) to the predicted accent structure; see Section 4.1.3.

[5] In practice it was found that a large number of the Roget cross-categorial combinations were not useful and often in fact detrimental: this is because Roget categories are usually topical rather than taxonomic classifications. However, 19 fairly taxonomic combinations were retained, because of their substantial predictive utility (Sproat, 1994, page 89).

[6] If one counts by token rather than by type, then the judges agree at 0.93, and the baseline/judge and statistical-algorithm/judge agreement scores are respectively 0.74 and 0.89. Interestingly, only 174 (15%) of the cases were actually handled by the hand-built

further details of the nominal accent system and its evaluation.

4.1.3 Models of accentuation within gentex

To date none of our non-English systems has an accentuation model as complete as that for our American English system. This section is therefore not a report on how one would build an accentuation module within the gentex model, but rather a sketch of how such a module could be built.

Accenting decisions that are based purely upon lexical class are clearly straightforward to handle using WFST's: one merely has to provide a set of accenting rules that are sensitive to the lexical classes in question. Similarly, one can straightforwardly handle positional information, such as that relevant for the accentuation of cue phrases, preposed adverbials, and fronted PP's: here one needs rules that are sensitive to lexical properties, as well as to position within the phrase.

More difficult to handle would be premodified nominals. The treatment of accentuation in nominals described above depends upon a large amount of lexical information, including information on the relationships between words. A naive implementation of this information in terms of WFST's would result in impractically large machines. However, it should be possible to factor the problem in a way similar to that discussed for homograph disambiguation in Section 4.2. For example, one could implement a finite-state tagger that tags each element of a nominal according to its "semantic" class: so *kitchen* would be tagged (among other things) as ROOM, and *towel* would be tagged (among other things) as HOUSEWARE. A very small finite-state grammar stated over these semantic classes could then be used to determine the preferred accenting pattern.[7]

This would be satisfactory for binary nominals, where one is merely interested in the relationship between two words, but it is not enough for larger nominals, where one is also interested in constructing an analysis, usually in the form of a tree, of the input. In the NP system, nominals longer were analyzed using a context-free grammar, and parsed with a CKY (chart) parser. A nominal such as *city hall parking lot*, would be analyzed as shown in Figure 4.1. Noun nodes would be constructed over the (lexically listed) items *city hall* and *parking lot*. Then a rule that sanctions the combination of two (possibly complex) noun compounds into a larger compound would allow one to construct another noun node over the entire construction. Concomitant with this syntactic analysis is the accentual analysis of the input. In isolation the two main components of *city hall parking lot* would be accented thus: *city háll* and *párking lot.* In combination, the first accent shifts back following the so-called rhythm

---

rules of NP: the 964 remaining cases were assigned accent on the basis of the trained models described above, suggesting that corpus-trainable models may be more extensible than hand-built systems in this domain.

[7] See also Section 3.6.7 for some discussion of the implementation of accentual phrase boundary placement in Japanese, which is similar in many respects (including complexity) to English nominal accenting.

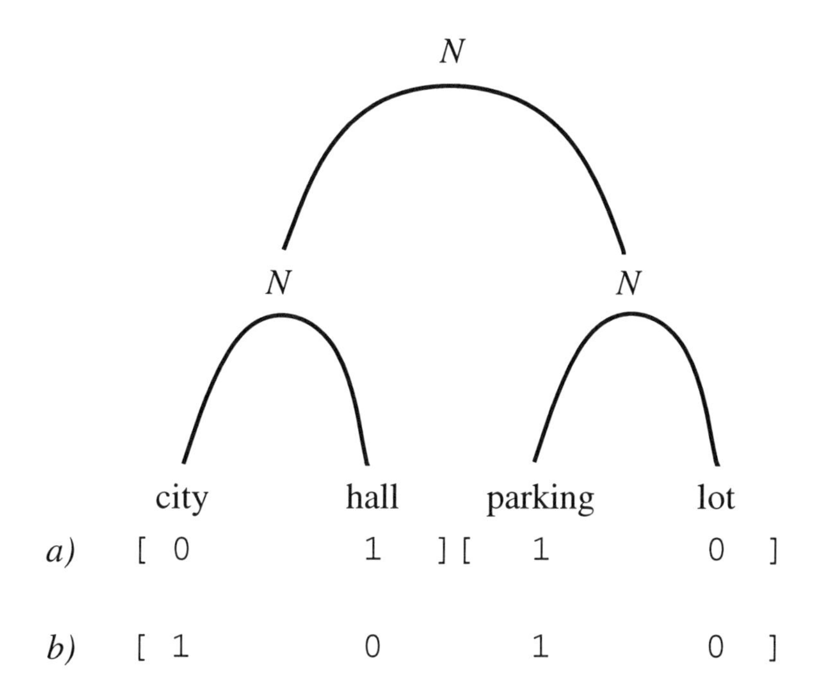

Figure 4.1: The analysis of *city hall parking lot* in NP. The 1's and 0's denote accent patterns, with 1 denoting the more prominent, and 0 the less prominent member: (a) represents the accent of the independent elements *city hall* and *parking lot*; (b) represents the accentuation of the entire construction, after the application of the rhythm rule.

rule (see, e.g., (Liberman and Prince, 1977)): *cíty hall párking lot*. Within NP, the accent shift is handled dynamically as the nominal is constructed: *city hall* and *parking lot* would be constructed with their default accent pattern 01 and 10 — where 1 represents the most prominent word. By rule, a combination of the form 01 + 10 will be reanalyzed as 10 + 10. One approach to implementing this kind of model within the gentex framework would be to adopt a proposal along the lines discussed in (Roche, 1997) for implementing context-free grammars using successive compositions of FST's. For the example at hand, one could imagine a transducer corresponding to the lexicon in the NP system that simply tags and brackets lexical entries such as *city hall* and *parking lot*. The output of such a transducer might look as follows, where the lexically prominent words are here starred, and the lexically non-prominent words are preceded by a minus sign: $[_{\mathrm{N}}$ −*city* **hall* $_{\mathrm{N}}]$ $[_{\mathrm{N}}$ **parking* −*lot* $_{\mathrm{N}}]$. A subsequent transducer could be used to group the two nouns into a larger unit, and reverse the accenting in *city hall* given the context to the right. Needless to say, the details of such a proposal remain to be worked out.

Accenting decisions that are sensitive to discourse structure are the most tricky to implement in a purely finite-state approach. To see this, consider the deaccenting of *given* words. Clearly it is not possible to construct a priori a finite-state device that remembers every word or lemma that has been seen in the preceding discourse: such a machine would be of astronomical size. On the other hand, it is clearly possible to implement something akin to the discourse stack that forms part of the accenting module of the American English TTS system, and then implement the deaccenting of given items using finite-state transducers which constructed "on the fly". In the finite-state toolkit on which gentex is based, there is, underlying the notion of a finite-state machine, the more general notion of a *generalized* state machine (GSM). For a GSM it is not necessary to know beforehand which arcs leave a given state; rather one can construct just the arcs one needs on the fly as one is using the machine, for example in a composition with another machine. Thus one could envision dynamically building states that "remember" that a particular word has been seen in the prior discourse.

4.2 Homograph Disambiguation

We return in this section to the problem of lexical disambiguation, a topic that was touched upon to some extent in the previous chapter. More specifically, we will discuss a solution to the particular problem of homograph disambiguation, and its implementation within the computational framework introduced in the last chapter. Work on homograph disambiguation for the Bell Labs TTS system has been most extensively carried out for English, and so we will discuss the general methodology with examples from English. We will end this section with a discussion of some preliminary work on Mandarin and Russian.

In addition to the rule-based methods for homograph disambiguation such as those discussed in the previous chapter, we use a *decision-list* based method due to Yarowsky, and discussed extensively in his published work; see, for example, (Yarowsky, 1994b; Yarowsky, 1994a; Yarowsky, 1996; Yarowsky, 1997).[8] Before we turn to the description of homograph disambiguation, we would like to start by situating this topic within the broader topic of sense disambiguation.

For written language, sense ambiguities can be classified into two broad categories, namely "natural" ambiguities and "artificial" ambiguities. Natural ambiguities consist of words that are polysemous, an example being the word *sentence*; see (54a). Examples like *sentence* which do not involve a difference of pronunciation, are of most crucial importance in applications where a mistake in detecting the intended sense of a word would have potentially disastrous or amusing consequences. Machine translation is one such application: in an English-to-French translation system, for example, we would want to translate

[8] David Yarowsky also implemented the homograph disambiguation portion of our American English Text-to-Speech system.

the word *sentence* in 1 as *peine*, but the instance in 2 as *phrase*. While such cases clearly have potential importance for TTS too,[9] of more immediate importance are the subcategory of polysemous words that are homographs, such as *bass* in the examples in (54b).

(54) (a) 1. She handed down a harsh **sentence**.
2. This **sentence** is ungrammatical.

(b) 1. He plays **bass**. (/beis/)
2. This lake contains a lot of **bass**. (/bæs/)

Artificial polysemy is created when words which would under most conditions be orthographically distinct, are rendered identical by one or another means. So, *polish* and *Polish* are normally distinguished by capitalization, but in sentence-initial position or in case-free contexts (e.g. in headers where all of the text may be capitalized), the distinction is lost. Similarly, the French words *côte* 'coast' and *côté* 'side' are typically distinguished by accent diacritics. But in cases where all letters are capitalized, or in some 7-bit renderings of French text (see Appendix A), the distinction is once again lost. Artificially polysemous examples are also introduced in the case of spelling or typing errors. Spelling correctors start with the assumption that the correct word is within a certain *edit distance* — measured in terms of insertions, deletions and substitutions — from the correct word. Of course, it is usually the case that there is more than one possible actual word within a given edit distance of the misspelled word. For example, the typo *vrain* could in principle be either *brain* (1 substitution), or *vain* or *rain* (1 deletion), and one would typically have to consider the context in which the word occurs in order to decide which correction is appropriate.[10]

What kind of evidence is useful for disambiguating polysemous words? As one might expect, the answer depends upon the particular example being considered. In some cases, a polysemous word may be ambiguous between two grammatical categories; for example, the word *project*, which may be a noun (*prójject*) or a verb (*projéct*). In such cases, local grammatical information (e.g., grammatical part-of-speech sequences) may be useful in disambiguation. At the other extreme are cases where the polysemy is within a single grammatical category, but there is nonetheless a clear semantic distinction between the senses: *bass* (fish or musical range) offers an obvious example from English. In such cases, grammatical information for the most part gives little leverage: rather lexical information on words in the immediate or wider context is generally much more useful. This suggests the need for a model which allows for both local and "wide" context evidence, something which the decision-list model provides.

[9] For example, the accenting algorithm discussed in Section 4.1 should not generally deaccent an instance of the 'phrase' sense of *sentence*, given a previous occurrence of the 'punishment' sense.

[10] See (Golding, 1995) for an example of context modeling for spelling correction, and see (Yarowsky, 1996) for some discussion.

Decision List for *lead*		
Log Likelihood	Evidence	Pronunciation
11.40	*follow/V* + lead	⇒ lid
11.20	*zinc* ↔ lead	⇒ lɛd
11.10	lead *level(s)*	⇒ lɛd
10.66	*of* lead *in*	⇒ lɛd
10.59	*the* lead *in*	⇒ lid
10.51	lead *role*	⇒ lid
10.35	*copper* ↔ lead	⇒ lɛd
10.28	lead *time*	⇒ lid
10.16	lead *poisoning*	⇒ lɛd
⋮	⋮	⋮

Table 4.2: Partial decision-list for the homograph *lead*.

The decision-list approach to homograph disambiguation is best understood by example. Consider the partial list in Table 4.2, which represents a decision list for the word *lead*, one of the homographs that is modeled in our American English TTS system. The second and third columns of this table represent, respectively, the particular piece of evidence, and the sense — 'foremost' or 'metal' — predicted when that evidence occurs. So, in the V+N construction *follow ... lead*, for example, the most likely sense is 'foremost'. On the other hand, if the word *zinc* occurs somewhere in the context, then the more likely interpretation is the metal reading. The list is ordered by log likelihood — the numbers in the left column — defined as follows, where *Collocation* denotes the particular piece of contextual evidence:

$$Abs(Log(\frac{Pr(Pron_1|Collocation)}{Pr(Pron_2|Collocation)})) \tag{4.1}$$

The largest log likelihood values will thus be associated with those pieces of contextual evidence that are most strongly biased towards one of the senses.

A decision list such as the one in Table 4.2 is used in the order given: that is, when attempting to disambiguate the word *lead* one would first compute whether the word was likely to be in a Verb-Object relation with *follow*; next one would look for the word *zinc* anywhere in the context; next one would see if a form of the noun *level* occurs to the right; and so forth. If none of the contexts match, then an empirically determined default is chosen.

Decision lists are constructed automatically from labeled data as follows:[11]

[11] See the various cited papers by Yarowsky for full descriptions of the algorithm; for example see (Yarowsky, 1996, pages 83–91). Note that the method described here, and the one used

1. Construct a training corpus containing examples of the homograph to be disambiguated, tagged with the correct pronunciation in each case.

2. Count instances of various *collocations*, where by collocation we mean simply a juxtaposition of words or their properties; see (Yarowsky, 1996, Chapter 4). To clarify further, the algorithm looks for certain features of the context, including:

   (a) The word, lemma, or part of speech one place to the right of the target;

   (b) The word, lemma, or part of speech one place to the left of the target;

   (c) Words or lemmata within a pre-defined $\pm k$ words of the target;

   (d) The verbal head word/lemma to the left of the target. (This is clearly an English-specific feature.)

   For example, in a given training corpus one might notice that a form of the word *level* occurs 192 times in the position one to the right of the target in case it is /lɛd/, and 0 times in case it is /lid/; or that the word *narrow* occurs 61 times one to the left of the target if it is /lid/, and 1 time if it is /lɛd/; or, finally, that the word *zinc* occurs 212 times somewhere within $\pm k$ words of the target in case it is /lɛd/, and 0 times if it is /lid/.

3. Compute the log likelihoods as previously defined; see Yarowsky (Yarowsky, 1996) for details on how the probabilities are estimated from raw counts, and for details on smoothing in the case of undersampled events.

4. Sort the list of collocations by log likelihoods into a preliminary decision list.

5. Further smooth and prune the preliminary list to remove rules that are completely subsumed under higher ranking rules, and to provide better estimates for probabilities of rules that are partially subsumed by higher ranking rules.

As described above, the decision list is used on new data by looking for the first rule in order of decreasing log likelihood that matches the context, and using that rule alone for prediction. In cases where more than one rule matches, one could conceive of other ways of combining the multiple evidence. However, perhaps surprisingly, simply using the first rule rarely performs worse than combining the evidence from all matching rules, and often outperforms a more complicated approach.

Table 4.3 (see also (Yarowsky, 1996, Table 5.9)) lists some results for some common English homographs. The important columns are the last two, which

---

for homograph disambiguation in the Bell Labs TTS system is a *supervised* learning algorithm, since it is trained on hand-labeled data. Yarowsky also discusses various unsupervised techniques for sense disambiguation.

Word	Pron1	Pron2	Sample Size	Prior	Performance
lives	laivz	lɪvz	33186	.69	.98
wound	waʊnd	wund	4483	.55	.98
Nice	nais	nis	573	.56	.94
Begin	bɪgín	béigɪn	1143	.75	.97
Chi	tʃi	kai	1288	.53	.98
Colon	koʊlóʊn	kóʊlən	1984	.69	.98
lead (N)	lid	lɛd	12165	.66	.98
tear (N)	tɛɚ	tɪɚ	2271	.88	.97
axes (N)	æksiz	æksɪz	1344	.72	.96
IV	ai vi	fɔrθ	1442	.76	.98
Jan	džæn	jɑn	1327	.90	.98
routed	rutɪd	raʊtɪd	589	.60	.94
bass	beis	bæs	1865	.57	.99
TOTAL			63660	.67	.97

Table 4.3: Sample results for various English homographs.

list the prior probability (how often one would be right by simply guessing the most frequent reading), and the performance of the disambiguation method sketched here.

The method is applicable to TTS, of course, not only in the case of ordinary homographs, but to many other classes where the orthographic form is ambiguous between several possible readings. These include such cases as:

- Abbreviations like *St.*
- Three and four digit numbers, which can be read as dates (*983 = nine eighty three*, *1935 = nineteen thirty five*) or as ordinary numbers (*nine hundred eighty three*, *one thousand nine hundred thirty five*).
- Roman numerals, which usually have more than one possible interpretation: *IV* could be *four* or *IV* (for *intravenous*).
- Expressions of the form *number/number* could be dates or fractions.

Yarowsky (Yarowsky, 1996; Yarowsky, 1997), reports excellent results on these cases of disambiguation. Note that one could view the decision-list approach as being a more general solution to the rule-based disambiguation techniques for cases like Russian percentage expressions discussed in Section 3.6.5.

How can the decision-list framework fit into the WFST-based text-analysis model described in Chapter 3? On the face of it, there should be no particular problem in doing this since the questions that are asked about the context

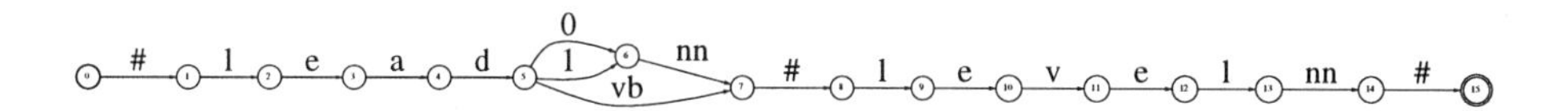

Figure 4.2: An ambiguous lattice with various possible lexical analyses of *lead level*.

would appear for the most part to be easily stateable as ordered rewrite rules: for example, one could write a pair of rules for *lead* that could be roughly stated as follows:

(55) lead $\rightarrow$ /lɛd/ / zinc Σ^* ___
lead $\rightarrow$ /lɛd/ / ___ Σ^* zinc

A moment's reflection will reveal that this approach is hopeless: even on an unrealistically small symbol alphabet of ninety symbols, the pair of rules for *zinc* in (55) compile into a transducer with about 25 states and nearly 2000 arcs. The composition of a realistic set of such rules will in general be too large to be practical. In the case of Chinese or Japanese, where, the text-analysis alphabet numbers into the tens of thousands of symbols, even compiling a single rule of this kind would be prohibitive. We therefore need to consider an alternative factorization of the problem.

Recall from Chapter 3 that prior to the application of "language models", of which homograph disambiguators would be an instance, the lexical analysis consists of a lattice of possible analyses. For the sake of the current method, we will assume that the various possible lexical renditions of a homograph are equally weighted, meaning that it will be entirely up to the disambiguation model to decide which one is correct. For example in Figure 4.2, various possible lexical analysis of the phrase *lead level* are given. For simplicity of exposition, we will dispense for the present discussion with the more richly annotated lexical representations used in the previous chapter. The lexical analyses for *lead* assumed here are:

	Current representation	Full represenation	Pronunciation
(56)	lead vb	l'{ea|i}d [V]	/lid/
	lead 0 nn	l'{ea|i}d [N]	/lid/
	lead 1 nn	l'{ea|ɛ}d [N]	/lɛd/

Given an input lattice such as that in Figure 4.2, we can then factor the problem into the following stages:

1. Find instances of a particular homograph in the input string, and tag it with a tag H0, H1 ..., depending upon which of the n (usually two) possible pronunciations such a lexical analysis would imply. We will adopt the convention of assigning H0 to the default pronunciation: in the example we are considering, H0 will be assigned to *lead vb* and *lead 0 nn*,

whereas H1 will be assigned to *lead 1 nn*. If there are no such instances then proceed no further. We will refer to this operation as the *homograph tagger*.

2. For each piece of contextual evidence, tag it for (a) the *type* of evidence it constitutes, and (b) the predicted pronunciation from the set $0, 1, \ldots$ *In addition, associated with each tag will be a cost that indicates the evidence's place in the decision list.* We will notate the different types of evidence as follows:

type of evidence	code
V	head verb
C	occurs somewhere in context
L	occurs immediately to the left
R	occurs immediately to the right
[	left part of a three-element collocation surrounding the target
]	right part of a three-element collocation surrounding the target

The numbers $0, 1, \ldots$ will be used to denote the pronunciation, so that C0, for example, will tag an element that, if it occurs somewhere in the context, will trigger the zeroeth pronunciation of the target. Note that any given word may belong to more than one category. For example, the third entry in the list could state that if w_i occurs one to the left then the zeroeth pronunciation is predicted; and the tenth entry could state that if w_i occurs one to the right the first pronunciation is predicted. In this case w_i would be tagged as both L0$< 3 >$ and R1$< 10 >$.

We will term this stage the *environmental classifier*. The only features that will be relevant for subsequent classification are the environmental tags, the homograph tags and the word boundary markers '#' (for the purpose of counting distance). All other symbols can therefore be transduced to a dummy symbol Δ.

3. Optionally change the homograph tags H0, H1, ... into the tag *ok*, when they occur in an appropriate environment: for example, H1 will optionally be tagged as *ok* if the preceding word was tagged as L1. H0 — the default — can always be optionally rewritten as *ok*, but with high cost. We will term this phase the *disambiguator*.

4. Filter untransformed H0, H1 .... This will leave a lattice of possible analyses with the tag *ok*. The lowest cost path of this lattice will be the analysis that corresponds to the disambiguation conditioned by the first element in the decision list that matches the environment.

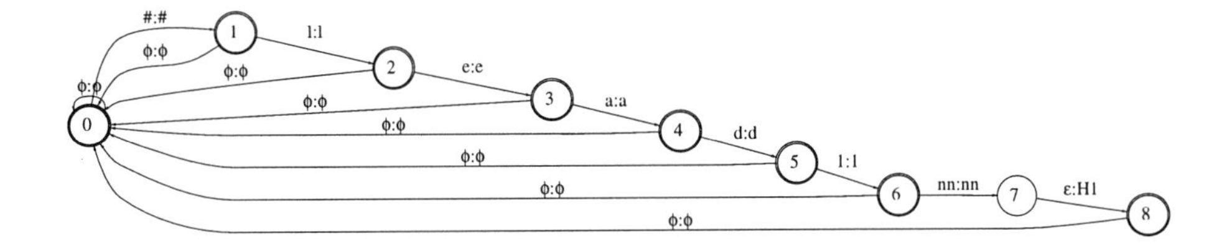

Figure 4.3: A failure function transducer for the H1 pronunciation of *lead*.

These operations can be more precisely defined in terms of transductions as follows:

1. The homograph taggers need to mark each instance of particular strings with a designated tag. One could envision doing this with a rewrite rule, but over any realistically sized alphabet this would result in a very large set of transducers. Fortunately, there is a straightforward solution in terms of failure function transducers (cf. (Mohri, 1994)). An example tagger, which inserts the tag H1 after *lead 1 nn* is shown in Figure 4.3.[12] The transducer in Figure 4.3 we will call H_1; the equivalent transducer for H0 we will call H_0.

2. The environmental classifier consists of two transducers. C_1 optionally rewrites any symbol except the word boundary or the homograph tags H0, H1 ..., as a single dummy symbol Δ, and C_2 classifies contextual evidence from the decision list according to its type, and assigns a cost equal to the position of the evidence in the list; and otherwise passes Δ, word boundary and H0, H1 ... through. A partial table of the mappings performed by C_2 based on the decision list in Table 4.2 is given in Table 4.4. Note that for the context *of___in* the costs sum to 4, which is the position of that evidence in the list (cf. Table 4.2)

3. The disambiguator D implements a set of optional rules as shown in Table 4.5.

4. The filter F simply removes all paths containing H0, H1, ... This can simply be stated as a filter of the form $\neg(\Sigma^*(\text{H0}|\text{H1})\Sigma^*)$, where the Σ

[12] The arcs labeled $\phi : \phi$ specify failure transitions, that is transitions which are taken from an arc when no other arc matching the input symbol is available at a given state. In the particular implementation of failure functions assumed in Figure 4.3, we will continue to fail *consuming no input* until we reach a state which either has an arc that matches the input symbol, or else has a self-loop on $\phi : \phi$ (i.e., state 0 in this example); in the latter case one symbol of the input is transduced to itself, and we attempt again to match the input from that state.

# follow vb #	$\rightarrow$	# Δ^{+} V0 # <1>
# zinc nn #	$\rightarrow$	# Δ^{+} C1 # <2>
# level(s?) nn #	$\rightarrow$	# Δ^{+} R1 # <3>
# of pp #	$\rightarrow$	# Δ^{+} [1 # <2>
# in pp #	$\rightarrow$	# Δ^{+} 1] # <2>
	$\vdots$	

Table 4.4: Some mappings performed by the environmental classifier transducer C_2, for the English homograph *lead*.

H0	$\rightarrow$	*ok*	/	V0 Σ^* ___
H1	$\rightarrow$	*ok*	/	C1 Σ^* ___
H1	$\rightarrow$	*ok*	/	___ Σ^* C1
H0	$\rightarrow$	*ok*	/	___ # Δ^* R0
H1	$\rightarrow$	*ok*	/	___ # Δ^* R1
H0	$\rightarrow$	*ok*	/	[0 # Δ^* ___ # Δ^* 0]
H1	$\rightarrow$	*ok*	/	[1# Δ^* ___ # Δ^* 1]
	$\vdots$			
H0	$\rightarrow$	*ok* $< 20 >$		

Table 4.5: Rules for the disambiguator D. Note that the final rule freely maps H0 to *ok* with the (arbitrary) high cost 20.

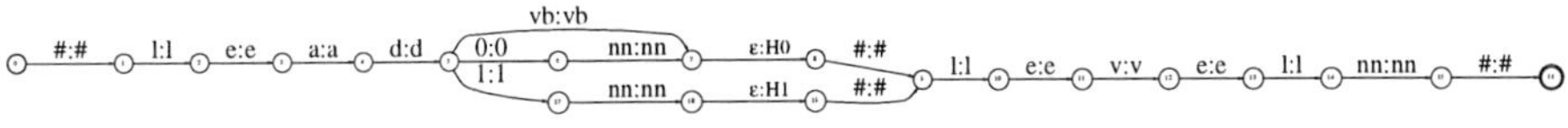

Figure 4.4: The result of $L \circ H_0 \circ H_1$.

consists of the small alphabet: H0, H1, C0, C1, L0, L1, R0, R1, [0, 0], [1, 1], Δ, #.

To understand how this system is used, consider again the input *lead level* as depicted in the lattice in Figure 4.2, which we will call L. The output of composing L with the homograph taggers H_0 and H_1 is shown in Figure 4.4. After composition with C_1 and C_2 we have the transducer in Figure 4.5. Finally, the result of the composition $L \circ H_0 \circ H_1 \circ C_1 \circ C_2 \circ D \circ F$ is shown in Figure 4.6. The correct analysis (*lead 1 nn*) can be computed by first computing the cheapest path in the transducer in Figure 4.5, then computing the left projection, and finally intersecting this back with L:[13]

$$L \cap \pi_1[\ BestPath\ [\ L \circ H_0 \circ H_1 \circ C_1 \circ C_2 \circ D \circ F\ ]\ ]$$

This will have the effect of removing all but the optimal analysis from the lattice L.

The algorithm just sketched for implementing decision lists within the gentex framework may seem overly baroque, but as was pointed out before, this factorization is justified by the large reduction in size that it achieves over any attempt to implement decision lists as a single transducer. To quantify this claim somewhat consider Table 4.6, which gives the sizes (in terms of the numbers of states and arcs) of H_0, H_1, C_1, C_2, D and F; vs. the size of $H_0 \circ H_1 \circ C_1 \circ C_2 \circ D \circ F$. The sizes are based on an unrealistically small alphabet consisting of 47 symbols: for a larger alphabet the effects would be even more striking, particularly because the sizes of some of the factorized transducers — H_0, H_1, C_2, D, F — would not change with a larger alphabet. One other point that may not immediately be obvious is that whereas $H_0 \circ H_1 \circ C_1 \circ C_2 \circ D \circ F$ is particular to a given homograph, in the factorized representation some of the transducers can be shared among all homographs, and thus only need to be represented once. The shareable transducers are C_1, D and F.

The model just described has been implemented and forms part of gentex. To date it has only been applied to a few limited examples in Mandarin and

[13] In order for this to work properly we actually need to remove the costs from the best analysis of $L \circ H_0 \circ H_1 \circ C_1 \circ C_2 \circ D \circ F$: the costs assigned by this transducer bear no relation to the costs in the lexical analysis L.

Figure 4.5: The result of $L \circ H_0 \circ H_1 \circ C_1 \circ C_2$.

	States	Arcs
H_0	9	17
H_1	9	16
C_1	1	90
C_2	40	57
D	28	228
F	1	11
$H_0 \circ H_1 \circ C_1 \circ C_2 \circ D \circ F$	260	5984

Table 4.6: Sizes of the component transducers for homograph disambiguation, compared with the size of the result of composing all of these transducers together.

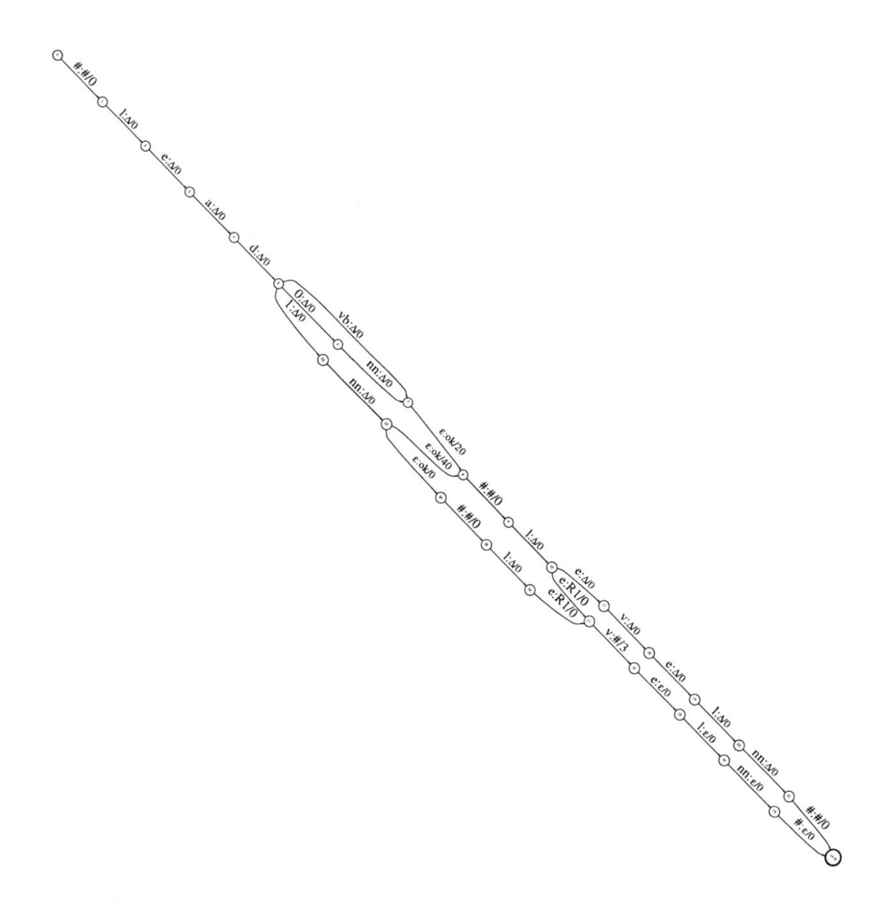

Figure 4.6: The result of $L \circ H_0 \circ H_1 \circ C_1 \circ C_2 \circ D \circ F$.

Language	Homog.	Pron 1	Pron 2 (3)	Gloss 1	Gloss 2 (3)
Mandarin	行	*xíng*	*háng*	'travel'	'line'
	率	lǜ	*shuài*	'rate'	'command'
Russian	берег	/bʲérʲik/	/bʲirʲók/	'shore'	'cared for'
	черт	/čert/	/čort/	line (gen pl)	'devil'
	дорогой	/dʌrógəi/	/dərʌgói/	'way' (instr)	'dear'
	душе	/dúši/	/dušé/	'shower' (prep)	'soul' (prep)
	села	/sʲilá/	/sʲélə/ /sʲólə/	'village's'	'sat' (fem) 'villages'
	стоит	/stóit/	/stʌít/	'cost'	'stand'
	тома	/tómə/	/tʌmá/	'volume's'	'volumes'
	все	/fʲsʲe/	/fʲsʲo/	'everyone'	'everything'

Table 4.7: Homographs treated by the Mandarin and Russian TTS systems.

Russian.[14] Note that Russian is a particularly rich source of homographs, due to lexical variation in stress patterns, and the failure to distinguish ë /jo/ from e /je/ in normal writing: as we noted in Section 3.7, in a corpus of only two million words, we have collected close to six hundred such potential homographs, each of which occurs at least fifty times.

At the time of writing we have implemented the examples in Table 4.7. We hope to report fully on these and other instances of multilingual sense disambiguation in future work.

4.3 Prosodic Phrasing

It is typical for human readers to break long sentences up into smaller prosodic phrases. Appropriate places for prosodic phrase breaks are occasionally indicated by punctuation marks, especially commas, but this marking is far from consistent, and in any case there is some justification in the view that marking prosodic boundaries is not the primary function of punctuation (Nunberg, 1995). For longer utterances where appropriate breakpoints are not consistently marked with punctuation it is necessary to automatically infer such breakpoints.[15]

[14] Again, the implementation of homograph disambiguation in our American English system is done in a completely different way.

[15] We will, of course, have nothing to say here about the intonational consequences of phrasing: this issue will be addressed in Chapter 6.

Various techniques have been proposed for computing appropriate phrase break locations, ranging from fairly simple local syntactic analysis of the input text (O'Shaughnessy, 1989), to the use of a full-blown syntactic parser (Bachenko and Fitzpatrick, 1990). (See also Section 3.9.4.) The method used in the Bell Labs American English TTS system[16] is a corpus-based statistical method, based on labeled corpora of read speech. Classification and Regression Tree models — CART (Breiman et al., 1984; Riley, 1989) — were trained on a subset of the ATIS corpus (Hemphill, Godfrey, and Doddington, 1990). Input to the tree-growing algorithm is data that mark, for each space between adjacent words w_i w_{i+1}, various features of the context that were judged a priori to be of likely interest to a phrasing prediction model. These features include:

- The length of the utterance in words.
- The distance in syllables, lexically stressed syllables, and words between the putative boundary point and the beginning and end of the utterance.
- The accentuation derived from the accent prediction model for w_i. and w_{i+1}.
- Stress level of the last syllable in w_i.
- Part-of-speech for a four word window around w_i w_{i+1} .
- Whether w_i w_{i+1} is part of a premodified nominal; if it is, the distance of w_i from the beginning of that nominal.
- Mutual information scores for a four-word window around. w_i w_{i+1}.

The overall cross-validated rates for multi-speaker read speech are 81.9% for a three-way distinction between utterance boundaries, intonational phrase boundaries (in the sense of (Pierrehumbert and Beckman, 1988)), and word boundaries. The most successful predictors are part-of-speech, some constituency information, and mutual information; each can predict a large percentage of observed boundaries. However, since it is not very practical to include a large table of mutual information scores in the TTS system, that was dispensed with, and the runtime system makes use only of part-of-speech information and information on whether the word sequence in question is part of a premodified nominal. The runtime system is a straightforward use of the tree in that for each word boundary, the decision tree is walked starting at the root, and questions are asked about the context at each node, until a leaf node is reached and a decision is arrived at on whether to interpret the word boundary as a higher phrase boundary. Besides the work for American English, an application of this model to Spanish is reported in (Hirschberg and Prieto, 1994).

How would a model of phrasing of this kind fit into the gentex text-analysis model? Phrase boundary prediction can be viewed as a species of sense disambiguation. Each word boundary is considered to be ambiguous between being

[16] Due to Hirschberg: see (Wang and Hirschberg, 1992) for a more extensive discussion.

an actual word boundary, or some higher level phrase boundary. The problem is then to disambiguate between these possibilities, making use of the context. Two possible approaches then suggest themselves. The first would be to model the problem using the decision-list based approach previously discussed in the context of sense disambiguation; a prerequisite for this approach would be the conversion of a decision-tree model into a decision-list model. A second would be to keep the tree-based models as is, and compile directly from trees into WFST's using the algorithm reported in (Sproat and Riley, 1996).

We briefly describe this latter option using the toy example in Figure 4.7.[17] In this toy example, we will consider the labeled path on the left-hand side of the tree descending from the root to leaf node 16. The interpretation of this path is as follows:

- Question [1] in this simplified example relates to the distance between the word boundary in question and the previous punctuation mark (*dpunc*): if it is less than four words away, go left, otherwise go right.
- Question [2] relates to the grammatical part of speech of the word to the left (*lpos*): if it is a noun, verb, adjective, adverb or determiner go left, otherwise go right.
- Question [3] similarly asks about the part of speech of the word to the right (*rpos*): if it is a noun or an adjective go left, otherwise right.
- Finally, question [4] further refines question [1], asking if the distance to the previous punctuation mark is fewer than three words. If it is then we arrive at node 16, where in 133 out of 200 examples in the training data, the answer to the question "is this word boundary also a phrase boundary", is "no".

The basic insight of the tree compilation algorithm reported in (Sproat and Riley, 1996) is that if the question asked at each node can be expressed in terms of regular expressions, then the decision at each leaf node can be modeled as rewrite rule that can be compiled into a WFST using the compilation algorithm for weighted rewrite rules presented in (Mohri and Sproat, 1996); by extension the entire tree can be modeled as the intersection of all of the WFST's corresponding to all of the leaf nodes.[18] To see this informally with the current example, consider that each of the questions asked in traversing the tree from the root to node 16 can be modeled as a regular expression relating either to the left or right context. For example, question [1] relates to the left context

[17] A real decision-tree model of phrase boundary prediction would be somewhat too complex to use as an illustration here.

[18] As we mentioned in Section 2.5.2, regular relations, and thus WFST's are *not* in general closed under intersection. However, as also previously mentioned, certain sub-classes of regular relations, including 'same-length' relations *are* closed under this operation. So long as the relations expressed by leaf nodes of the tree are restricted to the classes of relations for which closure under intersection holds, there is no problem in modeling the entire tree as an intersection of WFST's.

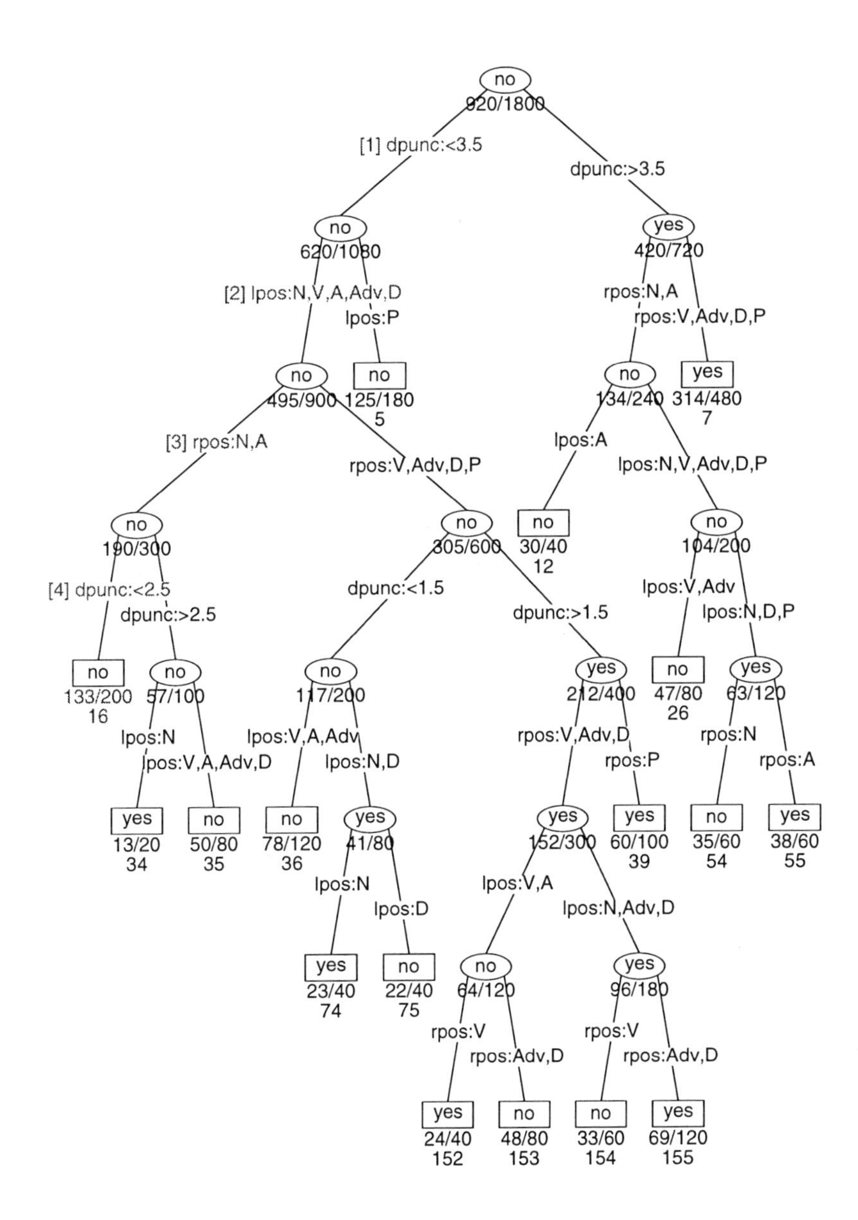

Figure 4.7: A toy example of a decision-tree-based model of phrase boundary prediction.

Node 16	λ	ρ
	[1] $(\Sigma^*(p\omega\|p\omega\#\omega\|p\omega\#\omega\#\omega))$ $\cap$	[3] $N\|A$
	[2] $(\Sigma^*(N\|V\|A\|Adv\|D))$ $\cap$	
	[4] $(\Sigma^*(p\omega\|p\omega\#\omega))$	

Table 4.8: Regular expressions for the left (λ) and right (ρ) contexts for labeled path in Figure 4.7.

(λ) (see Section 2.5.4 and (Mohri and Sproat, 1996)) and can be stated as the regular expression $(\Sigma^*(p\omega|p\omega\#\omega|p\omega\#\omega\#\omega))$, where p represents the previous punctuation, ω stands for any word, and # represents a word boundary. The entire left context λ for the rule is constructed by intersecting together, in this case, the expressions for questions [1], [2] and [4]. The sole right-context (ρ) restriction is question [3]. See Table 4.8 for the entire set of regular expressions for these questions. The final rewrite rule for node 16 can be expressed as follows:

(57) $\# \Rightarrow (\# < 0.41 > |p < 1.09 >) \ / \ p(\omega\#)?(N|V|A|Adv|D)$ ____ $N|A$

Simply put, a word boundary is rewritten as itself with cost 0.41 and as a phrase boundary with the higher cost 1.09, when the lefthand context consists of one or two words, where the immediately preceding one is N, V, A, Adv or D; and where the immediately following one is N or A. The costs are, again, simply the negative logs of the estimated probabilities: $\frac{133}{200} = .67$, and $\frac{67}{200} = .33$. For more extensive details on the tree compilation algorithm, the interested reader is referred to (Sproat and Riley, 1996).

Assume that the lexical analysis has every word boundary marked as potentially ambiguous between being a word boundary and a phrase boundary. Assume further that we have a transducer P that deletes all labels except those needed for the phrase boundary prediction. Then, assuming that we have a WFST T representing the phrase-boundary decision tree, the predicted phrasing for a given lexical analysis L will be given by:

$$L \cap \pi_1[\ BestPath\ [\ L \circ P \circ T\ ]\ ]$$

This is, of course, essentially the same as the modeling scheme described in the previous section on homograph disambiguation.

4.4 Some Final Thoughts on Text Analysis

Needless to say, there are many other topics that one could cover in a treatment of text analysis for synthesis, which we have not covered here. For example, we have had little to say about grammatical part-of-speech assignment, or syntactic parsing models.

As we saw in Section 3.8.1, our English TTS system incorporates a trigram-based model of part-of-speech assignment, originally due to Church (Church, 1988), and widely cited in the literature. Such a model is trivially implementable using WFST's. Other approaches to part-of-speech assignment (e.g. Brill's tagger (Brill, 1992)), are probably also implementable as WFST's.

Full-blown syntactic analysis is probably less common in TTS systems than some of the "shallower" methods of syntactic analysis, such as part-of-speech assignment; see Section 3.9.4. (In part this may be because the arguments for having a complete syntactic analysis for the purposes of, e.g., prosodic phrasing, or lexical disambiguation, have never been overly compelling.) Nonetheless it is interesting to consider possible ways of integrating full syntactic parsing into the general text-analysis framework we have presented in these two chapters. The problem, of course, is that interesting syntactic grammars are usually assumed to require (at a minimum) context-free power. Given a finite-state lattice representing a set of lexical analyses, how would we use syntactic models to disambiguate those lattices — i.e., by weeding out some analyses as syntactically ill-formed?

Several approaches suggest themselves, of which the following constitute a subset:

1. Use a full-blown syntactic parser to parse the lexical lattice (rather than just a lexical string), weeding out those paths that do not contribute to a well-formed parse of the input.

2. Construct a finite-state approximation of a context-free syntactic grammar, along the lines of (Pereira and Wright, 1997). Intersect the resulting FSA with the lexical lattice.

3. Mimic a context-free syntactic parser by iterative composition of transducers, along the lines discussed in (Roche, 1997).

4. Eschew entirely building a parse tree. Rather, model syntax as a sequential application of local grammatical constraints which are easily implementable as WFST's (Mohri, 1994; Voutilainen, 1994).

We will not choose among these alternatives here, but rather leave it as a topic of future research.

As we can see from the discussion in this chapter, the gentex model is still incomplete, and there is much future work to be done to incorporate into the model various kinds of information necessary for the accurate analysis of input text. Nonetheless, despite the present incompleteness and limitations, we believe that gentex is virtually unique among TTS text-analysis models: it provides a completely uniform treatment of a variety of problems that are normally treated separately and in quite different ways; and it has been applied in a similarly uniform fashion to a variety of typologically quite different languages and writing systems.

Chapter 5

Timing

Jan van Santen

5.1 Introduction

The synthesis component produces output speech with a prosodically appropriate temporal and intonational structure. The task of the timing component is to compute this temporal structure from symbolic input such as phoneme symbols, stress, and accent markings.

There are several reasons why this a hard task. One is that many factors affect timing, and that the joint effects of these factors are complex. Another reason is that the number of factorial constellations — "cases" that can occur in the language — is vast. The final reason is that very little is known about the underlying processes responsible for speech timing.

The timing component of our system is conventional in that it specifies temporal structure in the form of *segmental durations*. This is by no means the only way to characterize temporal structure. For example, one could specify temporal structure in the form of sub-segmental parameterized time warping functions, parameterized articulatory trajectories, or durations of sub-segmental speech intervals.

The timing component is unusual in that it is neither purely rule based (as in (Allen, Hunnicutt, and Klatt, 1987)), nor purely statistically based (e.g., (Riedi, 1995)). Instead, it borrows from both of these opposing approaches. And, of course, the component is also multilingual: all information pertinent to a given language is stored in tables that are external not only to the run time modules but also to the analysis software.

In this chapter, we first reflect on these basic design choices, and then proceed to describe our approach in detail.

5.2 Representation of Temporal Structure

Whereas the phonetic representation of intonation is reasonably unproblematic (in the form of fundamental frequency, F_0), the phonetic representation of temporal structure is remarkably controversial. Focal points in the controversy are the status of the phonetic segment and what we shall call the *concatenative assumption*. Although it would appear more appropriate to discuss the latter in Chapter 7, we have to address it here because there is obviously an intimate connection between computation of timing (by the duration component) and what entities this temporal structure will be imposed on (by the synthesis component).

5.2.1 The concatenative assumption

The concatenative assumption is at the heart of concatenative synthesis. It states that a realistically sized acoustic unit inventory exists (i.e. a collection of stored recorded speech intervals), from which one can generate natural-sounding speech for any input text. By "realistically sized", we mean that the required natural speech data base can be recorded in a practically feasible time frame, without putting absurd demands on the speaker. The term "realistic" is deliberately vague; however, in practice it is quite clear that 5,000 is realistic, and 500,000 is not. And by "generate" we refer to the usual operations performed by the synthesis component: concatenation of successive units, adjusting their temporal structure, and imposing a pitch contour.

Path Invariance

The essence of the concatenative assumption is that there are limits on the amount and variety of contextual influences on a given speech interval, including effects from both coarticulatory influences from adjacent speech sounds and prosodic factors such as word stress.[1] This can be made more precise using the concept of *path-invariance*, which we illustrate with results from a study in which the effects of postvocalic voicing on the spectral trajectory of the preceding sonorant region were analyzed (van Santen, Coleman, and Randolph, 1992). Utterances produced by a male speaker were of the form *Please say* *<monosyllabic word>*. The monosyllabic words formed minimal pairs, differing only in the final segment (/d/ vs. /t/). Figure 5.1 provides some examples of these minimal pairs. We only analyzed the sonorant regions of the utterances.

Consider the segment sequence /m-ɛ-l/ (as in *meld* and *melt*). At the articulatory level, this sequence is produced by a complex sequence of articulatory events involving the lips and several dimensions of tongue movement. At the acoustic level — specifically in formant space — there is a sharp increase in F_1

[1] This subsection touches on aspects of speech production that are discussed more fully in Chapter 7, but should be comprehensible without consulting this later chapter.

starting at the onset of the vowel, followed by a sharp decrease in F_2 as the /l/ approaches.

The overall duration of the sequence /m-ɛ-l/ in the word *meld* is much longer than in *melt*, due to well-known effects of postvocalic voicing in utterance final position (e.g., (Klatt, 1973; van Santen, 1992a)). Given this large temporal difference, given the complexity of the underlying articulatory events, and given the non-simultaneity of the changes in F_1 and F_2, it would appear surprising if the two occurrences could be generated by a single, common template. Instead, we would expect the changes in the formants to be *asynchronous* in the sense that the dynamic relations between F_1 and F_2 change when a voiceless post-vocalic consonant is replaced by a voiced post-vocalic consonant.

Yet, when we inspect Figure 5.1 (top left panel), we see that the F_1, F_2 trajectories are remarkably close in terms of their *paths*, i.e. the curves generated by smoothly interconnecting the successive points of the trajectories.[2] Figure 5.1 shows five additional examples. Of course, this does not necessarily mean that the same would be found if we also plotted other formants, spectral tilt, and still other acoustic dimensions. Nevertheless, the degree of *path equivalence* is striking.

When two trajectories approximately traverse the same path, then, by definition, one trajectory can be approximately generated from the other by first producing its corresponding path using smooth interpolation and then sampling this path at points closest to those of the second trajectory. In other words, two trajectories share a path if an only if they are related by a time warp. If we now consider a collection of path-equivalent trajectories, we can single out one of them as a template, and generate all other trajectories with time warping.

The concatenative assumption redefined

We now restate the concatenative assumption as the statement that *a realistic number of speech intervals exists that are each path-equivalent over the entire range of contexts defined by the input domain and jointly span the entire language* (i.e., combinations of phone label sequences and prosodic markers.)

It is clear, of course, that path equivalence is a matter of degree, and that microscopic analysis of speech will ultimately show that no perfect path equivalences exist. Our point of view is that, while the concatenative assumption is hence fundamentally untrue, it is at the same time sufficiently accurate that in the context of current text to speech systems problems caused by this assumption are overwhelmed by problems in other synthesis components. But in due

[2] These centroid trajectories were computed as follows. The cepstral trajectories of four of the five tokens of one word (e.g., *meld*) were time-warped onto the fifth token (the "pivot"), and for each frame of the latter the median vector was computed for the five cepstral vectors mapped onto that frame. The same was done with each of the other four tokens playing the role of pivot. Subsequently, the process was repeated, with now the median vector trajectories taking the place of the original cepstral trajectories. The process was continued until convergence was reached.

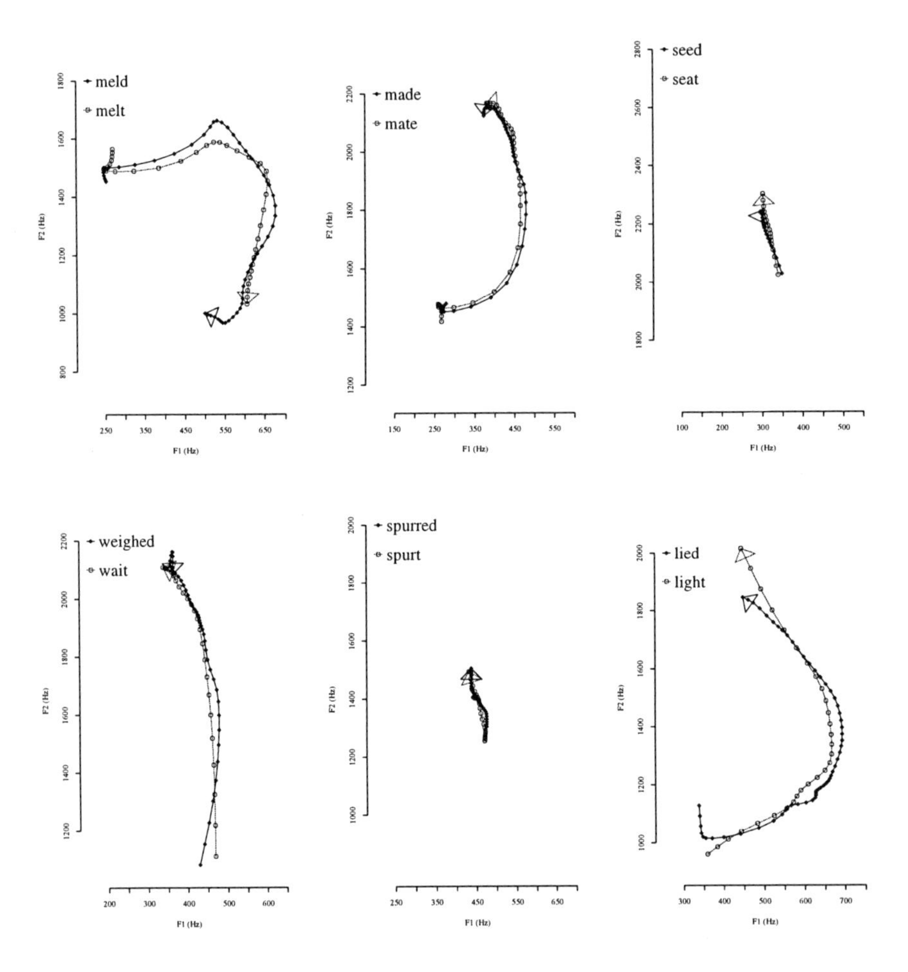

Figure 5.1: F_1, F_2 trajectories (centroids) for minimal word pairs. Open and closed symbols are placed at 10 ms. intervals. Only sonorant portions are shown. Arrow heads indicate the end of the trajectories. See text for computation of centroids.

time, as these other problems are addressed, the concatenative assumption will have to be dropped, and entirely new synthesis methods have to be adopted.

We emphasize that the concatenative assumption is solely concerned with the acoustical level. It is not an assumption about underlying phonological structure, or about underlying speech processes. We elaborate on this point in the following subsection.

5.2.2 Segmental duration, phonetic segments, and the concatenative assumption

The concept of phonetic segment has been discredited over the years, making the use of segmental duration suspect. We want to clarify here the relation between phonetic segments, the concatenative assumption, and segmental duration; in passing, we argue that the concept of an acoustically invariant phonetic segment — which is the discredited concept — is irrelevant to our approach.

First, the concatenative assumption does not require the existence of acoustically invariant phonetic segments. It only requires that there be path-equivalent speech intervals sufficient to cover the input domain.

Second, the concatenative assumption does not require that we use the phonetic segment as the unit of timing. This results from the fact that the concatenative process whereby two acoustic units are combined to generate a larger chunk of speech can modify temporal structure in many ways. For example, it may allow only for a linear stretch or compression of an entire unit; or it may allow the same for sub-intervals of units that — by some rule — correspond to phonetic segments; or, as illustrated in the preceding subsection, it may allow for non-linear time warps, resulting in some parts of segments being stretched or compressed more than other parts. In other words, there is considerable freedom in terms of the time interval whose duration is modified.

Third, segmental duration does not require the existence of acoustically invariant phonetic segments. The reason is that duration rules typically incorporate information about the phonemic context of a segment — for example whether a vowel is followed by a nasal or a voiceless stop. Hence, although the same phonological symbol is used to refer to the vowel region in both contexts, nothing is assumed about the acoustic invariance of this region. To elaborate, consider unflapped /t/ and flapped /t/. Our system has separate duration rules for predicting how these two allophonic variants are affected by their respective contexts; using the same symbol /t/ for both cases is merely a labeling convenience (or inconvenience) and has no theoretical or empirical implications.

Fourth, segmental duration (or sub-segmental duration) *does* require that we can reliably and meaningfully detect *boundaries* in the speech signal. Most boundaries are relatively straightforward — both acoustically and articulatorily. For example, transitions of obstruents to vowels or semi-vowels involve sudden energy increases in specific frequency bands; these increases are typically produced by a sudden change in the position of the obstructing articulator. The same is true for transitions between voiceless fricatives and between voiced fricatives. This is not to say that some of these boundaries, although acoustically well-defined, do not have a certain arbitrariness. For example, why not use onset of the nasal murmur instead of oral closure as the boundary between vowels and nasals? The truly problematic cases mostly involve transitions between vowels, or between vowels and glides. Here, conventions based on formants can lead to reliable segmentation (as measured by inter-judge reliability), but the resulting boundaries do not necessarily correspond to well-defined articulatory events (such as oral closure, voicing onset, or opening of the velar flap). Note that the nature of a given boundary is affected by coarticulation. For example, the /p/-/æ/ boundary looks different in *pat* than in *spat* because the /s/ removes the aspiration part of the /p/. Thus, segmentation rules themselves are context-dependent, do not assume the existence of acoustically invariant boundaries, and hence do not assume the existence of acoustically invariant phonetic segments.

Note, however, that as one migrates to a time-warp based approach, boundaries become less of an issue. In fact, one of the main goals of the time warp study (van Santen, Coleman, and Randolph, 1992) was to circumvent the definition of boundaries. But in the context of the current TTS system, boundaries are important.

5.2.3 Timing of underlying features vs. timing of speech intervals

The key distinction here is whether to specify timing in terms of the speech signal itself, or in terms of underlying parameters that are then used to generate a synthetic speech signal (e.g., the work by Stevens and his coworkers (Stevens and Bickley, 1991; Bickley, Stevens, and Williams, 1997), Coleman (Coleman, 1992; Dirksen and Coleman, 1997), and Local (Local and Ogden, 1997)). Given that speech is produced by quasi-independent articulators, it would seem that the latter approach is more theoretically sensible.

In Coleman's work (Coleman, 1992; Dirksen and Coleman, 1997), for example, a hierarchy is proposed consisting of feet, syllables, sub-syllabic entities (onset, rhyme, coda, nucleus), and features (e.g., heaviness, height). Notably absent are phonemes. Within syllables, features have their own quasi-independent time courses. During synthesis, these features are mapped onto acoustic parameter trajectories that control a formant synthesizer. This framework is effective in representing key coarticulatory phenomena, such as assimilation (e.g., vowel nasalization), deletion (as in *mem'ry*), addition (as in *Chompsky*), substitution (/ŋ/ in *congress*), and reduction (of unstressed vowels into schwa).

The issue at stake here is not the validity of this account, but whether it is at variance with the concatenative assumption. There need not be a contradiction between a multivariate dynamic conception of the speech production process on the one hand, and the concatenative assumption on the other hand. The reason is that the parameters involved may be quasi-independent, in particular across different phoneme sequences, yet for a given phoneme sequence they may be strongly coupled, even as this phoneme sequence is embedded in a range of contexts. Strong coupling of articulatory parameters results in path equivalence at the acoustic level.

We are not convinced that synthesis based on these concepts (e.g., York-Talk (Coleman, 1992), and synthesis produced by Ken Stevens' group (Stevens and Bickley, 1991; Bickley, Stevens, and Williams, 1997)) is currently superior to the best concatenative systems. We think this is due, however, more to the inherent difficulties of synthesis-by rule (see Chapter 7) than to flaws in the theoretical concepts. And even if it turns out that these inherent difficulties are unresolvable, we believe that in the attempt to build these systems extremely valuable knowledge is accumulated that is also helpful for other synthesis approaches, including concatenative synthesis. For example, our acoustic inventory design is based significantly on this type of articulatory knowledge.

5.2.4 Timing of supra-segmental units

We already mentioned our belief in the importance of manipulating the durations of sub-segmental intervals. In this subsection, we discuss a viewpoint that takes strong exception to this belief. This viewpoint holds that even segments are too small a unit.

First, we should discuss a point of confusion that often enters in discussions of supra-segmental units: units as *phonological entities* vs. units as *speech intervals*. As will be discussed below, phonological units larger than the phoneme play a critical role in our duration component. We predict segmental duration from location of the phone in the syllable, the location of the syllable in the word and phrase, its stress status, and the location of the phrase in larger phrases, etc. In some of these discussions, it is often surprisingly unclear whether the "importance of the syllable" refers to the obvious fact that as a phonological entity it plays an important role in predicting temporal structure, or to the syllable as a speech interval.

An important contribution to this discussion was made in (Campbell, 1992), where the syllable is proposed as the key speech interval, and in (Barbosa and Bailly, 1994), where the *inter-perceptual center group* is championed. The contribution of these papers is to provide a precise mathematical model of how syllable duration or inter-perceptual center group duration is related to segmental duration.

According to Campbell's model, the duration of a syllable depends on the prosodic context of the syllable (e.g., stress, location in the phrase), and also — but only very coarsely — on which segments it contains. Specifically, what matters is how many segments a syllable contains, and whether the nucleus is a vowel vs. a diphthong vs. a syllabic consonant. We call this relative independence of syllable duration from segmental contents *segmental independence*.

The second fundamental assumption (which we call *syllabic mediation*) is that the duration of a segment depends only on the duration of the syllable, the identity of the segment, and the identities of the other segments in the syllable.

Thus, when the same syllable has the same duration in two different contexts, then all its constituent segments should also have the same duration in the two contexts.

The inter-perceptual center group based model by Bailly and Barbosa is the same as Campbell's syllabic timing model, except that the model uses intervals between vowel starts instead of syllables.

We are in complete disagreement with these two proposals. First, how are we to resolve the conflicting demands of syllables vs. inter-perceptual center groups? For example, if syllables are the relevant unit then the boundary between /r/ and /t/ in *important* is more important than the boundary between /t/ and /ə/; if inter-perceptual center groups are the relevant unit the opposite would be true.

Since segment-based approaches that predict *both* the /r/-/t/ and /t/-/ə/ boundaries accurately would be acceptable according to both large unit approaches, the question can be asked whether large-unit timing modules have some unique property that allows them to predict large-unit boundaries more accurately than segment-based approaches.

The answer appears to be negative, for the simple reason that the segmental independence assumption is wrong. Consider syllables having the same structure (e.g., *consonant – vowel – consonant*, such as *sit* and *knit*. The syllabic model predicts identical durations for *sit* and *knit*, because it largely ignores the segmental makeup of syllables (it considers these two syllables segmentally equivalent because they have the same structure). There are recent data indicating that the durations of syllables having the same structure, and confined to a fixed context (e.g., word-initial, phrase-medial, stressed) depend strongly on the segments that make up the syllable (van Santen and Shih, 1997). For example, a monosyllable such as *sit* is 65 ms. longer than *knit*, matching a similar difference in average duration between /s/ and /n/ in other contexts (61 ms.). So this model would systematically produce syllabic durations that are too short for *sit* or too long for *knit*.

Thus, unless one takes into account such segmental effects on syllable duration, the prediction of syllable duration may be *less* accurate than the prediction provided by segmental approaches. However, once these detailed segmental effects are taken into account, a syllabic timing model ceases to be a syllabic timing model.

In summary, proposals for units larger than the segment claim that boundaries between large units are perceptually more important than boundaries between unit-internal segments, and hence must be predicted with particular accuracy. Current proposals differ sharply in what the large units are, provide little evidence for the perceptual importance of their own units, and have not shown that their implementations can predict these boundaries more accurately than segmental timing modules. Although the idea that some boundaries — or certain intra-segmental markers — are perceptually more important than others seems plausible in principle, until more evidence is available we must focus on attempting to accurately predict *all* segment boundaries; in addition we must start to pay attention to segment-internal timing, as discussed in the preceding section.

5.2.5 Summary

Our view on temporal structure can be summarized as follows.

- The concatenative assumption, while theoretically false, has not been shown to be responsible for audible acoustic flaws in current synthesis systems. Hence, for now we view it is a useful working hypothesis.
- One can reliably measure boundaries in concatenative speech intervals that correspond to phonetic segments. This does not assume that the acoustic properties of phonetic segments are invariant. In fact, the set of acoustic units in which a given segment occurs should capture all the major coarticulatory effects on that segment.
- It is important to move towards time-warp-based temporal control; controlling timing via segmental duration should be viewed as a convenient but coarse approximation to time-warp-based control.
- Controlling durations of time intervals longer than the segment does not add to the accuracy of temporal control. However, the role of phonological entities larger than the phoneme is quite important for prediction of timing.

5.3 General Properties of Segmental Duration

Now that we have explained our focus on segmental duration, we discuss the general properties of segmental duration that have led us to our current approach.

5.3.1 Factors known to affect segmental duration

We use the term "factor" to refer to a categorization such as syllabic stress, which has as *levels* primary stressed (third syllable in *regulation*), secondary stressed (first syllable), and unstressed (second and fourth syllables). In text-to-speech synthesis, these levels are computed from text, joined into feature vectors, and then given to the timing module. Several studies have shown that the following factors affect duration in American English (Crystal and House, 1982; Crystal and House, 1988a; Crystal and House, 1988b; Crystal and House, 1988c; Crystal and House, 1990; Klatt, 1976; Umeda, 1975; Umeda, 1977).
Typical numbers of levels are between parentheses.

1. *Phonetic segment identity (30).* For example, holding all else constant /ai/ as in *like* is twice as long as /ɪ/ in *lick*.

2. *Identities of surrounding segments (10).* For example, vowels are much longer when followed by voiced fricatives than when followed by voiceless stops.

3. *Syllabic stress (3).* Vowels in stressed syllables and consonants heading stressed syllables are longer than vowels and consonants in unstressed syllables.

4. *Word "importance" (3).* This factor is quite complex because the importance of a word is highly context dependent. The factor used in our work on timing (van Santen, 1992a) is *predicted word accent*, which refers to whether the word is predicted to have a major pitch movement. This is predicted from features such as parts of speech of the word and its neighbors, its lexical identity, and whether the word occurred in sentences preceding the current sentence (*given* vs. *new* information; see Section 4.1.1). An alternative way of predicting word importance is from word frequency (Umeda, 1975).

5. *Location of the syllable in the word (3).* Syllables at the end of a word (word-final syllables) are longer than syllables that start words (word-initial syllables), which are in turn longer than syllables in the middle of words (word-medial syllables).

6. *Location of the syllable in the phrase (3).* Phrase-final syllables can be twice as long as the same syllable in other locations in a phrase.

Effects of additional factors have been claimed, but the empirical status of these effects is less certain than for the above factors. For example, it has been claimed that syllables are shorter in long stretches of unstressed syllables than in short stretches (e.g., (Lehiste, 1977)), but evidence to the contrary exists (van Santen, 1992a). To further illustrate this point, (Crystal and House, 1988b) contains a list of effects that have been claimed in the literature but which these authors were unable to replicate.

There are at least two reasons why the effects of many factors are unclear. First, there may exist systematic differences between studies in terms of speakers, textual materials, and speaking modes. Second, in typical text most factors are confounded (e.g., in American English, syllabic stress and within-word position are confounded because in most polysyllabic words the first syllable receives stress).

Although it is quite likely that research will show the effects of additional factors, there is evidence that the factors on the consensus list capture a large percentage of the durational variance. For example, in a study of American English, 94.5% of the variance of vowel duration that can be predicted *from text* can be predicted by processing text in terms of these factors (van Santen, 1992a). Nevertheless, some of these new factors may have large effects in specific sub-cases, such as phrase-final syllables or function words. Their small contribution to the overall percentage of variance explained does not mean that they have no perceptual effect, but that the frequency with which these sub-cases occur is relatively small.

5.3.2 Interactions

The effects of a given factor can be modulated by other factors (See Section 2.3.1 in Chapter 2). For example, the effects of stress levels are different for vowels and consonants. For vowels, the stress level of the syllable containing the vowel has an effect, but the effect on consonants depends on whether the consonant occurs in the onset or the coda of the syllable; in the latter case there is little effect. Similarly, when a voiceless stop in the onset of a stressed syllable is preceded by an /s/ its duration is dramatically decreased (in English); the same is not true for other consonants, for voiceless stops in other positions, or in unstressed syllables.

As we pointed out in Section 2.3.1 in Chapter 2, the term "interaction" is not always used consistently. The standard definition of interaction is that the effect of a factor measured in milliseconds is altered by a change in some other factor. This is what is tested in the analysis of variance (Winer, Brown, and Michels, 1991): *additive interactions*. For example, syllabic stress might increase the average duration of /ai/ by 42 ms. from 140 to 182 ms. and the average duration of /e/ by only 24 ms. from 80 to 104 ms. However, in this example the factors of phoneme identity and syllabic stress do not interact in the *multiplicative* sense because the percentage change is 30% in both cases.

The concept of interaction can be further broadened. Klatt (1973) found that vowels are longer when they are followed by a voiced consonant than by a voiceless consonant. Moreover, this effect — whether measured in ms. or as a percentage — was much larger in phrase-final syllables than in phrase-medial syllables; thus, the interaction was both additive and multiplicative. This interaction has been often documented since (e.g., (Crystal and House, 1988c; van Santen, 1992a)). This interaction does not involve *reversals*: Holding all else constant, phrase final vowels are longer than phrase-medial vowels, and vowels followed by a voiced consonant are longer than vowels followed by a voiceless consonant. In other words, the factor of phrasal position does not *reverse the direction* of the effect of the postvocalic consonant, it only *amplifies* it.

Such "reversal interactions" do exist, however. For example, when we compare words such as *butter*, *return*, *finer*, and *beneath*, we find that the /t/-burst in *return* is longer than the /n/ in *beneath* while the opposite holds contrasting the /t/-burst in *butter* with the /n/ in *finer* (van Santen, 1993c). This is a reversal of the effects of the segmental identity factor brought about by a change in the stress levels of the surrounding vowels.

There are many factors with additive or multiplicative interactions, but there are not very many reversal interactions. We shall see that this has important implications for timing module construction.

5.3.3 Sparsity

In open-domain text-to-speech synthesis any sequence of words — including neologisms and strange names — must be correctly synthesized. The combinatorial possibilities are astronomical, even when one imposes restrictions on sentence length and vocabulary size. Since the input domain for the timing module consists of feature vectors and not of text, this does not necessarily mean that this module also faces a myriad of combinatorial possibilities. For example, if the only factors affecting timing were *syllabic stress* (stressed, unstressed) *word accent* (accented, deaccented) *phrasal position* (final, non-final) and *phoneme identity*, then there would be only a few hundred possibilities, which could easily be covered by a small amount of training data. But if many more factors play a role and if few restrictions apply to their combinations, then the feature space may be too large for complete coverage by the training data base: the training data base will be *sparse* with respect to the full feature space.

The issue of coverage is not straightforward, as the following three subsections demonstrate. Note that sparsity problems occur regardless of the timing representation used because sparsity has mostly to do with the input domain of timing modules, not with their output.

Coverage and statistical models

Coverage cannot be defined independently of the type of statistical model employed for prediction of timing. Sixty percent coverage — as measured by the type count[3] of the feature vectors in the training sample divided by the corresponding count in the language — may be adequate for estimating the parameters of one class of models, yet be completely inadequate for parameter estimation for a different class of models.

On the one extreme, consider a lookup table that lists the average duration for each feature vector in the training data base — the simplest statistical model imaginable. Constructing this timing module obviously requires complete (100%) coverage of the feature space.

On the other hand, if one were to apply the *additive model*, then far less training data is needed (see Section 2.3.1). According to the additive model, the duration $\mathrm{DUR}(\vec{f})$ for a feature vector $\vec{f} = \langle f_1, \cdots, f_N \rangle$ is given by

$$\mathrm{DUR}(\vec{f}) = A_1(f_1) + \ldots + A_N(f_N). \tag{5.1}$$

Here f_i represents a value on the i-th factor; e.g., if the i-th factor is syllabic stress, then f_i could be "stressed" and f_i' "unstressed". $A_i(f_i)$ is a parameter whose value reflects the contribution of factor i when it has level f_i.

To show that the additive model requires less coverage than the lookup table, again suppose that the feature space consists of the factors *syllabic stress*,

[3] The *type count* of a set of feature vectors is the number of *distinct* vectors; the *token count* is the total number of vectors in the set.

word accent, *phrasal position*, and *phoneme identity*, so that with 30 phonemes there would be 360 possible feature vectors. Since the number of independent parameters is only 34, a training sample as small as 34 could be sufficient for estimating the parameters.[4] If the additive model is valid, then accurate predictions can be made for the remaining 326 feature vectors; that is, the additive model can be used to *fill in missing data*. For this model 9.44% coverage is adequate.

Between these two extremes — the lookup table and the additive model — is a continuum of statistical models, ranging from models that make very few assumptions but have limited capacity for filling in missing data to models that make strong and possibly incorrect assumptions but can handle large fractions of missing data.

Inherent variability of speech

Speech timing can be predicted from text only up to a point. For example, an analysis of durations of the same vowel spoken in identical contexts by the same speaker corrected for per-utterance speaking rate showed that at least 8% of the total variance is text-independent (van Santen, 1992a). Apparently, speech accelerates and decelerates in the course of even a brief utterance in a way that cannot be predicted from text. This implies that if we were to construct a lookup table of feature-vector - segmental duration pairs, multiple observations would be needed for each pair to reach sufficient statistical accuracy.

Methods other than the lookup table approach are also vulnerable to this intrinsic variability because it directly affects the reliability of the parameter estimates, but to a lesser extent than lookup tables because for a given amount of data the number of data points per parameter is much larger. Thus, as with the coverage issue discussed in the preceding subsection, the impact of inherent durational variability on predictive performance depends on the statistical model.

Nonuniform feature vector frequency distribution

Two remaining questions concerning coverage must be answered. First, do we have any idea how many feature vectors might exist in a given language? Second, if this number turns out to be very large, what are the drawbacks of being able to accurately predict duration only for the most frequently occurring vectors?

A tentative answer to the first question was provided in an analysis of 750,000 sentences of English (van Santen, 1993c). This text contained 22 million phonetic segment occurrences. Corresponding feature vectors were computed by a particular system — the text-analysis portion of the Bell Labs

[4] For an additive model for N factors where factor i has n_i levels, the total number of free parameters (Winer, Brown, and Michels, 1991) is $1 + \sum_{i=1}^{i=N} (n_i - 1)$.

TTS system — but there is little reason to doubt the generality of these findings. Feature vectors computed were based on a factorial scheme described in (van Santen, 1994), which is relatively coarse. For example, the factor of "phrasal location" had only three levels ("phrase-final", "phrase-penultimate" and "other").

For subsets of various sizes of these 750,000 sentences we computed the type count of the feature vectors contained in the subset. Of course, the type count increased as a function of sample size. What we did not expect, however, was that there was no sign of the type count tapering off. By extrapolation, the type count would be roughly 30,000 for sets with a size of 22 *billion*. Using less coarse schemes would further increase this count.

The answer to the second question is that the number of types with low occurrence probabilities is sufficiently large that the probability that a randomly selected sentence contains at least one low probability type is high. For example, in the same analysis it was found that *the probability that a randomly selected 50-phoneme sentence contains a vector that occurs less than once in a million segments was more than 95%*.

We conclude that the number of feature vector types that can occur in a language is on the order of tens of thousands, and that rare feature vectors cannot be ignored by the timing module because the chance of encountering *some* rare vector in even a small amount of text is surprisingly large.

More generally, we have found that frequency distributions of many other entities — such as tetraphones, feature vectors relevant for predicting phrase boundaries, etc. — have this property. Apparently, the combinatorics of American English — and undoubtedly that of many other languages as well — is such that text-to-speech synthesis schemes unprepared for the unusual are doomed to failure, because, in language, the unusual is the rule. This captures the experience many have had who work often with text-to-speech synthesis systems: one notes a flaw in a sentence, brushes it aside because it involves a rare constellation of contextual factors, only to be confronted with a flaw involving another rare constellation in the next sentence.

5.3.4 Directional invariance

In section 5.3.2 we remarked that relatively few factors affecting duration exhibit reversal interactions, but that many have additive or multiplicative interactions. The influential Klatt model (Klatt, 1973) captures the interaction between postvocalic voicing and phrasal position as follows:

$$\mathrm{DUR}(V, C, P) = S_{1,1}(V)S_{1,2}(C)S_{1,3}(P) + S_{2,1}(V) \qquad (5.2)$$

Here, V denotes the vowel identity factor, C the class of the postvocalic consonant (*voiced* vs. *voiceless*) and P the phrasal position factor (*phrase-medial* vs. *phrase-final*). In the usual formulation, $S_{2,1}(V)$ is the minimum duration of vowel V, *MINDUR(V)*; $S_{1,1}(V)$ is the *net duration* defined as the difference between the inherent duration, *INHDUR(V)*, and the minimum duration;

$S_{1,2}(C) = K_C$; and $S_{1,3}(P) = K_P$. The latter two are constants tied to the postvocalic consonant and to phrasal position.

To clarify, each $S_{i,j}$ is a parameter vector, each parameter corresponding to a level on the j-th factor. The subscript i (having values 1 and 2) refers to the fact that Equation 5.2 is a sum of two product terms, $S_{1,1}(V)S_{1,2}(C)S_{1,3}(P)$ and the trivial product $S_{2,1}(V)$.

Klatt's model illustrates the possibility that interactions can be modeled using equations that are neither additive nor multiplicative, yet are not much more complicated because they involve few parameters and use well-behaved mathematical operations (addition and multiplication). This raises hopes that equations may indeed exist that can handle both interactions and sparse training data.

The problem with Klatt's model, unfortunately, is that it is not an accurate description of some of the interactions that have been observed. For example, evidence exists that the following model is significantly more accurate (van Santen, 1994):

$$\text{DUR}(V, C, P) = exp[S_{1,2}(C)S_{1,3}(P) + S_{2,3}(P) + S_{3,1}(V)] \qquad (5.3)$$

Aside from modeling the *logarithm* of the duration, the model in Equation 5.3 differs from that in Equation 5.2 in that there are three product terms (vs. 2), the P factor occurs in two terms (vs. 1), and the vowel factor occurs in one term (vs. 2). Because the P and C factors have only two levels each and the V factor at least 15 levels, the model in Equation 5.3 has far fewer parameters (21 vs. 34), although it has more terms.

Klatt's model and the model in Equation 5.3 are merely instances of a very large set of possible models, which, for obvious reasons, we call *sum-of-products models* (van Santen, 1993a). Below, in section 5.4, we will show that this class of models has properties that make them uniquely suitable for describing the types of interactions often observed in duration: near-absence of reversal interactions, but ubiquity of additive and multiplicative interactions.

5.4 The Sum-of-Products Approach: Theory

In our system, we describe the input domain of the duration module as a factorial space. We then model duration in two phases. First, we split the space along some standard distinctions such as vowels vs. consonants, ultimately producing a tree. Second, we model the cases subsumed under each terminal node of the tree by a sum-of-products model.

5.4.1 What is a sum-of-products model?

According to sum-of-products models (see Section 2.3.2), the duration for a phoneme/context combination described by the feature vector $\vec{f}$ is given by:

$$\mathrm{DUR}(\vec{f}) = \sum_{i \in K} \prod_{j \in I_i} S_{i,j}(f_j). \qquad (5.4)$$

Here, K is a set of indices, each corresponding to a product term. I_i is the set of indices of factors occurring in the i-th product term. For example, in Equation 5.3 there are three product terms with index sets {2,3}, {3}, and {1}. The factors are indexed as 1 (V), 2 (C), and 3 (P). Note that the concept *product* refers to "product of 1 or more"; here, two of three products are trivial — one-item — products.

We already provided some examples in Equations 5.2 and 5.3. The additive model (Equation 5.1) is another example. For the additive model, $K = \{1, ..., N\}$, and $I_i = \{i\}$, and for the multiplicative model, $K = \{1\}$, and $I_1 = \{1, ..., N\}$.

It is shown in (van Santen, 1993a) that the structure of sum-of-products models (i.e., the index sets I_i) can be inferred from data by subtracting certain marginal means. Similar to the analysis of variance, where the absence or presence of each interaction term can be tested separately from the other interaction terms, the proposed analysis scheme for sum-of-products models allows determining for each individual product term whether it is needed to account for the data. This is important for practical reasons, because the number of distinct sum-of-products models grows extremely rapidly with the number of factors (roughly given by $2^{2^{N-1}-1}$), so that it is in practice not possible to exhaustively fit each model to a given data set.

Directional invariance

As mentioned, most of these models predict directional invariance (see Section 2.3.1). Consider, for example, in Equation 5.3. Suppose that for $P = \mathit{final}$ and $C = \mathit{voiced}$, we find that

$$\mathrm{DUR}(/\mathrm{ai}/, \mathit{voiced}, \mathit{final}) > \mathrm{DUR}(/o/, \mathit{voiced}, \mathit{final})$$

It is easy to show that this implies

$$S_{3,1}(/\mathrm{ai}/) > S_{3,1}(/o/)$$

so that

$$\mathrm{DUR}(/\mathrm{ai}/, \mathit{unvoiced}, \mathit{medial}) > \mathrm{DUR}(/o/, \mathit{unvoiced}, \mathit{medial}).$$

This is so independently of the actual values of the parameters $S_{1,2}(\mathit{voiced})$, $S_{1,2}(\mathit{unvoiced})$, $S_{1,3}(\mathit{final})$, $S_{1,3}(\mathit{medial})$, $S_{2,3}(\mathit{final})$, $S_{2,3}(\mathit{medial})$, $S_{3,1}(/\mathrm{ai}/)$, and $S_{3,1}(/\mathrm{Ai}/)$.

This non-reversal prediction is *inherent in the structure of the model*, as represented by the choice of the index sets {2,3}, {3}, and {1}. For the effects of the other factors, this prediction is not made. For example, by allowing $S_{1,3}(final)$ to be negative, the effects of C may reverse.

The directional-invariance property of sum-of-products models may be responsible for recent results obtained by Maghbouleh (1996). Maghbouleh trained two segmental duration models, and then tested their generalizability on new data, both "somewhat new" data taken from portions held out from the training corpus and "very new" data taken from entirely different corpora. Generalizability was measured by the correlation between observed and predicted segmental durations. One model was the Classification and Regression Tree model (*CART*) (Riley, 1992), the other a sum-of-products model. In CART, in the training phase, a tree is formed by successively dichotomizing the factors (e.g., the stress factor is split into {1-stressed, 2-stressed} vs. {unstressed}) to minimize the variance of the durations under the two newly formed subsets of the speech corpus. For each node of the tree, the observed average duration of the associated subset of the speech corpus is listed. In other words, CART is a general-purpose statistical method that imposes little structure on the data. In effect, it is a condensed lookup table.

Results were straightforward: a training corpus of a few hundred data points was sufficient for the sum-of-products model to reach an asymptote at generalizability levels higher than reached by CART, even after training the latter on as many as 10,000 data points. Moreover, the difference was more pronounced for the "very new" data than for the "somewhat new". We claim that these results are due to the fact that the directional invariance property allow sum-of-products models to *extrapolate* from cases observed in the training data to new cases, whereas CART does not; in fact, the extrapolation concept is alien to CART.

5.4.2 Heterogeneity: sub-categorization

Obviously, not all interactions are directionally invariant. First, some factors are irrelevant for certain regions of the linguistic space. For example, the factor of combined stress, which is clearly important for intervocalic consonants, is irrelevant for vowels because English contains very few words with three or more successive vowels. It thereby becomes difficult to model vowels and intervocalic consonants with a single model. Second, even factors that are relevant in more than one region may have large effects in some of these regions and either no effects or differently ordered effects in other regions. For example, for voiceless stops in word onsets, being preceded by a tautosyllabic /s/ has the dramatic effect of deaspiration, but for word-final voiceless stops there is no such effect because they already are deaspirated. Again, using a single model to cover these two cases may not be possible.

What is required is to divide (and sub-divide) the feature space into categories that are *homogeneous* in the sense that within-category interactions are well-behaved and can hence be captured by sum-of-products models. In other words, one has to construct a *tree* with as leaves not actual durations — as in (Riley, 1992) and (Pitrelli and Zue, 1989) — but sub-categories characterized by a particular sum-of-products model together with estimated parameter values. Thus, where homogeneity is defined by these authors in terms of the variance of the durations of the cases subsumed under a leaf, here homogeneity is defined in terms of a goodness-of-fit criterion. As a consequence, factor levels that differ sharply in duration such as a short and a long vowel will be on different leaves in a variance-based tree, but will be on the same leaf in a goodness-of-fit based tree provided that these two vowels are *similarly affected by the same set of contextual factors*. There is strong evidence for this type of equivalence between vowels (van Santen, 1992a). This type of *parallel behavior* of different phonetic segments in the same phonetic class (e.g., vowels, voiced stop closures) is an important phenomenon that should be taken advantage of; variance-based trees do not do this.

5.5 The Sum-of-Products Approach: Practice

Here we describe how one in practice constructs a tree, selects sum-of-products models, and estimates parameters. The crux of the process is to divide activities between the human expert and the computer — leaving to either what they do best.

5.5.1 Overview

Three types of activity are involved:

Decisions based on the research literature

The categories in the tree involve conventional phonetic and phonological distinctions, and are based on known facts about segmental duration in a given language. Likewise, the decision on which factors to initially include in the analysis is not based on training data — although the factors ultimately included do involve training data (see next item).

It should be noted that any duration system construction — whether fully automatized or not — critically involves decisions of this type, because all approaches need a description of the context of a segment that leaves out certain pieces of information (e.g., the identity of the segment eight segments removed from the target segment). Which information is left out is an a priori decision that is not based on the training data at hand but is based on broader knowledge about segmental duration. Thus, the distinction between data based vs. knowledge based approaches is not an absolute one — all approaches use prior knowledge and data. The key distinction is how much prior knowledge is used, how sophisticated it is, and the degree to which a model can incorporate or represent this knowledge. Perhaps the key strong point of our approach is that it makes good use of both prior knowledge and data. Conventional rule based approaches such as the Klatt system are not very good at incorporating data (because model parameters are not estimated in a cogent, globally optimal manner), whereas standard data based approaches such as CART do not incorporate enough knowledge (such as directional invariance, and why certain categories must be analyzed separately).

Exploratory data analysis

This involves applying a variety of tools to answer the following questions:

- Which levels should be distinguished on a given factor for each category?
- Which factors interact?
- What equation best describes the observed pattern of interactions?

Parameter estimation

For a given leaf on the tree, once an equation has been selected, parameters are estimated with standard least-squares methods.

5.5.2 Exploratory analysis of duration data

"Holes" in training data have immediate effects on statistical analysis. Standard tools for analysis of multi-factorial data (i.e., data where some dependent variable — such as duration — depends on multiple factors) fare best when each factorial constellation occurs *equally often* (this is, of course, diametrically the opposite of the sparsity discussed earlier). One can then ascertain which distinctions to make on a factor simply by computing the marginal means (i.e., for each level in a factor one computes the mean duration over all observations where the factor has this level). However, when some constellations occur more often than others, these marginal means can be deceptive. For example, in two-syllable words, vowels in the first syllable are on average slightly longer than vowels in the second syllable (van Santen, 1992a). However, this is caused by the fact that initial syllables are more often stressed than second syllables. If one measures vowel duration in stressed syllables only, vowels in the second syllable are substantially longer than in the first syllable. So, if on the basis of the marginal means one had decided to collapse the within-word position factor, this would have lead to poor prediction of duration.

There are methods that can handle factor confounding (van Santen, 1992a) (see Section 2.3.1). In one of these methods (quasi-minimal sets analysis; see Section 2.3.2), for a given critical factor (in the example above, within-word position) one finds a subset of the data where all constellations are matched on a number of *matching* factors[5] and differ only on the critical factor. In that way, one removes the confounding effects of the matching factors and can get

[5] Factors one has reason to believe cannot be ignored in the analysis; in the example above, syllabic stress, but probably also vowel identity and phrasal position.

a cleaner look at the genuine effects of the critical factor. *However, when the training data have large holes, this analysis can often not be performed because no subsets matched on important factors can be found.*

A second method allows for bigger holes in the data. Here, we assume that the effect of the critical factor and the *joint effects* of the remaining factors combine multiplicatively. If we denote the j-th factor as F_j (where F_j has levels f_j, f_j', f_j'', ...), the critical factor as F_1, and define DUR(f_1, ..., f_n) to be the "true" duration for the cell defined by F_1 having level f_1, F_2 having level f_2, etc., then we are assuming that

$$\mathrm{DUR}(f_1, ..., f_n) \quad = \quad \mathrm{A}(f_1) \times \mathrm{B}(f_2, ..., f_n). \tag{5.5}$$

Here, $A()$ and $B()$ represent unknown parameters to be estimated from the data. They reflect the contributions of the *critical factor* F_1 and the *corrective factors* F_2 ,..., F_n, respectively. One could also have assumed an additive combination rule, where in Equation 5.5 the $\times$-sign is replaced by the $+$-sign. The key point is that estimates of $A()$ are estimates of the mean durations of the levels of factor F_1 from which we have removed the effects of the confounding factors. Thus, these estimates can be considered to be *corrected marginal means* (See Section 2.3.2 in Chapter 2).

This method can handle bigger holes than the quasi-minimal pairs approach because it requires only as many constellations as there are parameters to be estimated. The number of parameters can be as small as the number of levels on the critical factor. Most importantly, the method does not require a subset where all levels on the critical factor are matched on all matching factors. It only requires that for most pairs of levels there is such a subset.

But even this method has limits in terms of how large a hole it can accept. And these holes can be very large indeed. For example, in English it is known that vowel duration depends on the identities of the preceding two and the following two segments. Thus, /ɛ/ is longer in *bend* than in *bent* or *bet*, and /a/ is longer in *lot* and *slot* than in *plot* or **splot* (van Santen, 1992a). Now, if one encodes each segment in terms of the identities of the surrounding four segments (with 42 possibilities each), one has created a space of 3,111,696 possibilities[6] — to be further multiplied by the remaining factors. Yet, in a fairly large (54,336-segment) data base (van Santen, 1992a) only 30,120 of these possibilities where present; and of these only 1,315 occurred more than five times — a reasonable minimum for reliable statistical estimation given the inherent instability of timing in speech. These 1,315 constellations constitute less than 0.05 percent of the total space. In other words, we must generalize from this 0.05 percent to the remaining 99.95 percent.

[6] Unless one confines phoneme strings to be contained inside words, there are remarkably few segment sequences that can *not* occur in a language. For example, with 42 phones, 74,088 is the theoretical maximum number of possible triphones. Of these, 48,092 actually occur in high-frequency English names. However, only 717 can occur as the initial triphone in first names.

Under these circumstances, the parameter estimation involved in multiplicative correction (or any other modeling effort) runs into two problems:

1. Parameter estimability is not merely a matter of comparing the number of parameters with the number of distinct constellations. The distribution of non-empty "cells" must have a particular pattern (Dodge, 1981). For example, when — by the type of bad luck all too common in sparse data — some factor level occurs exactly once and occurs in a constellation with some level on another factor that also occurs exactly once, then the parameters cannot be estimated.

2. Even when parameters can be estimated, this does not mean that the estimates are acceptable. In particular, estimates of parameters for rare levels of factors can be unreliable. This is further compounded by the inherent instability of speech timing.

As a consequence, *any statistical analysis requires sparsity reduction*, i.e., reduction in the number of factors or distinct levels on factors. The Catch-22 is that reducing sparsity requires the very statistical analyses that are made impossible by sparsity.

The only solution lies in using judgment based on the research literature or other prior knowledge. For example, in English the /ž/ phoneme (as in *measure*) is quite rare. However, /ž/ is a voiced fricative, and it is known that all other voiced fricatives (/z/ as in *rose*, /v/ as in *love*, /ð/ as in *bathe*) cause preceding vowel duration to be longer than any other consonant class. Thus, on the postvocalic consonant factor, one may decide to pool /ž/ with the other voiced fricatives.

This type of judgment plays also a critical role in constructing the text used for the training data. Some of the sparsity problems can be avoided by carefully constructing this text; e.g., by increasing the number of /ž/ occurrences. But to be able to do that, one has to be able to specify the factors and which distinctions are likely to be relevant *prior to data collection*.

In summary, the role of prior knowledge is critical both for data collection and for data analysis. When the research literature contains little information about timing in the language of interest, it is advisable to conduct some carefully designed single-factor pilot experiments where one analyzes the effect of a factor while holding all other factors constant. Such experiments nonetheless cannot *replace* a large data base, because they do not allow for measurement of interactions and because, unless one is very careful, differences between single-factor pilot studies can produce parameter estimates that jointly do not provide a consistent picture of a speaker speaking in a single mode and rate.

5.5.3 Statistical package

We use an internally developed statistical package for exploratory data analysis and parameter estimation. We list here the capabilities of this package. Prior to analysis, the data are summarized in a data table which lists for each segment in the data base both descriptive information (the vector $\vec{f}$) and the measured duration.

During analysis, the user constructs a directory structure, representing the tree. Each subdirectory contains the relevant subset of the data table. It also contains "feature map files", specifying for a given factor which factor levels can be combined. E.g., the first column contains all phoneme labels, and the second column phoneme class labels such as "voiced stop closure". The effect is that in the ensuing statistical analyses, all voiced stop closures will be treated the same.

The capabilities of this package are the following:

- **Feature Map:** Edit specific feature map.
- **Select:** Create subtree.
- **Raw Marginals:** Raw marginal means.
- **Raw 2-way Marginals:** Raw 2-way marginal means.
- **Corrected Marginals:** Marginal means corrected for other factors.
- **Corrected 2-way Marginals:** Corrected 2-way marginals.
- **SoP Structure:** Explore space of sum-of-products models.
- **SoP Estimation:** Estimate and display sum-of-products model parameters.
- **Addmodel Estimation:** Estimate and display additive model parameters.
- **Multmodel Estimation:** Estimate and display multiplicative model parameters.
- **Residuals:** Residuals from SoP model.
- **Variance Explained:** Variance explained by models.
- **Extremes:** Find extreme durations in selected data subset.
- **Aov** N-way analysis of variance.

The user typically starts by creating a few subtrees based on conventional distinctions. Next, an iterative process follows where feature maps are continually changed based on results from the various "Marginals" analyses (see Section 2.3.2). In this process, phonetically informed judgment is critical. When data are quite sparse, it is often impossible to use sum-of-products models other than the additive or multiplicative models. The items "Residuals", "Variance Explained", and "Aov" are used to evaluate results.

As we have noted previously, the system is truly multilingual in that all information pertinent to a given language is stored in tables that are external not only to the run time modules but also to the analysis software. Finally, we note that to create run-time tables, a compiler program takes the results from the analysis phase and automatically converts them into tables. Thus, one can instantaneously listen to the perceptual consequences of different analyses.

5.5.4 Applications to multilingual synthesis

The Bell Labs Text-to-Speech System has employed this procedure for all covered languages. Here we describe as examples Mandarin Chinese (Shih and Ao, 1997), French (Tzoukermann and Soumoy, 1995), and German (Möbius and van Santen, 1996).

The system has a group of *core factors*, which is the union of all factors used in any of our languages. Naturally, this group expands when some highly language-specific factors such as Mandarin tone are encountered, or when new theoretical insights are gained (see below). However, as the number of covered languages grows, we expect the group of core factors to become reasonably complete. For example, for some languages (Spanish, German) no additional factors were required.

Among the core factors are the following. First, *phone identity factors*: identity of the (1) current segment, (2) previous segment(s), and (3) next segment(s). Second, *stress-related factors*: (4) degree of discourse prominence, and (5) lexical stress. And third, *locational factors*: location of the (6) segment in the syllable, (7) syllable in the word, (8) word in the phrase, (9) phrase in the utterance.

These particular factors imply certain theoretical choices. For example, even though the internal data structure of our synthesizer contains information about the stress levels of syllables surrounding a given syllable, this information is not extracted in the form of a durational factor. Thus, no information about stress feet can be used by the timing component. However, precisely because the information is available in the data structure, only small changes in the source code are required to include factors representing stress feet.

However, in another sense, these factors are reasonably theory-neutral. For example, we can fully express in terms of the factors at hand a phonological theory which claims the importance of syllable-openness in conjunction with the number of syllables in a word and the location of the stressed syllable in the word. More generally, while some theories may require new factors, others can be expressed by combining (and collapsing levels on) available factors.

Mandarin. Training text was created by greedy text selection methods applied to a text corpus of 15,620 newspaper sentences.

We had to add three *tone-related factors* to the group of core factors: identity of the current, previous, and following tone. A second Mandarin-specific factor, Syllable type (*CG-vowel-C, CG-vowel, C-vowel-C, CG-*

	Mandarin	French	German
Number of Cases	46,265	7,143	24,240
Number of Sub-classes	6	16	30
Number of Parameters	298	782	674
Correlation	0.872	0.847	0.896
Rms	0.026	0.025	0.019
Obs Mean	0.076	0.074	0.060
Obs Std Dev.	0.053	0.047	0.043

Table 5.1: Number of phonetic segments, terminal nodes, estimated parameters, correlation and root mean squared deviation of observed and predicted durations (Rms, in seconds), and mean and standard deviation of the observed durations (both in seconds).

diphthong, C-vowel, C-diphthong, vowel-C, vowel, and *diphthong*; here, C is a non-glide consonant, and G a glide), was computable from the core factors.

A classification scheme was constructed with six subclasses: *vowels*, *fricatives*, *plosive bursts and aspirations*, *plosive closures*, *sonorants in syllable onsets*, and *sonorants in syllable codas*. This classification is exhaustive, because codas can only contain sonorants in Mandarin.

Among findings that we typically have not seen in other languages were the following. First, absence of utterance-final lengthening. Second, a compensatory effect where coda consonants were shortened by intrinsically long nuclei. Some of the overall statistics are shown in Table 5.1.

French. Similar to Mandarin, the training text was created by greedy text selection methods applied to a text corpus of 15,000 newspaper sentences.

A factor specific to French is whether a consonant is created by *liaison* (see Section 3.6.2). Here, *les autres* (the others) must be pronounced /le zotr(ə)/ and not /le otr(ə)/, with the addition of an intervocalic consonant /z/. If the second word does not start with a vowel, the /z/ is not pronounced.

Issues of special interest are complete *schwa deletion* in words such as /guvɛrnmɑ̃/ *gouvernement* 'government', and the high degree of confounding of lexical stress (always on word final syllable unless its nucleus is a schwa) with intra-word location.

It was found that consonants created by liaison are shortened relative to consonants in other contexts. Also, certain stressed vowels are shorter in open than in closed syllables. This is of particular interest because it contradicts the hypothesis that speakers tend to keep syllable durations constant.

German. Our study of German duration used the Kiel Corpus of Read Speech, recorded and manually segmented at the Kiel Phonetics Institute and published on CD-ROM (Institut für Phonetik und digitale Sprachverarbeitung, 1994).

No German-specific factors or factor levels were found to be necessary. Likewise, the sub-classification scheme followed standard distinctions — the same as used for English (van Santen, 1994) — except that vowels are split into central vowels (schwa), diphthongs, and full (non-central) monophthongs.

Among the key findings were that, for consonants in syllable onsets, stops were more than doubled in length by syllabic stress, whereas fricatives were hardly affected. In contrast to English (van Santen, 1992a), the proportional effect of stress on vowel duration was significantly larger in utterance final position than in utterance medial position.

5.6 Future Developments

The durations obtained with our method are quite acceptable both statistically and perceptually (van Santen, 1994). Also, the procedure is quite efficient: once a segmented data base is available, the analyses can be performed in a few days. The amount of manual labor is proportional to the sparsity of the data, because it consists mostly of catching the resulting statistical artifacts. Sparsity will be reduced in the future by better automatic segmentation systems that will provide us with much larger data bases. It is quite likely that ultimately the entire procedure can be automatized, with human supervision focusing entirely on data base design.

However, we want to remind the reader of three inherent limitations in our approach. First, the predicted durations can only be as good as the veracity of the information provided to the duration component by text analysis components. The prediction of prosodic factors such as pitch accent placement and prosodic phrase boundary location (Chapter 4) is still only partially adequate, and much more work needs to be devoted to these fundamental problems in text analysis.

Second, we must in the future do a better job of modeling sub-segmental timing. This is particularly clear in the case of utterance-final lengthening, where it is primarily the latter half of the utterance-final that syllable must be stretched.

Third, there is a sizable amount of durational variability that cannot be predicted from text. We conjecture that this durational variability is not random, and has systematic dynamic patterns that one should attempt to describe mathematically.

Chapter 6

Intonation

Jan van Santen, Chilin Shih, Bernd Möbius

The task of the intonation module in the TTS system is to compute an F_0 contour from the same input as is used by the duration component, with phoneme durations added.

Intonation research is extremely diverse in terms of theories and models, much more than is the case with duration. On the phonological side, there is little consensus on what the basic elements should be — tones, uni-directional motions, multi-directional gestures, etc. Reflecting the situation on the phonological side, modeling at the phonetic level is also quite diverse, including interpolation between tonal targets (Pierrehumbert, 1980), superposition of underlying phrase and accent curves (Fujisaki, 1983; Fujisaki, 1988), and concatenation of line segments ('t Hart, Collier, and Cohen, 1990).

The status of our own work on intonation reflects this diversity. Traditionally, we have used a tone sequence and interpolation method for English, for tone languages, and for certain other languages. However, recently we have moved to a different approach, which is in the superposition tradition. We use this approach now for all languages except Chinese, Japanese, and Navajo. Although we conjecture that this approach can also be used for the latter three languages, we do not have proof of that yet.

In this chapter, we first give a bird's eye overview of some of the major controversies in intonation research. We then describe our work in the tone sequence approach. Finally, we describe the new method — with more emphasis on the empirical results, because very few publications on this method exist.

6.1 Issues in Intonational Theory

The fundamental problem for intonation analysis and synthesis is that the variation of fundamental frequency (F_0) as a function of time is the acoustic correlate of a number of linguistic prosodic features with quite diverse time domains: from utterances or even paragraphs through accentual phrases or accent groups to words and syllables. Whether F_0 movements are caused by, for instance, word accentuation or phrase intonation, cannot easily be decided by means of acoustic measurements or by perceptual criteria. A separation of these effects, however, can be achieved on a linguistic, i.e. more abstract, level of description. Here rules can be formulated that predict accent or phrase related patterns independent of, as well as in interaction with, each other.

Two major classes of intonation models have evolved in the course of the last two decades. There are, on the one hand, hierarchically organized models that interpret F_0 contours as complex patterns resulting from the superposition of several components (*superposition models*). On the other hand, there are models that claim that F_0 contours are generated from a sequence of phonologically distinctive tones, or categorically different pitch accents, that are locally determined and do not interact with each other (*tone sequence models*).

The main difference between the tone sequence models and the superpositional models can be characterized by how they define the relation between local movements and global trends in the intonation contour. The competing points of view are illustrated by the following two quotations:

> The pitch movements associated with accented syllables are themselves what make up sentence intonation ... there is no layer or component of intonation separate from accent: intonation consists of a sequence of accents, or, to put it more generally, a sequence of tonal elements. (Ladd, 1983a, page 40)

> Standard Danish intonational phenomena are structured in a hierarchically organized system, where components of smaller temporal scope are superposed on components of larger temporal domain ... These components are simultaneous, parametric, non-categorical and highly interacting in their actual production. (Thorsen, 1988, page 2)

We will now describe the features of the prevalent intonation models of both categories. In the subsequent section we will discuss two other approaches, namely the intonation models developed at IPO (Instituut voor Perceptie Onderzoek [Institute for Perception Research], Eindhoven) and the Kiel model, respectively, which defy a categorization as being either of the superpositional or the tone sequence type.

6.1.1 Tone sequences or layered components?

Tone sequence models

Arguably the most influential work based on the tone sequence approach is Janet Pierrehumbert's dissertation (Pierrehumbert, 1980). Pierrehumbert's intonation model builds upon metrical (Liberman and Prince, 1977) and autosegmental phonology (Leben, 1976). The starting point for her model is a metrical representation of the sentence that yields, first, rule-based information on strong and weak syllables and, second, a sequence of high and low tones that are associated with stressed syllables and prosodic boundaries by way of context sensitive association rules.

This formal description is a continuation of the tone level tradition (Pike, 1958; Trager and Smith, 1951), but Pierrehumbert avoids the theoretical and methodological pitfalls of the earlier approaches by casting her model within the more constrained theory of autosegmental phonology. As Bolinger (1951) pointed out, the most essential shortcomings of Pike's four level theory are the lack of principled separation between the tone levels in the paradigmatic domain as well as the lack of explicit criteria for the transition from one tone to another in the syntagmatic domain. Autosegmental phonology provides two or more parallel tiers, each of which consists of a sequence of phonological segments, and whose mutual relations are determined by association rules; segments are defined as "the minimal unit of a phonological representation" (Goldsmith, 1990, page 10) and do not necessarily correspond to phonetic segments. In the case of intonation, specific rules associate the tone tier with syllabic, phonemic or subphonemic tiers.

In Pierrehumbert's model, an *intonational phrase* is represented as a sequence of high (H) or low (L) *tones*. H and L are members of a primary phonological opposition. The tones do not interact with each other but merely follow each other sequentially in the course of an utterance. There are three types of accents:

1. *Pitch accent*, marked by the "*" symbol (H*, L*); pitch accents can be bi-tonal. (H*+L, H+L*, L*+H, L+H*).

2. *Phrase accent*, marked by the "-" symbol (H-, L-). It is used to represent pitch movement between a pitch accent and a boundary tone.

3. The *boundary tone*, marked by "%" (H%, L%), is aligned with the edges of a phrase.

The combination of the three accent types are constrained by a finite-state grammar which generates well-formed intonational representations (Pierrehumbert, 1980) and can be formulated as follows:

$$\left\{ \begin{array}{c} H* \\ L* \\ H*+L \\ H+L* \\ L*+H \\ L+H* \end{array} \right\}^{+} \left\{ \begin{array}{c} H- \\ L- \end{array} \right\} \left\{ \begin{array}{c} H\% \\ L\% \end{array} \right\}$$

The regular expression stipulates that an English intonation phrase consists of three parts as indicated by the three sets of curly braces: one or more ($^+$) pitch accents, followed by one phrase accent, ending with one boundary tone. In each set of curly braces there is a list of possible choices at each state: six types of pitch accent, two types of phrase accent, and two types of boundary tones.

The following example (from (Pierrehumbert, 1980, page 276)) illustrates the tonal representation of a sentence consisting of one intonational phrase, and the association of the tonal tier with the segmental tier, here represented by orthography:

```
That's a remarkably clever suggestion.
           |                |
           H*               H* L- L%
```

The abstract tonal representation is then converted into F_0 contours by means of a set of *phonetic implementation rules*, which also takes care of the temporal alignment of the tones with the stressed syllables. The phonetic rules determine the F_0 values of the H and L tones, based on information such as the prominence level and the preceding tones. Calculation of the F_0 values of tones is performed strictly from left to right, depending exclusively upon the already processed tone sequence and not taking into account the subsequent tones.

Pierrehumbert's intonation model is predominantly sequential; what is treated in other frameworks as the correlates of the phrase structure of a sentence or the global trends such as question or declarative intonation patterns, is conceptualized as elements of the tonal sequence and their (local) interaction. The framework stipulates, for example, that the English question intonation is embodied in the tonal sequence L* H- H%, and that there is no separate phrase-level "question intonation contour" that these targets ride on. Similarly, the observed downtrend in some types of sentences, particularly in list intonation, is accounted for by the downstep effect triggered by downstepping accents, such as H*+L, rather than being attributed to a phrase level intonation which affects all pitch accents equally.

There are a few aspects of the model that are hierarchical or non-local. According to Pierrehumbert's own assessment, the model is situated at the interface between intonation and metrical phonology, and inherits the hierarchical organization of the metrical stress rules. Another element whose effect is global is declination, onto which the linear sequence of tones is overlaid. Given these properties, Ladd (1988) characterized Pierrehumbert's model as a hybrid between the superposition and the tone sequence approach. Furthermore, discourse structure is hierarchically organized, and the information is used to control F_0 scaling so that the pitch height of discourse segments reflect the discourse hierarchy (Hirschberg and Pierrehumbert, 1986; Silverman, 1987).

A later paper (Liberman and Pierrehumbert, 1984) took a stronger position regarding the local nature of tone scaling, in that the authors concluded that most of the observed downtrend is attributed to downstep and that there is no evidence of declination in English. This position is reflected in the implementation of this model for TTS (Anderson, Pierrehumbert, and Liberman, 1984), as discussed in more detail in Section 6.2.1. The declination effect was reintroduced in the intonation model of Japanese (Pierrehumbert and Beckman, 1988).

D. Robert Ladd's phonological intonation model (1983b) is based on Pierrehumbert's work but integrates some aspects of the IPO approach ('t Hart, Collier, and Cohen, 1990) and the Lund intonation model (Bruce, 1977; Gårding, 1983) as well (see below). Like Pierrehumbert, Ladd applies the framework of autosegmental and metrical phonology. He attempts to extend the principles of feature classification from segmental to suprasegmental phonology, which would also facilitate cross-linguistic comparisons. In Ladd's model, F_0 contours are analyzed as a sequence of structurally relevant points, i.e., accent peaks and valleys, and boundary endpoints, each of which is characterized by a bundle of features. Acoustically, each tone is described in terms of its height and position relative to the segmental chain. Tones are connected by straight-line or smoothed F_0 transitions.

We will now turn to a discussion of superpositional intonation models.

Superposition models

Nina Grønnum (Thorsen) developed a model of Danish intonation (see (Grønnum, 1992) for a comprehensive presentation) that is conceptually quite different from the tone sequence approach. Her intonation model is hierarchically organized and includes several simultaneous, non-categorical components of different temporal scopes. The components are layered, i.e., a component of short temporal scope is superimposed on a component of longer scope.

Grønnum's model integrates the following components. The highest level of description is the text or paragraph, which requires a discourse-dependent intonational structuring (*text contour*). Beneath the text there are influences of the sentence or the utterance (*sentence intonation contour*) and of the prosodic phrase (*phrase contour*). The lowest linguistically relevant level is represented by *stress group* patterns. The four components are language-specific and actively controlled by the speaker. The model also includes a component that describes microprosodic effects, such as vowel intrinsic and coarticulatory F_0 variations. Finally, a Danish-specific component models the *stød*, a creaky voice phenomenon at the end of phonologically long vowels, or on the post-vocalic consonant in the case of short vowels; the *stød* is usually interpreted as a synchronic reflection of distinctive tonal patterns on certain syllables, which is lost in modern Danish but still exists in Norwegian and Swedish (Fischer-Jørgensen, 1987).

All components of the model are highly interactive and jointly determine the F_0 contour of an utterance. Therefore, for the interpretation of observed natural F_0 curves, a hierarchical concept is needed that allows the analytical separation of the effects of a multitude of factors on the intonation contour.

An important element of Grønnum's model is the notion of *stress group*. A stress group consists of a stressed syllable and all following, if any, unstressed syllables. In Danish, stressed syllables are tonally marked by local F_0 minima. The precise pattern of the stress group depends upon its position in the utterance and the sentence intonation contour onto which it is superimposed. The global slope of the sentence intonation contour, which is construed as a line connecting the stressed syllables in the utterance, correlates with sentence mode. Declaratives exhibit the steepest declining slope, whereas syntactically and lexically unmarked interrogatives (*echo questions*) show almost flat sentence intonation contours (also cf. (Möbius, 1995) for German).

Similar to Grønnum's work, the Lund intonation model (Bruce, 1977; Gårding, 1983) analyzes the intonation contour of an utterance as the complex result of the effect of several factors. A *tonal grid*, whose shape is determined by sentence mode and by pivots at major syntactic boundaries, serves as a reference frame for local F_0 movements. Thus, implicitly it is assumed that the speaker pre-plans the global intonation contour. At first glance, the Lund model also includes elements of the tone sequence approach in that it represents accents by sequences of high and low tones. But the difference with Pierrehumbert-style models is that the position and height of the tones is determined by the tonal grid, which is, by definition, a non-local component. Still, the Lund model indicates that it is possible to integrate aspects of the superpositional and the tone sequence approaches.

The classical superpositional intonation model has been presented by Hiroya Fujisaki (Fujisaki, 1983; Fujisaki, 1988). It can be characterized as a functional model of the production of F_0 contours by the human speech production apparatus, more specifically by the laryngeal structures; the approach is based on work by Öhman and Lindqvist (1966). The model represents each partial glottal mechanism of fundamental frequency control by a separate component. Although it does not include a component that models intrinsic or coarticulatory F_0 variations, such a mechanism could easily be added in case it is considered essential for natural-sounding synthesis.

The model additively superimposes a basic F_0 value (F_{min}), a *phrase component*, and an *accent component*, on a logarithmic scale. The control mechanisms of the two components are realized as critically damped second-order systems responding to impulse commands in the case of the phrase component, and rectangular commands in the case of the accent component. These functions are generated by two different sets of parameters: the timing and amplitudes of the phrase commands as well as the damping factors of the phrase control mechanism on the one hand, and the amplitudes and the timing of the onsets and offsets of the accent commands as well as the damping factors of the accent control mechanism on the other hand. The values of all these parameters are constant for a defined time interval: the parameters of the phrase component within one prosodic phrase, the parameters of the accent component within one accent group, and the basic value F_{min} within the whole utterance.

The F_0 contour of a given utterance can be decomposed into the components of the model by applying an analysis-by-synthesis procedure. This is achieved by successively optimizing the parameter values, eventually yielding a close approximation of the original F_0 curve. Thus, the model provides a parametric representation of intonation contours.

Adequate models are expected to provide both predictive and explanatory elements (Cooper, 1983). In terms of prediction, models have to be as precise and quantitative as possible, ideally being mathematically formulated. Fujisaki's model exploits the principle of superposition in a strictly mathematical sense, and succeeds in analyzing a complex system in such a way that both the effects of individual components and their combined results become apparent.

6.1.2 Other intonation models in speech synthesis

The IPO model

The intonation model developed at IPO shares some basic assumptions with the tone sequence approach (see ('t Hart, Collier, and Cohen, 1990) for a comprehensive presentation). The model postulates the primacy of (sentence) *intonation* over (word or syllable) *accenting*. Accenting is subordinate to intonation in the sense that the global intonation pattern of an utterance imposes restrictions for the sequence and shape of F_0 movements. However, intonation and accenting are not seen as two distinct levels of prosodic description; rather, the intonation contour of an utterance consists of consecutive F_0 movements.

The reason for not including the IPO model in the tone sequence category is that it is an essentially perception based, *melodic* model. One of its starting points is the observation that certain F_0 movements are perceptually relevant for the listener, while others are not.

The IPO method of intonation analysis consists of three steps. First, the perceptually relevant movements are *stylized* by straight lines. The task is "to draw a graph on top of the F_0 curve which should represent the course of pitch as a function of time as it is perceived by the listener" ('t Hart, 1984, page 195). The procedure results in a sequence of straight lines, a *close copy contour*, that is perceptually indistinguishable from the original intonation contour: the two contours are *perceptually equivalent*. The reason for stylizing the original intonation, according to the IPO researchers, is that the enormous variability of raw F_0 curves presents a serious obstacle for finding regularities.

In a second step, common features of the close copy contours, expressed in terms of duration and range of the F_0 movements, are then *standardized* and collected as an *inventory* of discrete, phonetically defined types of F_0 rises and falls. These movements are categorized according to whether or not they are *accent lending*; for example, in both Dutch and German, F_0 rises occurring early in a syllable cause this syllable to be perceived as stressed, while rises of the same duration and range, but late in the syllable, are not accent lending. Similarly, late falls produce perceived syllabic stress but early ones do not. The notion of accent lending adds a functional aspect to the otherwise purely melodic character of the model.

In the third and final step, a *grammar* for possible and permissible combinations of F_0 movements is written. The grammar describes both the grouping of pitch movements into longer-range contours and the sequencing of contours across prosodic phrase boundaries. The contours must comply with two criteria: they are required to be perceptually similar to, and as acceptable as, naturally produced contours. Thus, the complete model describes the melodic possibilities of a language.

There are problematic issues at each of the three steps. First, the stylization method can yield inconsistent results when the same original contour is stylized more than once, either by the same or by different researchers, which may yield different parameter values. It is claimed, however, that in practice any inconsistencies are below perceptual thresholds (Adriaens, 1991, page 38) and therefore negligible. Second, a categorization of F_0 movements is not achieved by applying objective criteria but rather by heuristics (Adriaens, 1991, page 56). Even though the procedure is nowhere described in full detail, it is clear that it is guided by one overriding goal, namely a melodic model whose adequacy can be perceptually assessed. Third, the grammar is not prevented from generating contours that are acceptable but have not been observed in natural speech, thus violating the perceptual similarity criterion — which may be too restrictive anyway. More problematic, however, is the possibility of the grammar to generate contours that are perceptually unacceptable.

The IPO model was originally developed for Dutch, but it was later applied to English (de Pijper, 1983), German (Adriaens, 1991), and Russian (Odé, 1989). It has been implemented in speech synthesis systems for Dutch (Terken, 1993; van Heuven and Pols, 1993), English (Willems, Collier, and 't Hart, 1988), and German (van Hemert, Adriaens-Porzig, and Adriaens, 1987).

The Kiel model

The Kiel intonation model (KIM, (Kohler, 1991)), developed for German by Klaus Kohler, is in the tradition of the British school of intonation and in particular of Halliday (1967). However, the actual connection with the British tradition of intonation research is not so much the analysis of utterances in terms of prenucleus-nucleus structures and the pertinent prosodic elements (*tones*, cf. (Kohler, 1977)). Kohler favors this approach for another reason too: it allows for an alignment of intonation contours, their F_0 peaks and valleys, with the segmental structure, whereas intonation models in the American (Pike) tradition atomize the contours into targets or levels.

KIM is designed to be a generative, functional model of German intonation, based on results of research on F_0 production and perception. According to Kohler, KIM differs from most models in that it does not ignore microprosodic F_0 variations, and integrates syntactic, semantic, pragmatic, and expressive functions (*meaning functions*). The model applies two types of rules. Input information to the *symbolic feature rules* is a sequence of segmental symbols annotated for stress and pragmatic and semantic features. The rules convert this input into sequences of binary features, such as $\pm$*late* or $\pm$*terminal*. Finally, *parametric rules* generate duration and F_0 values and control the alignment of the F_0 contour elements with the segmental structure of the target utterance. The parametric rules include rules for the downstepping of accent peaks during the course of the utterance as well as microprosodic rules.

One of the starting points in the development of KIM was the study of accent peak shifts (Kohler, 1987; Kohler, 1990), which discovered three distinct locations of the F_0 peak relative to the segmental structure of the stressed syllable: *early* peaks signal established facts that leave no room for discussion; *medial* peaks convey new facts or start a new argument; *late* peaks put emphasis on a new fact and contrast it to what already exist in the speaker's (or listener's) mind. Thus, shifting a peak backwards from the *early* location causes a category switch from *given* to *new* information.

The functional relevance of F_0 peaks crucially depends upon the ability of the speaker to deliberately and consistently produce the pertinent contours, and of the listener to identify the peak location. Kohler found that in his experiments, untrained speakers were usually unable to consistently produce an intended intonation type, and that they had difficulties in the realization of certain contours, especially early peaks (Kohler, 1991, pages 126–127). Given the claimed communicative relevance of the peak location, this finding is unexpected. However, the functional opposition between early and non-early peaks can be considered as being well established for German, and it is postulated to also exist in English and French (Kohler, 1991, page 126).

KIM has been implemented in the German version of the INFOVOX speech synthesis system (Carlson, Granström, and Hunnicutt, 1989).

6.1.3 Synopsis

Although the tone sequence and the superposition models of intonation diverge in formal and notational terms, they nevertheless may be more similar from a descriptive or implementation point of view.

The strongest position of the tone sequence model is presented in Liberman and Pierrehumbert (1984), where it is claimed that tone scaling is a strictly local operation without global planning, without even declination. There is nonetheless one aspect that is superpositional in nature, which is the additive effect of downstep and final lowering. Most variants of the model, such as the ones presented in (Pierrehumbert, 1980; Pierrehumbert and Beckman, 1988), and several implementations for TTS discussed in Section 6.2, actually incorporate a global declination component and can be seen as a hybrid of the tone sequence model and the superposition model.

Furthermore, Pierrehumbert's and Ladd's models build on autosegmental phonology by assigning a tonal tier whose alignment with syllabic and phonemic tiers is defined by association rules. Autosegmental theory also allows for various independent levels of suprasegmental description and their respective effects on the intonation contour by an appropriate phonological representation. According to Edwards and Beckman (1988), the most promising principle of intonation models is seen in the capability to determine the effects of each individual level, and of their interactions. Although probably not intended by the authors, this can also be interpreted as an argument in favor of a hierarchical approach and of superposition models of intonation. Thus, the conceptual gap between the different theories of intonation does not seem to be too wide to be bridged. In fact, Ladd more recently proposed a metrical approach that incorporates both linear and hierarchical elements (Ladd, 1990).

It is important to note, though, that the notion of *hierarchy* is not necessarily an appropriate criterion for differentiating tone sequence and superposition models, especially since its meaning is ambiguous. Both types of models contain hierarchical elements in the sense that utterances consist of prosodic phrases which in turn consist of accent groups or pitch accents; and even the most influential tone sequence model (Pierrehumbert, 1980) provides a non-local element, namely declination. There is another meaning of hierarchy: to make choices in various components of the prosodic system of a given language, higher levels having priority over, and setting constraints for, lower levels (cf. (Thorsen, 1988)). While Grønnum's model is hierarchical in this sense, there is no such preponderance of one component over another in Fujisaki's model.

Most of the models discussed here assume (explicitly or implicitly) a mechanism of pre-planning in speech production; therefore, the difference between the models should rather be seen in terms of how they represent this mechanism. It is possible for tone sequence models to provide a higher F_0 onset in longer utterances, but the relations between the individual pitch accents are not affected. In Grønnum's model (Grønnum, 1992), utterance length determines the slope of declination, short utterances having a steeper baseline, but it has no effect on the utterance-initial F_0 value.

Even among researchers representing different types of intonation models there is widespread agreement that the F_0 contour of an utterance should be regarded as the complex result of effects exerted by a multitude of factors. Some of these factors are related to articulatory or segmental effects, but others clearly have to be attributed to linguistic categories.

In contradiction to the strong assumption of the tone sequence model (Liberman and Pierrehumbert, 1984) that intonation is determined exclusively on a local level, there is actually ample evidence for non-local factors. In a study of utterances containing parenthetical clauses, Kutik, Cooper and Boyce (1983) show that the intonation contour is interrupted by the parenthetical and resumed right afterwards such that the contour is similar to the contour in the equivalent utterance without parenthesis. Also, Ladd and Johnson (1987) demonstrate how the first accent peak in an utterance is adjusted in height depending upon the underlying syntactic constituent structure. Furthermore, there is evidence that the speaker pre-plans the global aspects of the intonation contour, not only with respect to utterance-initial F_0 values but to phrasing and inter-stress intervals as well (Thorsen, 1985).

The preceding considerations favor models that directly represent both global and local properties of intonation. These models also provide a way of extracting prosodic features related to the syntactic structure of the utterance and to sentence mode. Generally speaking, the analytical separation of all the potential factors considerably helps decide under which conditions and to what extent the concrete shape of a given F_0 contour is determined by linguistic factors (including lexical tone), non-linguistic factors, such as intrinsic and coarticulatory F_0 variations, and speaker-dependent factors.

Superpositionally organized models are particularly appropriate for a quantitative approach: contours generated by such a model result from an additive superposition of components that are in principle orthogonal to, or independent of, each other. The components in turn can be related to certain linguistic or non-linguistic categories. Thus, the factors contributing to the variability of F_0 contours can be investigated separately. In addition, the temporal course pertinent to each individual component can be computed independently.

One of the intonation models currently used in several of the Bell Labs TTS systems (English, Spanish, French, German, Italian, Romanian, and Russian) follows the superpositional tradition (see Section 6.3), while in another version of English, Chinese, Navajo, and Japanese, a Pierrehumbert-style tone sequence approach is applied (see Section 6.2).

6.2 The Tone Sequence Approach

Before we describe the tone sequence intonation module, it is important to understand that several features that are often considered to fall under the rubric of intonation, have already been computed by previous modules of the system: these include lexical and phrasal accentuation, accent type as well as prosodic phrasing. Thus, the intonation module inherits a richly annotated TTS structure where accent types and phrasing are already determined, and phonemes are already parsed into a hierarchical prosodic structure including (at least some of the) following levels, from top to bottom:

1. *major phrase* (*utterance* in the sense of (Pierrehumbert and Beckman, 1988))
2. *minor phrase* (*intonational phrase*)

3. *intermediate phrase* (*Iphrase*)
4. *accentual phrase* (*Aphrase*)
5. word
6. syllable
7. segment (phone)

Each language may assign language-specific attributes to these levels.

One version (the older version) of the English intonation model, as well as the intonation of Chinese (Mandarin and Taiwanese), Navajo, and Japanese in the Bell Laboratories TTS systems, are modeled after the intonation framework of Pierrehumbert and Liberman (Pierrehumbert, 1980; Liberman and Pierrehumbert, 1984). The intonation contour of an utterance is represented as abstract tonal targets high (H) and low (L), which are predicted from text, and are aligned crudely with the text by the pre-intonation modules of the system. The actual height of the targets as well as their precise temporal alignment are controlled by phonetic implementation rules, taking into account the tonal specification, segmental information, various downtrend effects, and the prominence setting of each accent. Tonal targets may be sparsely placed, and the F_0 contour in between is generated by interpolation. Finally, a smoothing function is used to generate the actual F_0 contour.

In the implementation of tonal targets, three default F_0 values are first assigned to define the topline, the reference line, and the baseline, respectively. These numbers are modified in the course of the utterance. Note that in this framework, these three "lines" may not actually be lines, but might rather be step functions, depending on how and whether some of the lowering effects apply. The H and L targets are scaled with reference to these lines. Typically, H targets are scaled upwards from the reference line, and will land on the topline if it has a prominence setting of 1, while L targets are scaled downwards from the reference line. There are special cases in both English and in Japanese where some types of L tones are scaled downward from the topline. The baseline is the lowest pitch, which is typically realized only at the end of the utterance.

The lowering effects that modify the value of these lines include: *declination*, which denotes a gradual, time-dependent downtrend that is independent of accent choice (Pierrehumbert, 1980; Liberman and Pierrehumbert, 1984; Poser, 1984; Pierrehumbert and Beckman, 1988); and *downstep*, which denotes the lowering of F_0 values that is caused by specific tones (Liberman and Pierrehumbert, 1984). A common cause of downstep is the L tonal target, so that in a sequence H L H, the second H is lower than the first one. Although it has been claimed (Liberman and Pierrehumbert, 1984) that there is no residual declination effect in English after the downstep effect has been factored out, there is evidence of declination in Mandarin, which will be discussed in more detail in Section 6.2.2. *Final lowering* affects the end of the utterance: It is often observed that the last pitch accent/tone is lower than expected after declination and downstep effects have been accounted for (Liberman and Pierrehumbert, 1984).

In natural speech, the prominence of each accent is controlled by the speaker, reflecting the structure and the meaning of the utterance and the speaker's rendition of them. In the Pierrehumbert-Liberman framework, prominence settings can be interpreted only after effects such as declination, downstep, and final lowering have been factored out. Because these effects create an intonational downtrend in a declarative utterance, two adjacent peaks with identical F_0 values will be interpreted as having different prominence settings: the early peak is interpreted as weaker than the later peak, because all other things being equal, the early peak is subjected to less downtrend.

In the Bell Labs system, many attributes used in the modeling of the intonation contour can be controlled by commands — *escape sequences* — which are typically inserted into the text before the word or location to which they apply. Escape sequences in the current version of the Bell Labs system always begin with a backslash "\". For example,

```
\!r0.9 \!*L*+H0.8 Humpty \!*L*+H0.8 Dumpty
\!r0.85 \!*L*0.5 sat on a \!*H*1.2 wall.
```

will deliver the nursery rhyme with L*+H accent on the words *Humpty* and *Dumpty*, L* accent on *sat*, and the default H* accent on *wall*. The number following the accent is the prominence setting. The speaking rate is controlled by \!r and a number indicating the rate. The normal setting is 1, with a smaller number indicating a faster rate. The values of the topline, the reference line, and the baseline can also be changed by escape sequences.

6.2.1 English

The implementation of the tone sequence model of English has been described previously (Anderson, Pierrehumbert, and Liberman, 1984; Silverman, 1988). There are several major points of departure from the original framework (Pierrehumbert, 1980). Most notably, tones are scaled in relation to the topline and the reference line, not the baseline, and the declination effect has been excluded.

Some expansions of Pierrehumbert's intonation grammar in Section 6.1.1 occur more frequently than others. Figure 6.1 shows the two most common English intonation types, declarative and question, on the same text material *Anne*. F_0 values in Hz are plotted as a function of time. The three vertical lines mark the segmentation boundaries: the beginning of the utterance, the boundary between the vowel [æ] and the nasal [n], and the end of the utterance. The declarative intonation is shown in the left panel, with the tonal sequence H* L- L%. The F_0 peak on the vowel [æ] is represented in the framework by the pitch accent H*, and the falling contour after the peak is represented by a low phrase accent L- and a low boundary tone L%. The question intonation is shown in the right panel, with the tonal sequence L* H- H%. The low F_0

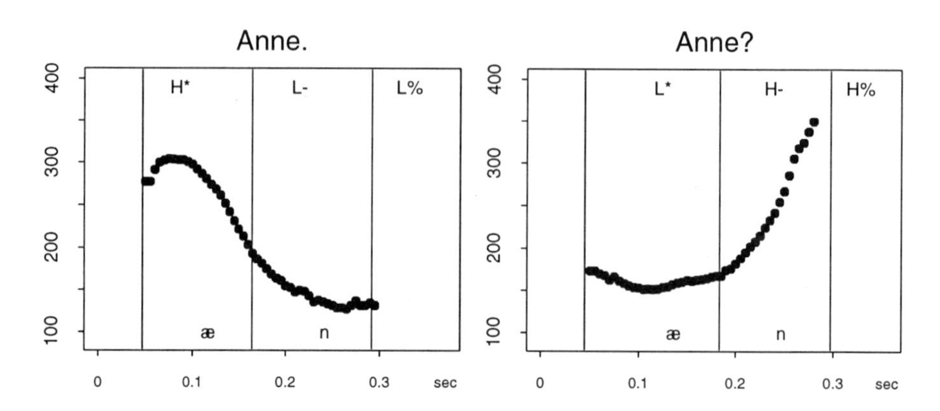

Figure 6.1: Tonal representation of English declarative intonation and question intonation

contour on the main vowel [æ] is represented by the pitch accent L*, and the rising tail is represented by a high phrase accent H- and a high boundary tone H%. In the implementation of the American English system the tone sequence (H%) H* L- L% is the default assignment of a declarative sentence, and the tone sequence L* H- H% is the default assignment for a question.

Figure 6.2 illustrates the target placement and scaling for the sentence *We were away a year ago.* The text-analysis modules determine that this is a declarative sentence with two accented words *away* and *year*, and the tone sequence assignment is H% H* H* L- L%. The two boundary tones are associated with the beginning and the end of the sentence, the first H* tone is associated with the stressed syllable *way* of *away*, and the second H* tone is associated with *year*. A default set of prominence values have also been assigned to each tone: the prominence value of the initial H%, the H* on *away*, and the phrase accent L- is 0.5; the prominence value of the H* on *year*, and of the final boundary tone is 1.

An abstract H and L target may be implemented with single or multiple *time*/F_0 points. In Figure 6.2, the initial H% tone is implemented with two points, the L- phrase accent and the L% boundary tone are each implemented with one point, while the H* tones are implemented with three points each. The points are scaled in relation to three values representing the topline, the reference line, and the baseline, respectively, which are shown as dotted lines. No declination is implemented, and no downstep effect is applicable in this rendition of the sentence, so the F_0 values of these three lines remain constant except at the end of the baseline, which is affected by the final lowering effect illustrated with the dashed line: the baseline value drops 10% in the final 500 ms.

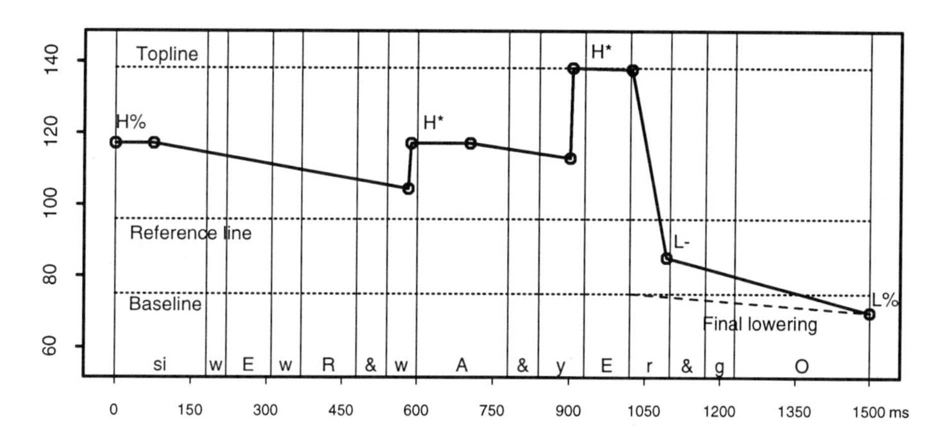

Figure 6.2: Implementation of English declarative intonation. Here, E=i, R=ər, &=ə, A=ei, o=oʊ.

The first point of the H% tone is placed at the beginning of the utterance which includes a region of silence marked as *si*. The value of this H% is halfway between the reference line and the topline, due to its prominence setting of 0.5. The second point of the H% assumes the same F_0 value, but is placed 76 ms. to the right. This interval is rate sensitive: longer for a slower speaking rate and shorter for a faster speaking rate. Note that both points are placed before the speech actually starts.

The H* accent is made up of a step with one initial lower point followed by two higher points. The temporal alignment of the third point is determined first, as a proportion of the rhyme duration plus an offset. The proportion and the offset values are estimated from experimental data and stored in a lookup table covering the interaction of five factors: whether the accented landing rhyme is word final, is a nuclear accent, is the final accent of a minor phrase, whether there is a stress clash and whether there is an accent clash. The H* on *away* is word final without stress and accent clash, and its location is 44.1% into the rhyme plus 25 ms. The H* on *year* is a word final nuclear accent, and its location is calculated as 40.5% into the rhyme plus 25 ms. The first point of the H* accent is then placed before the third point, at a interval that is speaking-rate dependent. At the default speaking rate, this interval is 122 ms. The second point is placed 5 ms. after the first point. The higher points in the H* accent (h) are calculated as follows, where p represents the prominence value, and *Top* and *Ref* represent the topline value and reference line value, respectively:

$$h = Ref + p(Top - Ref) \tag{6.1}$$

The low point of the H* accent (l) is scaled down from the high points by a fraction of the distance of the high points (h) and the reference line:

$$l = h - (h - Ref) * 0.6 \quad (6.2)$$

The L- phrase accent controls the duration and the slope of the fall from the preceding H* accent. Its placement depends on the distance between the previous accent and the end of the phrase. In this case it is placed at the end of the phone after the H* accent. Its value is calculated downwards from the (final-lowered) reference line. *Base* represents the baseline value:

$$l = Ref - p(Ref - Base) \quad (6.3)$$

The final L% accent is placed at the very end of the sentence. Its prominence setting is 1, so it falls on the final-lowered baseline.

Figure 6.2 does not contain a downstep accent, therefore no downstep effect is shown. A downstep accent such as H*+L in English (Liberman and Pierrehumbert, 1984) has the effect of lowering the topline, affecting all following H tones in the same phrase. The topline value at position i is scaled down from the previous H*+L tone at position $i-1$, as a fraction of the distance between the topline and the reference line. The downstep factor d ranges between 0 and 1, and the ratio 0.7 is used by default in the system.

$$Top_i = Ref + d(Top_{i-1} - Ref) \quad (6.4)$$

6.2.2 Mandarin Chinese

Mandarin Chinese distinguishes four lexical tones: high level (tone 1), rising (tone 2), low (tone 3), and falling (tone 4). The left panel of Figure 6.3 illustrates a set of tonal minimal pairs with the syllable *ma*: tone 1 *ma1* 'mother' is plotted in black diamonds, tone 2 *ma2* 'hemp' is plotted in black triangles, tone 3 *ma3* 'horse' is plotted in white diamonds, and tone 4 *ma4* 'to scold' is plotted in white triangles pointing downwards.[1] The four pitch tracks are aligned by the offset of the initial consonant [m], marked by the vertical line. On unstressed syllables the lexical tone may be lost completely, in that case the F_0 pattern on the syllable is largely determined by the preceding tone. Our notation for such a toneless syllable is tone 0. The right panel of Figure 6.3 contrasts the F_0 patterns of tone 0 on the syllable *le0* after four different lexical tones. Tone 1 and the following tone 0 are plotted in black diamonds, tone 2 and the following tone 0 are plotted in black triangles, tone 3 and the following tone 0 are plotted in white diamonds, and tone 4 and the following tone 0 are plotted in white triangles. The four pitch tracks are aligned by the offset of [l], marked by the long vertical line. Short lines on the pitch tracks mark the segment boundaries.

[1] For clarity we will use numerical representations of tone throughout this section, rather than the diacritic (¯ , ´ , ˇ , `) representation, used elsewhere in this book.

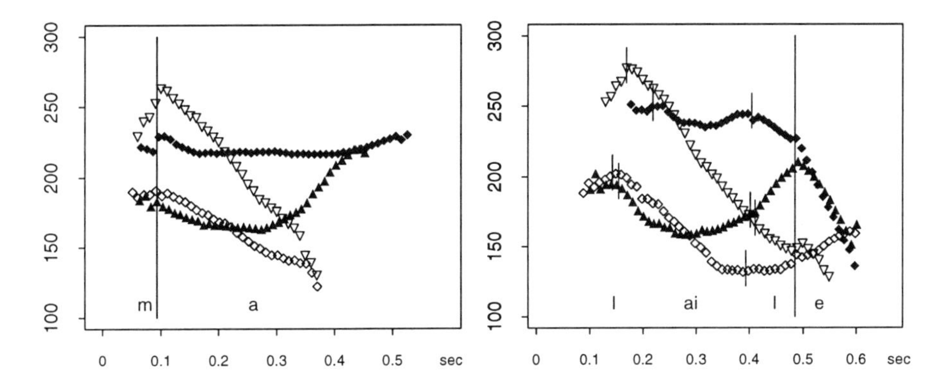

Figure 6.3: Pitch tracks of Mandarin tones. Four lexical tones are shown in the left panel: tone 1 *ma1* 'mother' is plotted in black diamonds, tone 2 *ma2* 'hemp' is plotted in black triangles, tone 3 *ma3* 'horse' is plotted in white diamonds, and tone 4 *ma4* 'to scold' is plotted in white triangles pointing downwards. Tone 0 following four lexical tones is shown in the right panel: tone 1 and the following tone 0 in black diamonds, tone 2 and the following tone 0 in black triangles, tone 3 and the following tone 0 in white diamonds, and tone 4 and the following tone 0 in white triangles.

A tone 1 starts in a speaker's high pitch range and remains high throughout the syllable. A tone 2 starts at a speaker's mid pitch range, remains level or declines slightly during the first half of the rhyme and rises up to the high range in the end. A tone 3 within an utterance starts at the speaker's mid range and falls to the low range, but in the utterance-final position some speakers produce this tone with a rising tail; hence it is also called a falling-rising tone. A tone 4 starts in a speaker's high pitch range, rises to form a peak, then falls to the low pitch range in the utterance-final position.

The beginning and end points of the four lexical tones fall on a few distinct levels rather than scattering across the board, so the tone shapes in Figure 6.3 can be captured with tone targets shown in Figure 6.4. The starting points of tone 1 and tone 4 are high, which are the same as the end points of tone 1 and tone 2. Tone 2 and tone 3 both have a mid target in the first half of the syllable, while tone 3 and tone 4 both fall to the low pitch range.

Tone 0 is best represented by a single target placed before the end of the syllable. The single target accounts for the lack of stability in the F_0 patterns of tone 0. The late position of the target allows for strong coarticulation effects with the previous syllable. However, if the target is placed at the very end of the syllable, we expect the ending F_0 values of tone 0 to be similar, which is not the case. A slightly earlier placement allows the F_0 trajectory to pass through the target and extend in the opposite direction of the previous targets, which

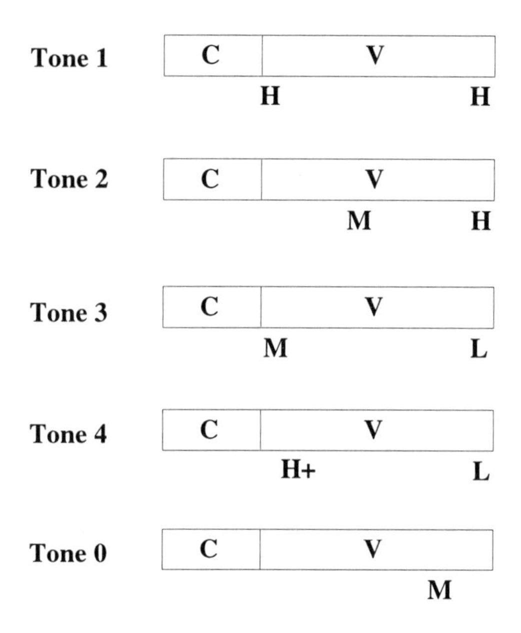

Figure 6.4: Target representation of Mandarin tones.

is a good approximation of the natural F_0 in Figure 6.3.[2]

Tonal transitional patterns are observed in the consonant region and the first half of tone 2, and these are accounted for by not assigning targets to these regions, allowing target interpolation to generate the desired shapes. Figure 6.5 shows the pitch variations in these regions in natural speech. The left panel shows the contrast in the pitch contours on the syllable initial consonant [m] in the second syllable *mao1* when the preceding syllable ends in H (mai1), plotted in black diamonds, and when the preceding syllable ends in L (mai3), plotted in white diamonds. The vertical lines mark the intervocalic [m] region, and the two pitch tracks are aligned by the offset of [m]. In the case of *mai1 mao1* (nonsense word) the targets before and after [m] are both H, and the pitch on [m] remains relatively flat. In the case of *mai3 mao1* 'to buy a cat', the target before [m] is L while the subsequent target is H, and the pitch on [m] rises abruptly from L to H. The right panel shows the contrasts in the pitch contours of a tone 2 syllable *mao2* 'hair' when the preceding syllable ends in H (after *yan1* 'smoke', plotted in black diamonds, or *yan2* 'salt', plotted in white triangles), and when the preceding syllable ends in L (after *yan3* 'eyes', plotted in white diamonds). The long vertical line marks the offset of [m], and

[2]The F_0 values of tone 0 after tone 4 are lower than the predicted value. This case is handled by a coarticulation rule; see below.

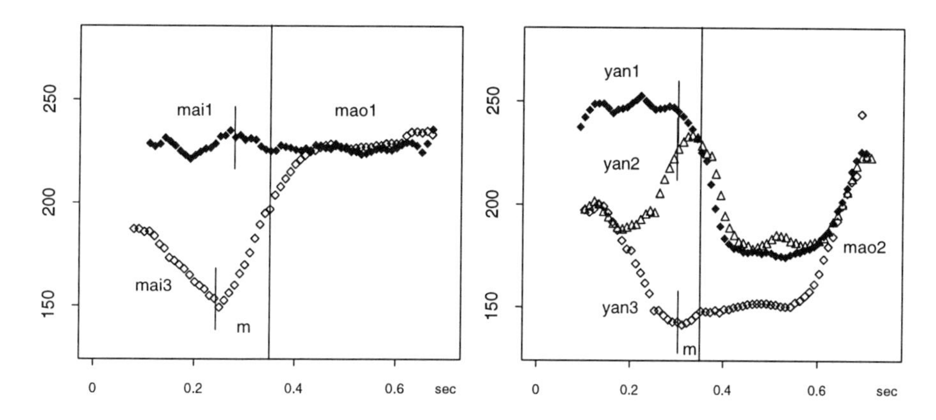

Figure 6.5: Tonal transition in Mandarin. The left panel compares tonal transition patterns on the [m]'s of *mao1* in two contexts: After a H tone in *mai1 mao1*, plotted in black diamonds, and after a L tone in *mai3 mao1*, plotted in white diamonds. The right panel compares the pitch tracks of *mao2* in three contexts: *yan1 mao2* in black diamonds, *yan2 mao2* in white triangles, and *yan3 mao2* in white diamonds.

short vertical lines mark the beginning of [m]. There are drastic but predictable variations in the F_0 trajectory of the first half of the tone 2 syllable *mao2*. The pitch rises from a previous L target, but falls from a previous H target. The pitch contour in the transitional region can be approximated rather closely by interpolating between the surrounding targets. The target in *mao2* is lower after a L target, which is consistent with the lowering effect that a L target has on subsequent tones.

Tones in connected speech deviate from their canonical forms in isolation considerably. The changes may come from tonal coarticulation (Shen, 1990; Xu, 1993) or phrasal intonation (Shen, 1985; Gårding, 1987; Shih, 1988), as well as various lowering effects. We use a set of tonal coarticulation rule to modify tonal targets in context. This set of rules are based on natural speech recordings of all disyllabic tonal combinations, and some trisyllabic tonal combinations in Mandarin. Both tonal target and target alignment may be changed by tonal coarticulation rules, but sentence-level scaling is accomplished by changing the value of topline, reference line, and baseline, thereby changing the value of H and L targets. The most important tonal coarticulation rules are listed below with brief descriptions.

1. *If tone 4 is not sentence-final, change its L target to M.*

   Our phonetic data shows that non-final tone 4 consistently falls to M, rather than to L. So the majority of tone 4 actually have the tonal targets

HM. This rule replaces a more limited rule that changes a tone 4 only before another tone 4 used in many systems (Zhang, 1986) apparently following the early report in (Chao, 1968). Listening tests show that restricting the rule to the tone 4 context results in over-articulation of tone 4 in other tonal contexts, and the result of our more general version of the rule is more natural.

2. *Given a tone 2, if the following tone starts with a H target, then delete the H of tone 2.*

   Comparing tone 2 before a H target and tone 2 before a L target, the rising slope before a H target is more gradual, and the pitch does not reach the expected H at the end of the tone 2 syllable. This is not a simple dissimilation effect. The desired pitch contour can be generated easily by assuming that the H target of tone 2 is absent. This effect can be observed in Figure 6.7 on the second syllable *wang2*.

3. *If a tone 0 follows a tone 4, change its M target to L.*

   Tone 0 should have a low falling pitch when it follows a tone 4. Since non-final tone 4 ends in M, assigning a L target on the tone 0 syllable will generate the desired falling contour.

4. *If the following tone starts with a H target, then the L target of tone 3 is raised:* $L = L + 0.15(M - L)$.

   The L target of tone 3 is raised when the following target is H. This is an assimilation effect.

5. *If tone 3 is sentence-final, add a M target, assigning its value to be* $M = M + 0.3(H - M)$.

   Tone 3 in the sentence-final position or in isolation may have a falling-rising shape, while other tone 3 have a low-falling shape. This rule adds the rising tail in the sentence-final condition. The pitch rises above the M point, but not as high as H.

In the implementation of these tone targets, we set the H target on the topline, the M target on the reference line, and the L target on the baseline, which can be fitted for an individual speaker or be controlled by escape sequences (see the description at the beginning Section 6.2) to reflect discourse functions. The topline and reference line values are modified in the course of the utterance to reflect the downstep effect and the declination effect. Downstep in English is caused by specific pitch accents such as H*+L, whereas in Mandarin, all non-high targets cause downstep, and the downstep ratios vary with the tones: the lower the target, the more downstep effect it has on the following tones. The low targets of tone 2 and tone 4 are both realized as an M target in non-final position, and they have a similar downstep effect. The low target of a tone 3 is realized as a L target in non-final position, and the

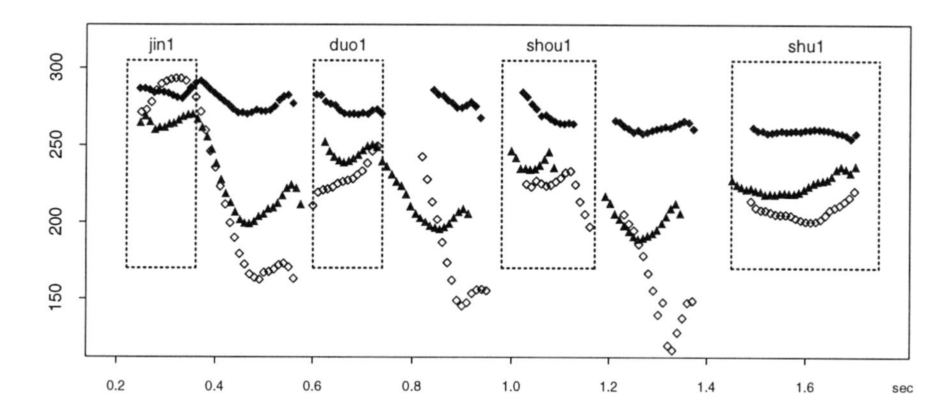

Figure 6.6: Downstep in Mandarin. Pitch tracks of three sentences showing the downstep effect in Mandarin. The sentence plotted in black diamonds consists solely of tone 1 syllables; the sentence plotted in black triangles has alternating tone 1 and tone 2, and the sentence plotted with white diamonds has alternating tone 1 and tone 3. The areas of interest — tone 1's on even-numbered syllables — are enclosed in boxes and the corresponding text is given on top of the boxes.

downstep amount is more than that of either tone 2 or tone 4. Figure 6.6 compares the F_0 patterns of three sentences: the sentence plotted in black diamonds consists solely of tone 1 syllables; since this sentence contains no low targets, there is no downstep effect. The sentence plotted in black triangles has alternating tone 1 and tone 2, and the sentence plotted with white diamonds has alternating tone 1 and tone 3.[3] The difference in downstep amount can be observed on the even-numbered tone 1 syllables enclosed in boxes, where the tone 1 after tone 3 is lower than the tone 1 after tone 2, which in turn is lower than the tone 1 after tone 1. This effect is clearest on the final syllable. Tone 1 syllables in the same box have the same segmental composition, which is given on top of the boxes.

Figure 6.7 shows a sentence with nine tone 1 syllables — all H tonal targets — following an initial two-syllable frame with a tone 3 and a tone 2. The initial tone 3 provides a reference point of the low pitch range of the speaker. In the nine-syllable stretch of H targets, the downstep effect is by definition absent, since there is no L target triggering the effect. So the observed lowering of subsequent H tones must be attributable to declination.

Data with Mandarin tone 1 offers a straightforward way to model the declination effect. Data are collected from tone 1 utterances two- to eleven-syllables

[3]The sentence with tone 4 is not shown, so as to avoid over-crowding of the plot. Suffice it to say that the downstep effect with tone 4 is nearly identical to that found with tone 2.

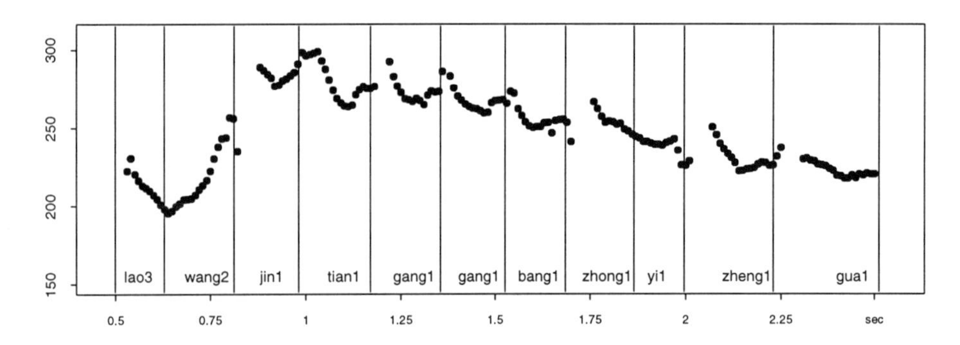

Figure 6.7: Declination in Mandarin shown in a sentence with nine tone 1 syllables — all H tonal targets — following an initial two-syllable frame with a tone 3 and a tone 2.

long. Regression models are fitted for two female speakers and two male speakers:

Female speaker A : $H_i = 0.89(H_{i-1}) + 23.52$
Female speaker B : $H_i = 0.83(H_{i-1}) + 39.36$
Male speaker A : $H_i = 0.79(H_{i-1}) + 24.10$
Male speaker B : $H_i = 0.80(H_{i-1}) + 25.94$

Contrary to previous claims (Pierrehumbert and Beckman, 1988) that declination is a linear effect where pitch drops a fixed amount over the same time frame, we find that the declination pattern is similar to downstep in that there is more pitch drop in terms of Hz value in the beginning of an utterance. Figure 6.8 shows the declination models fitted for female speaker B and male speaker A, with the initial value set to 280 Hz for the female speaker, and 137 Hz for the male speaker. The decline is steeper in the beginning, and asymptotes to 231 Hz for the female speaker and to 114 Hz for the male speaker.

6.2.3 Navajo

Navajo has two lexical tones: high and low. The implementation of Navajo tones is very similar to the Mandarin case. Tones are represented as H or L targets. The downstep effect is calculated tone by tone and the declination effect is calculated by the syllable position in the sentence. Navajo tones differ from Mandarin tones in one crucial way: Mandarin tones have fairly stable contours, which is accounted for by using two targets per syllable. The tonal contours in Navajo change depending on the tonal context, in a similar fashion as Mandarin tone 0. In general, Navajo tones are implemented with a single target placed at the end of the syllable to which the tone is associated. If the first syllable in an utterance has a H tone, the typical F_0 shape on that syllable is rising,

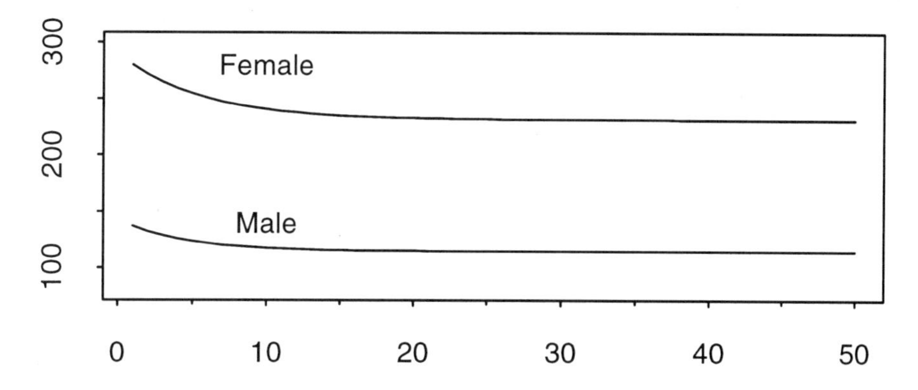

Figure 6.8: Declination models fitted for female speaker B and male speaker A, with the initial value set to 280 Hz for the female speaker, and 137 Hz for the male speaker. The decline is steeper in the beginning, and asymptotes to 231 Hz for the female speaker and to 114 Hz for the male speaker.

which points to the presence of a L% boundary tone in the beginning of an utterance. Figure 6.9 shows the natural pitch track of the word *diné bizaad* 'Navajo language', with a high tone on the second syllable *né*. Placing a single H target at the end of this syllable generates the rising contour on the high-toned syllable, as well as the falling contour on the following low-toned syllable.

One exception to the placement scheme suggested above involves HL tone

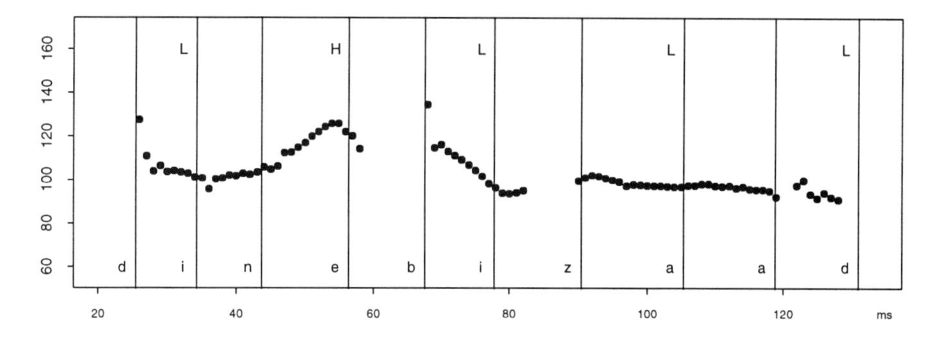

Figure 6.9: Natural F_0 contour of *diné bizaad* with the alignment of tonal targets.

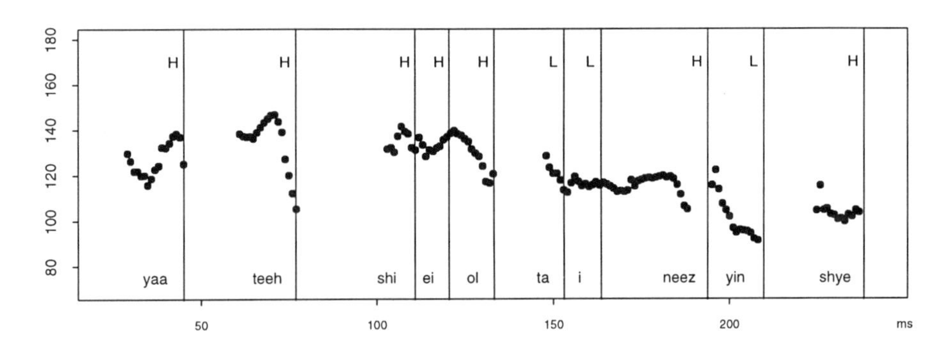

Figure 6.10: Declination and downstep effects in Navajo natural speech.

sequences on pairs of adjacent vowels. In this case, the pitch is falling throughout the first high-toned vowel, and remains low on the second low-toned vowel. To capture this pattern, the H target is shifted to the *beginning* of the first vowel, and the following low tone has two targets, one at the beginning and one at the end of the vowel.

Both the declination effect and the downstep effect can be observed in the natural speech shown in Figure 6.10: *Yá'at'ééh shí éí ólta i' nééz yinishyé* 'Hello I am called the Tall Learner'. Declination is present in the first half of the sentence, where there is a sequence of high tones spanning across five syllables. The pitch values of successive high tones drop slightly after a rise from an initial boundary tone. Downstep, a more drastic drop of high tone value after a low tone, is observed twice in the second half of the sentence: The lowering of *nééz* after low tones on *ta i'*, and the lowering of the final syllable *yé* after low tones on *yin(i)sh*.

Each minor phrase is given an initial L% boundary tone, which is placed on the *reference line*, the default value being 125 Hz. The low tone value is determined by the *baseline*, which is set at around 100 Hz. The high tone value is determined by the *topline*, which is initially set at 150 Hz. After a low tone, downstep is applied so the value of the topline is lowered. Downstep is calculated in the same way as English downstep. A downstep ratio of about 0.70 (about the same as the default in the English system) appears to be correct. In addition to downstep, a uniform declination of 3% per syllable is applied to both high and low tones in the course of a phrase. The values of topline, reference line, and baseline are reset to the default values at the beginning of each minor phrase.

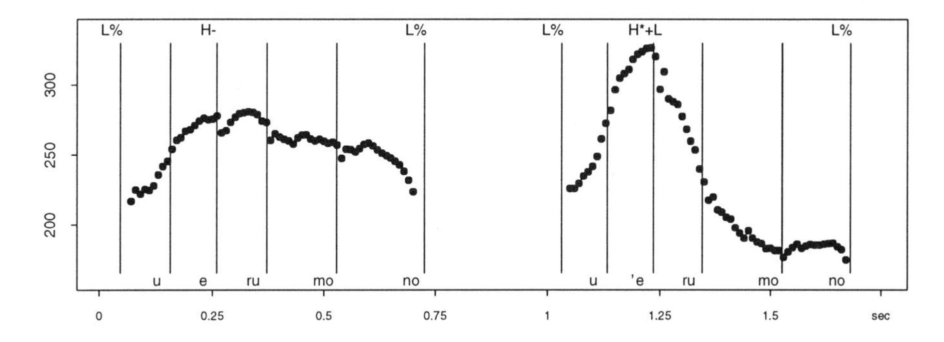

Figure 6.11: Two types of Japanese accentual phrases: on the left is an accentual phrase with two unaccented words, with the accent assignment L% H- L%. On the right is an accentual phrase with one accented word, with the accent assignment L% H*+L L%.

6.2.4 Japanese

The Japanese intonation model is an implementation of the model of Pierrehumbert and Beckman (1988), where Japanese intonation is represented as sequentially arranged tonal targets.[4]

The text analysis component described partially in Section 3.6.7 is responsible for the assignment of phrase breaks and applying deaccenting rules such as the Compound Stress Rule. The intonation module has two components. The first assigns accents, phrase accents, and boundary tones according to the phrasing and accent information annotated by the text analysis. The second is responsible for tone scaling and adjustment of target placement.

Japanese has a lexical distinction of accented and unaccented words, and for accented words, the location of the accent is also lexically determined. These two types of word have very different F_0 patterns. Accented words have a H*+L accent associated with the accented mora, while unaccented words have no accent. Figure 6.11 contrasts the F_0 patterns of these two types of words. On the left is a phrase of two unaccented words: *ueru mono*, 'something to plant'. The F_0 pattern shows an initial rise, then falls gradually to the end. The pattern is captured by the accent assignment L% H- L%, with no lexical pitch accent. On the right is a phrase with one accented word *u'eru mono*, 'the ones who are starved'. The intonation peak coincides with the accented syllable *'e*, followed by an abrupt fall. The movement of the intonation pattern is represented by L% H*+L L%.

[4]Example sentences in this section all come from the database collected by Jennifer Venditti.

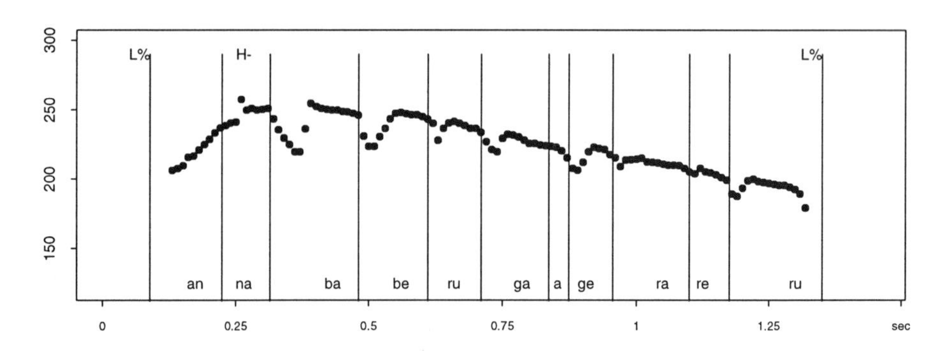

Figure 6.12: A long Japanese sentence with many unaccented words, with the accent assignment L% H- L%. The pitch values of the syllables between H- and L% are generated by tonal interpolation.

The sparsely placed targets for the accentual phrase with no accented words work equally well for a long sentence, shown in Figure 6.12, *Anna baberu ga agerareru* 'I can lift a barbell like that one', where the long and gradual downward slope after the initial rise is generated by interpolation between the phrasal H- tone and the final L% boundary tone.

The location of the accented syllable in the sentence has an effect on the overall intonation pattern. If the accented syllable is more than two morae away from the beginning of the utterance, there is an additional phrasal H- tone before the H*+L accent, creating a plateau before the H*+L peak. Furthermore, the L% boundary tones come in two levels of strength that can be determined reliably from context: if the initial syllable of the accentual phrase is accented, or when the initial syllable is long, the initial L% is weak. In the utterance-medial position, the weak L% corresponds to a shallower F_0 valley than the strong L% tone. We use the notation wL% for this tone in the TTS system. In this section the notation L% includes both types unless specified otherwise.

A declarative accentual phrase may have any of the following three types of target assignment:

1. L% H*+L L% is assigned to a phrase with an early accent.

2. L% H- H*+L L% is assigned to a phrase with a late accent.

3. L% H- L% is assigned to a phrase with no accented words.

Japanese has both declination and downstep effects. Declination applies within an utterance, affecting both the topline and the reference line, at a rate of about 10 Hz per second. Downstep is triggered by the H*+L tone, which causes the following H tone to be lower. The downstep effect is calculated as

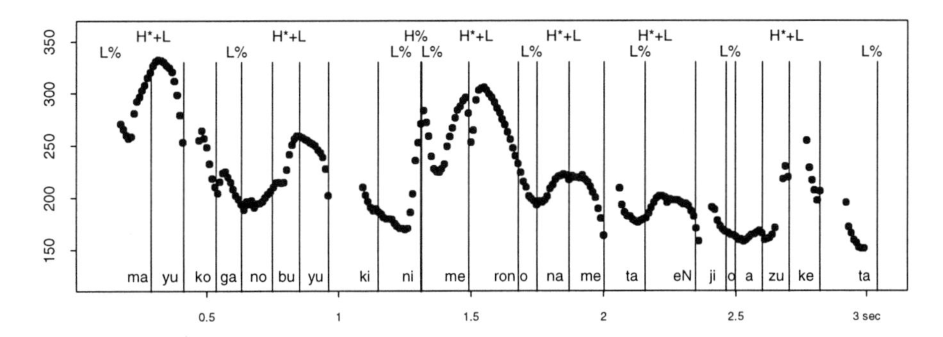

Figure 6.13: Downstep and declination in Japanese. There are two intermediate phrases (Iphrases), separated by the thick vertical line around 1.4 seconds. The downstep effect operates within each Iphrase, where each successive H*+L peak is lower than the previous one. The downstep effect does not apply across an Iphrase boundary: the pitch value of the first peak of the second Iphrase is reset to a value similar to, but not as high as the first peak of the first Iphrase. The difference is attributed to the declination effect.

in English. The downstep ratio is set to 0.7, which is similar to both English and Navajo. The downstep effect is restricted to the domain of an intermediate phrase (Iphrase), which is marked by short pause or the presence of phrase final boundary tones. The pitch value of the topline will be reset after an Iphrase boundary; however, the value is lower than the topline value at the beginning of the utterance because declination is still effective. Figure 6.13 shows the sentence *M'ayuko ga nob'uyuki ni m'eron o n'ameta 'enji o az'uketa* 'Mayuko left the child who licked the melon in Nobuyuki's care', where both downstep and declination is observed. There are two Iphrases, separated by the thick vertical line around 1.4 seconds. In each Iphrase each successive H*+L peak is lower than the previous one,[5] as a result of downstep. The first peak of the second Iphrase is reset, but it is not as high as the first peak of the first Iphrase. The difference is attributed to the declination effect.

In the case of question or continuation intonation, a H% boundary tone is attached to the end of the three declarative tonal strings. The value of the H% tone is not affected by downstep, declination, or final lowering and is set to the topline value at the beginning of the utterance.

The precise alignment of tonal targets with speech materials is affected by a combination of factors including accent type, tonal context, segmental or syllabic context, and the position of the accent. A few tonal targets have built-

[5] The very last pitch accent is higher than the previous peak. The explanation is that this accent has higher prominence setting than the previous one.

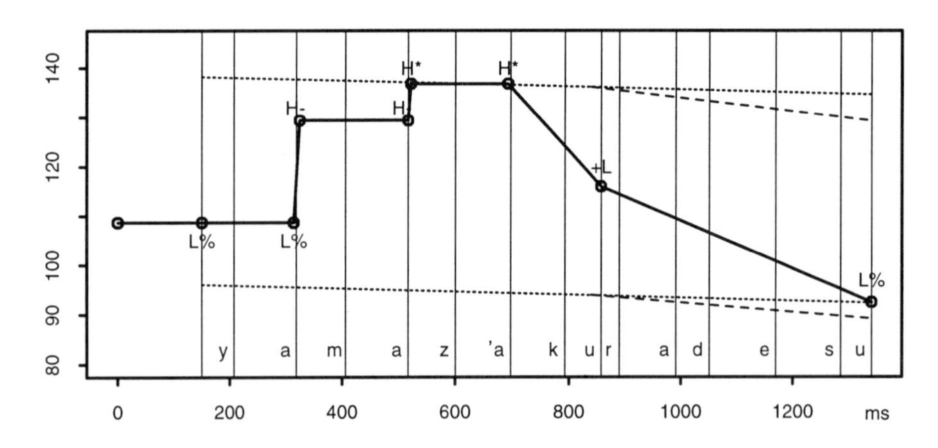

Figure 6.14: The placement of tonal targets in the Japanese TTS.

in duration that lasts one mora or one syllable long. Figure 6.14 illustrates the target placement of the sentence *Yama z'akura desu* 'It's a mountain cherry', with tonal sequence L% H- H*+L L%. The initial L%, the H- and the H* tones are implemented with two points each. In this sentence, the duration of each of these targets is one syllable long, so the first points of the targets are placed near the beginning of the syllable while the second points are placed near the end of the syllable. The placement of the L tone in H*+L depends on the number of following syllables in the phrase. If there are only one or two following syllables, the fall is abrupt and the L tone is placed very close to the H target. If there is further material in the phrase, the falling slope is more gradual and the L tone is placed one syllable later. This is the case in Figure 6.14.

The F_0 values of H and L are scaled with regard to two default values set at the beginning of each utterance: the topline value (138.24 Hz), and the reference line value (96 Hz). The top dotted line in Figure 6.14 shows the topline values transformed by declination , and the lower dotted line shows the reference line transformed by declination. Another lowering effect, final lowering, applies to the final 500 ms. of the utterance, which is shown with two dashed lines, one for the lowered topline and one for the lowered reference line. The lowering amount increases toward the end and reaches a 5% decrease in F_0 value at the end of the utterance.

Each accent type is assigned a default tonal prominence value. A phrasal H- tone has a prominence value of 0.8, while the H* tone is 1. The initial L% tone has a prominence setting of 0.8, the L of H*+L is 0.5, and the utterance final L% tone 0.9. The F_0 values are calculated as follows:

$$H = Ref + p(Top - Ref) \tag{6.5}$$

$$L = Top - p(Top - Ref) \tag{6.6}$$

Top and Ref are the topline and reference line transformed by declination, final lowering, and downstep, and p is the prominence setting of the tones. The H tone is scaled from the reference line up; the higher the prominence value, the higher the F_0 value. The L tone is scaled from the topline down; the higher the prominence value, the lower the F_0 value. The H* tone has a prominence value of 1, so it lies on the transformed topline. The final L% has a prominence value of 0.9, so its value is slightly higher than the transformed reference line represented by the dashed line. Downstep does not apply in this sentence. If there were another H tone following H*+L, the value of the second H would be lowered by 30% of the distance between the topline and the reference line.

6.3 The Superpositional Approach

6.3.1 Introduction

Currently, the intonation component used for English, French, German, Spanish, Romanian, Russian, and Italian is based on a different approach than the tone sequence approach presented above. This approach is in the superposition tradition, but is different in some key respects from the Fujisaki model. The model was developed as a natural result of a series of studies in English on the effects of the temporal and segmental structure of accent groups (i.e., syllable sequences led by a stressed syllable) on accent curves (i.e., local pitch excursions associated with accent groups). We note that currently we do not have particularly strong evidence in favor of this new approach and against tonal sequence and interpolation. In fact, we believe that without complete clarification of which assumptions are central and which ancillary, these two approaches cannot be empirically distinguished. For now, we prefer this new approach because it has turned out to be quite capable of capturing the essence of the data discussed next.

In this section, we first discuss the results of our studies on the effects of the temporal and segmental structure of accent groups on accent curves. We next sketch a general theory to account for these data, and finally we present the current intonation component — which implements a special case of this theory.

6.3.2 Studies on segmental alignment and height of accent curves

We use the word "accent curve" very loosely as some pitch excursion that corresponds with an accent group. At this point we do not subscribe to any more precise empirical or theoretical meaning (e.g., in terms of the Öhman-Fujisaki model).

These studies — earlier ones are described in (van Santen and Hirschberg, 1994) — were initiated largely because the temporal placement of pitch targets in our rendition of the Pierrehumbert-Liberman model, as applied to English, left much to be desired. We started out with attempting to predict peak location, but then also analyzed other "points" of the accent curves, as well as effects of consonant perturbation, intrinsic pitch, and overall duration on peak height. In the end, a fairly complex body of results could be summarized with a simple model, that then formed the basis of our new intonation component. However, we believe that this model has several interesting theoretical aspects, which we will discuss in Section 6.3.3.

Method

The results reported in this section were all based on speech recorded from a female speaker. Recorded, digitized speech was segmented manually and labeled for pitch accent, phrase accent and boundary tone type, using the system of intonational description presented in Pierrehumbert (1980). An epoch based pitch tracker (Talkin and Rowley, 1990) measured pitch at 10 ms. intervals.

In most recordings, the phrases were produced as single intonational phrases, with a single H* pitch accent, a low phrase accent, and a low boundary tone, and thus would be described as H*L-L% in Pierrehumbert's system. The accented syllable was either utterance final or was followed by one or more stressed syllables; correspondingly, the phrase final accent group had either one or more than one syllable.

Peak location and accent group duration

A basic question is whether peak location depends at all on the duration of accent groups. For example, in (Adriaens, 1991), rises start a fixed millisecond amount before vowel start; this would predict no correlation between peak location and duration of accent groups. Pierrehumbert (1981) assumes that peak location is a fixed percentage into the accented syllable. This would predict a positive correlation, but only within the set of monosyllabic accent groups; for polysyllabic accent groups, because variations in the duration of the post-accentual remainder affect accent group duration but not peak location, the expected correlation is low.

Figure 6.15 shows peak location (measured from the start of the accented syllable) as a function of accent group duration. Data are shown separately for monosyllabic accent groups and polysyllabic accent groups. In both cases there are significant correlations between accent group duration and peak location.

Yet, these correlations are not that strong. One interpretation is that peak location is only loosely coupled with segmental temporal structure. The other interpretation is that overall accent group duration is not the right predictor

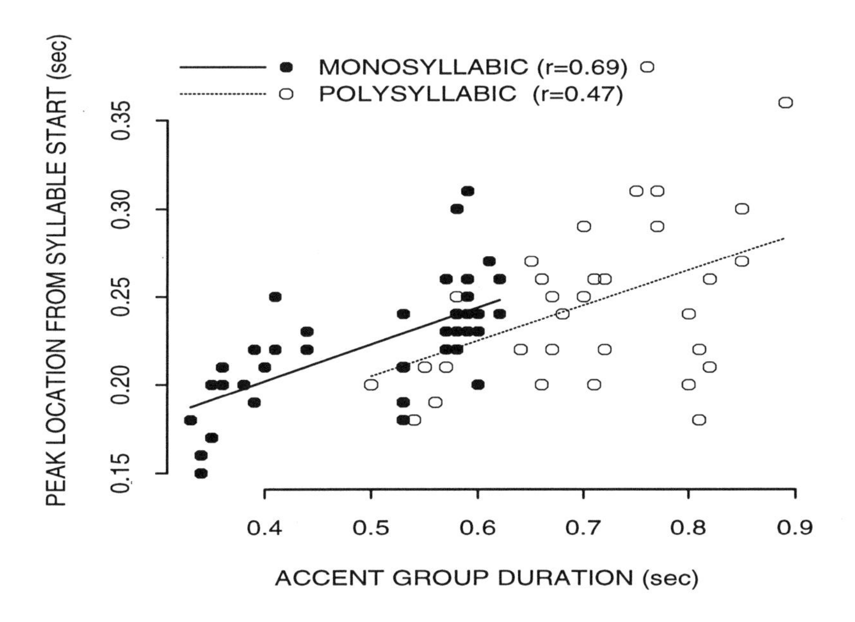

Figure 6.15: Effects of total accent group duration on peak location.

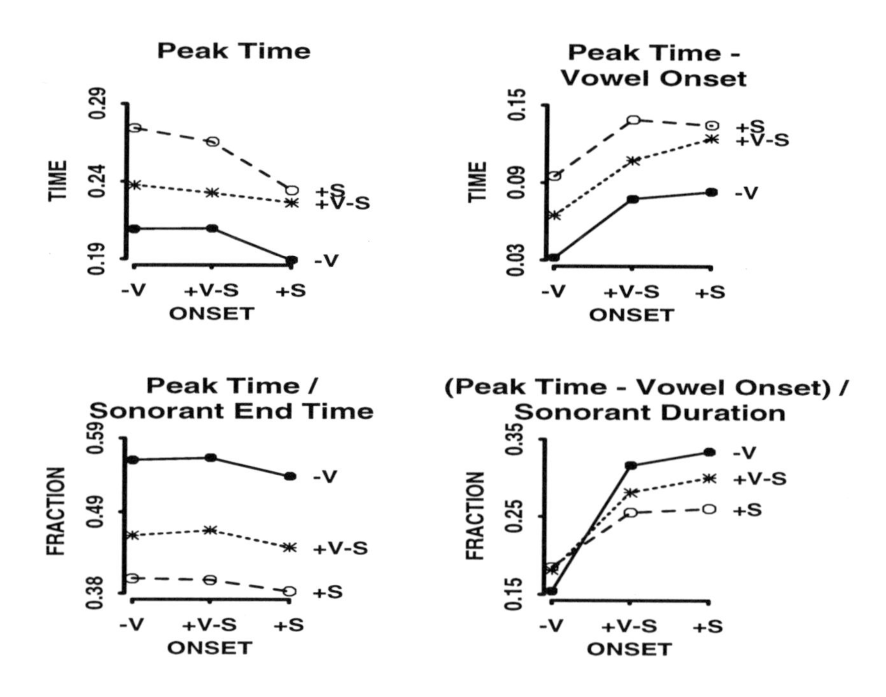

Figure 6.16: Effects of onset and coda (curve parameter) on peak time, measured in four different ways.

of peak placement. We will show that the second interpretation is the more accurate one.

Peak location and segmental composition

One way to show the inappropriateness of overall accent group duration is by investigating effects of the segmental composition of the accent group. We restrict ourselves here to monosyllabic phrase-final accent groups.

Figure 6.16 shows peak location as a function of the type of onset and coda of the accented syllable. We distinguish between voiceless (-V), voiced obstruent (-V+S), and sonorant (+S) onsets and codas. For example, the word *pan*, has a voiceless onset and a sonorant coda (e.g., the upper left point in the upper left plot in Figure 6.16). The word *name* has a sonorant onset and a sonorant coda. In the left top panel, peak location is measured from the syllable start; and in the right top panel, from vowel start. These measurements are normalized in the bottom two panels by division by the syllable duration and rhyme duration, respectively.

We can draw two inferences. First, there are effects of onset type and coda type on peak location, whether measured from syllable start or vowel start. Thus, mixing all syllable types together as was done in the preceding figure is not the right thing to do. This may have been one of the causes of the mediocre correlations between overall accent group duration and peak location.

Second, it is clear that no simple rule for peak placement works across all syllable structures. If we normalize peak location by dividing by syllable duration or by rhyme duration, there are still clear effects of onset type and coda type. This shows that the fixed-percentage-of-syllable-duration rule mentioned by Pierrehumbert (1981) is wrong. Of course, the rule developed by Adriaens (1991) for German of rises starting a fixed millisecond amount prior to vowel onset was already shown to be not applicable to our English data by the existence of systematic correlations between accent group duration and peak location.

Linear model for peak location

We just showed one failure of the hypothesis that overall accent group duration is the prime predictor of peak location, by demonstrating that peak location depends on the segmental makeup of the accent group. We now investigate a second implication of this hypothesis, which is that each "part" of the accent group affects peak location equally.

One way to measure the effects of different parts of an accent group is by splitting up an accent group into subsequences of phones, and measuring with a simple linear model how peak location depends on the durations of these subsequences.

For phrase-final accent groups, we split an accent group into *onset*, *sonorant rhyme*, and the unstressed *remainder* of the accent group. The *sonorant rhyme*, or *s-rhyme*, is defined as the phone sequence that starts with the first non-initial sonorant in the onset (or vowel start if the onset has no sonorants) and that ends at the end of the last sonorant (for monosyllabic accent groups) or of the last segment — whether sonorant or not — of the accented syllable (for polysyllabic accent groups). The onset consists of the consonants that make up the accented syllable onset, but without the sonorants assigned to the s-rhyme. Thus, *blank* is split into /b/ and /læn/, with the /k/ ignored; *seat* into /s/, /i/, and the empty remainder ϵ; *muse* into /m/, /ju/, and ϵ; *prım* into /p/, /rim/, and ϵ; and *practical* into /p/, /ræk/, and /tıkəl/.

For phrase-medial accent groups, we have the rhyme terminated by the end of the syllable instead of the end of the last sonorant in the syllable.

The somewhat unusual definition of s-rhyme was based on the following analyses of monosyllabic utterance final accent groups. First, the inclusion of coda sonorants and exclusion of coda obstruents in the s-rhyme was based on the fact that in monosyllabic accent groups the pitch value reached at the *sonorant end point* (i.e., the end of the last sonorant in the accent group) was extremely constant. It had only a 5-Hz range across coda types (137, 134, and 132 Hz for voiceless (-V), voiced obstruent (-V+S), and sonorant (+S) codas).

Moreover, this already small range must have been inflated by the fact that (1) the average slope of the line between the syllable start and the sonorant end point is negative, and (2) the sonorant end is much later for sonorant codas (644 ms.) than for voiced obstruent (460 ms.) or voiceless codas (362 ms.). By contrast, the value reached at the vowel end is far more variable (169 Hz for sonorant codas, 134 and 137 Hz for voiced obstruent and voiceless codas). In summary, the behavior of the sonorant end point appears to be highly invariant, which is a strong argument for having the s-rhyme terminate as defined.

Second, the inclusion of non-onset-initial sonorants in the s-rhyme was based on analyses using different versions of the linear model (described next) corresponding to alternative definitions of the s-rhyme. We compared parameter estimates based on obstruent-vowel-obstruent syllable data with parameter estimates based on obstruent-sonorant-vowel-sonorant-obstruent syllable data, and found that these estimates were similar when we included the sonorants in the s-rhyme, but significantly different when we excluded either sonorant from the s-rhyme.

The linear model states that peak location for an accent group with onset type C_o and coda type C_c (which we extend to include the "polysyllabic" type) is given by:

$$
\begin{aligned}
T_{peak}(D_{onset},\ D_{s-rhyme};\ C_o,\ v,\ C_c)\ =\ & \\
\alpha_{C_o,C_c} \times D_{onset}\ +\ & \\
\beta_{C_o,C_c} \times D_{s-rhyme}\ +\ & \\
\gamma_{C_o,C_c} \times D_{remainder}\ +\ \mu_{C_o,C_c}. & \qquad (6.7)
\end{aligned}
$$

According to this model, peak time for a syllable whose onset duration is D_{onset}, s-rhyme duration is $D_{s-rhyme}$, and remainder duration is $D_{remainder}$, is a weighted combination of these three durations plus a constant, which, like the Greek-lettered weights, may depend on C_o and C_c. We call these weights *alignment parameters*.

Figure 6.17 shows the results of fitting this model. Before discussing these results, we point out that the model provides a nice framework for representing a variety of simple rules about the invariance of peak location:

1. Measured from syllable onset: $\alpha = \beta = \gamma = 0$, $\mu > 0$.

2. Measured from vowel onset: $\alpha = 1$, $\beta = \gamma = 0$, $\mu > 0$.

3. Measured from syllable onset, divided by syllable duration: $\alpha = \beta$, $\mu = 0$.

4. Measured from vowel onset, divided by rhyme duration: $\alpha = 1$, $\mu = 0$.

In other words, these simple rules make predictions that can be tested by estimating the alignment parameters.

The intercept, μ, proved to be a statistically insignificant parameter whose presence added less than 2 percent to the amount of variance explained. The

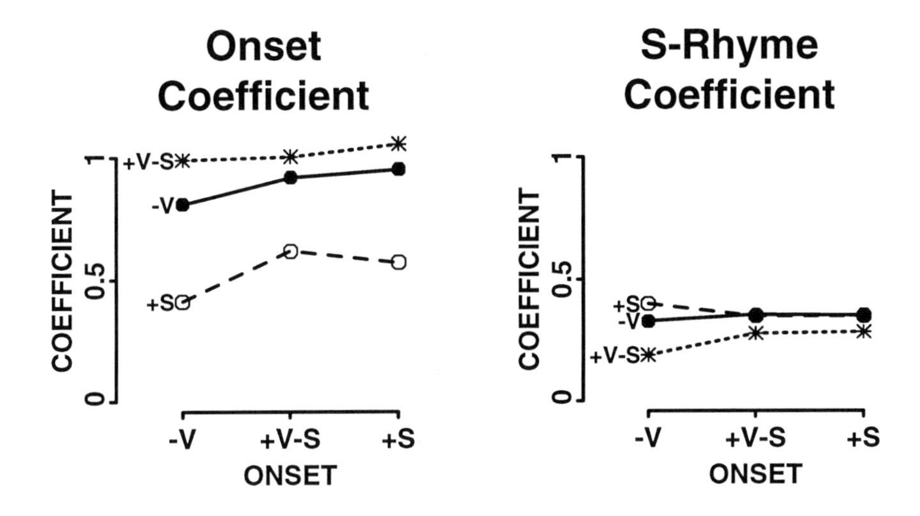

Figure 6.17: Regression coefficients for onset duration (α; left panel) and s-rhyme duration (β; right panel) as a function of onset and coda (curve parameter). These coefficients were obtained by a multiple regression analysis with peak location (measured from syllable start) as dependent measure and the durations of onset and s-rhyme as independent measures.

results (Figure 6.17) were based on the model altered by removing the μ parameters.

Correlations between observed and predicted peak locations varied between 0.61 and 0.87, with a median of 0.77. The median absolute deviation between predicted and observed peak location was 15 ms. Using an F test for testing the significance of added predictors (Morrison, 1967), these correlations proved significantly higher than the correlation between overall accent group duration and peak location.

Figure 6.17 shows that the α parameters were invariably larger than the β parameters, and that the latter were well above zero (all statistical significance levels are at $p < 0.001$ and will not be reported individually). These observations contradict the first three rules. The fourth rule fails completely for sonorant codas.

We now turn to the issue of whether the effects shown in Figure 6.16 are due exclusively to the durational differences between consonant classes. The parameters vary as a function of coda, and much less (if at all) as a function of onset. Somewhat to our surprise, this means that the effects of onset class shown in Figure 6.16 can be explained purely in terms of durational differences.

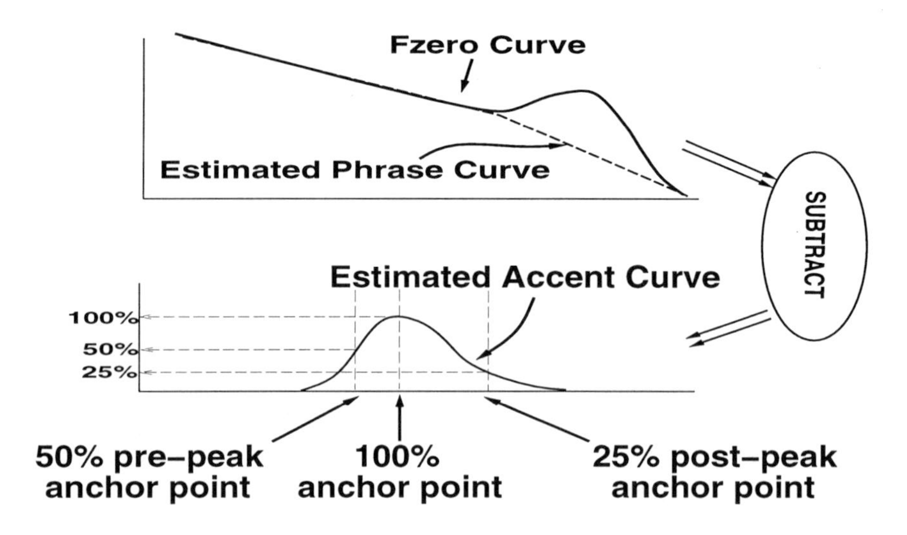

Figure 6.18: Estimation of accent curves.

But the effects of coda do involve differences other than durational ones.

We draw the following conclusions. First, for a given syllable structure; the onset weights are always much larger than the rhyme weights. In other words, increasing onset duration by 50 ms. pushes the peak farther to the right than increasing the rhyme duration by the same amount. Hence, it is indeed the case that durations of different parts of an accent group do not all have the same influence on peak location. In particular, this means that two syllables can have the same overall duration, yet the peak is systematically later in the one that has the longer onset and the shorter rhyme.

The second inference has to do with the curves being rather flat, yet clearly differing from each other, at least as far as the onset weight curves are concerned.

We infer that later peaks after voiceless onsets are purely the result of voiceless consonants being relatively long. However, onset duration has much less effect when the coda is a sonorant than when it is voiceless.

Anchor point location

We now discuss how we can extend this analysis of peak location to an analysis of the alignment of the entire accent curve. After all, the peak is only one point on that curve, and we are not convinced that it is the perceptually most important point (Bruce, 1990). The main reason, of course, that peak location has been studied so often is that it is relatively easy to measure its location.

What we will do is define points on the accent curve that will be called anchor points. The peak is one of those anchor points.

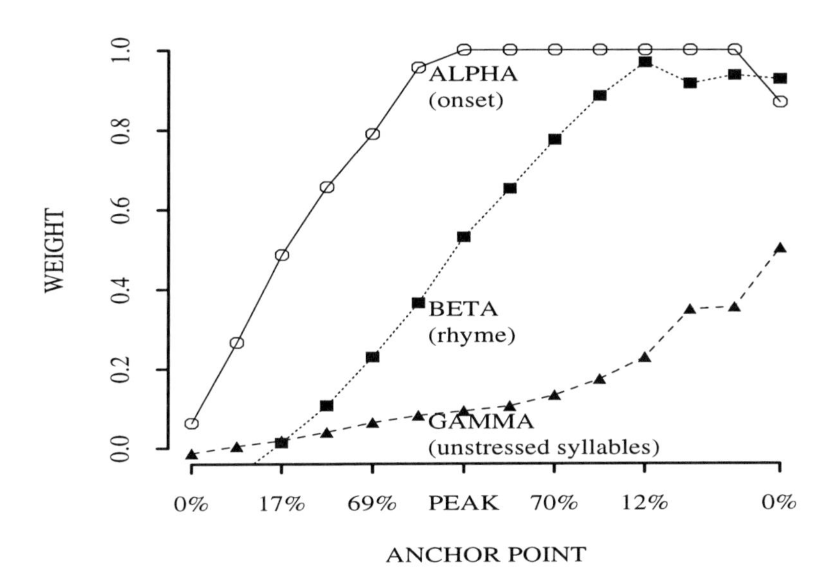

Figure 6.19: Linear weights as a function of anchor point, for each of the three sub-intervals of the accent group. Solid curve: onset; dotted curve: rhyme, dashed curve: remainder. Negative values are omitted.

Because of the extreme simplicity of the pitch contours that we studied up to this point, it is possible to reliably draw a line from the last frame of the sonorant that precedes the accented syllable (typically the vowel /oʊ/) to the last frame of the last sonorant in the accent group (See Figure 6.18). When we subtract this line from the observed F_0 curve, we obtained a curve that is roughly bell-shaped (where not interrupted by obstruents) and that starts and ends with a value of 0 Hz. For now we call this difference curve the *estimated accent curve*. There are many ways of defining characteristic points on the estimated accent curve — points whose locations can be reliably predicted from the accent group subsequence durations and that capture enough information about the estimated accent curve that synthesis of an entire accent curve can be generated from them. We chose a very simple solution: sampling the estimated accent curves at pre- and post-peak locations corresponding to 5%, 10%, 25%, etc. of peak height. By this definition, the peak is the 100% anchor point.

We now apply the same linear model to anchor point location as was used for peak location:

$$
\begin{aligned}
T_i(D_{onset},\ D_{s-rhyme};\ C_o,\ v,\ C_c) = \\
\alpha_{C_o,C_c,i} \times D_{onset} + \\
\beta_{C_o,C_c,i} \times D_{s-rhyme} + \\
\gamma_{C_o,C_c,i} \times D_{remainder} + \\
\mu_{C_o,C_c,i}.
\end{aligned} \tag{6.8}
$$

Here, the added subscript i refers to anchor point. Figure 6.19 shows the values of the regression coefficients.

The figure shows that the curves do not intersect. This means that all anchor points behave like the peak, in that the weights for the onset exceed the weights for the rhyme and the remainder of the accent group.

It is also clear that the curves are monotonically increasing. Moreover, they initially diverge, and then converge. Early anchor points mostly depend on onset duration and hardly at all on the durations of the rhyme and the remainder. But late anchor points depend more evenly on all three subsequence durations. The weights for the remainder never reach the value of 1.

Yes/no contours

We made recordings for the same speaker, with instructions to generate yes/no question contours. Only phrase-final monosyllabic accent groups were studied. Figure 6.20 shows three examples of this contour, confined to the final accent group. These contours consisted of a declining curve for the pre-accented region (not shown), an accelerated decrease starting at the onset of the accented syllable, and then a steep increase somewhere in the syllable nucleus. This steep increase was always preceded by a local minimum in the F_0 curve.

One question of interest about these contours is whether their time course is at all related to the temporal structure of the phrase-final accent group. After all, traditionally the yes/no feature is attached to phrases, not to syllables or feet.

Our data strongly suggest that there is a relationship. For example, the location of the point where the local F_0 minimum is reached can be predicted quite accurately ($r = 0.96$) from the onset duration and the sonorant rhyme duration, with respective coefficients of 0.94 and 0.75, with a non-significant intercept value. This means that the minimum point is not located at some fixed or random amount of time before the phrase boundary, in which case these correlations would have been zero and the intercept would have been significant. To the contrary, as measured from the end of the phrase, the minimum point occurs earlier as the s-rhyme becomes longer and as the onset becomes shorter. Similar results were obtained for the *rise point*, defined as the first point where the rise over a 10 ms. interval exceeds 10 Hz.

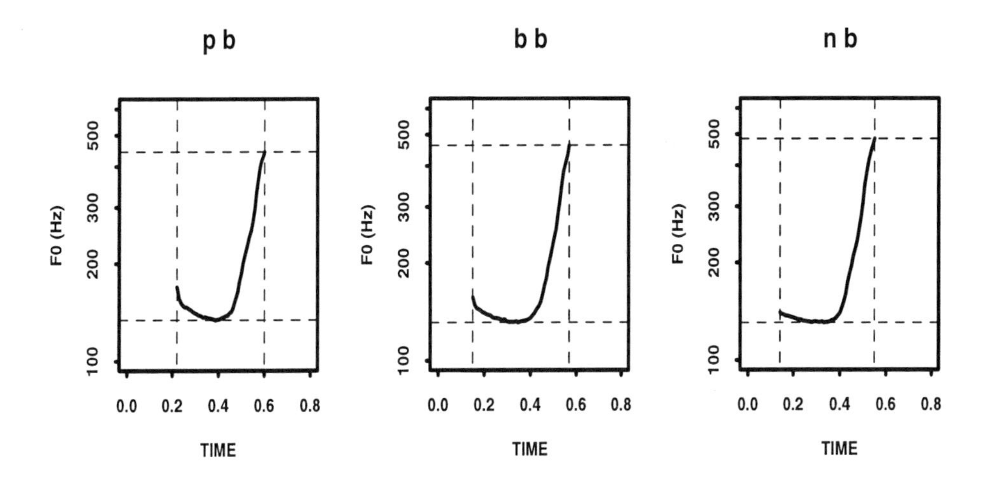

Figure 6.20: Three examples of yes/no question contours, for the words *pub*, *bib*, and *knob*. Contours are confined to the phrase-final accent group.

Continuation rises

Continuation rises were obtained from the same speaker. All recordings involved monosyllabic accent groups, all having the structure nasal-vowel-nasal. Figure 6.21 shows the centroid contour. We found an interesting phenomenon — a very low (in fact, slightly negative) correlation between the frequencies at the H* peak and the H% phrase boundary. Note that correlations between frequencies at similarly spaced anchor points for the standard contour and the yes/no contours are all positive, typically strongly so.

What could account for this correlation? When we compute the correlation between the frequencies at all anchor point locations, it turns out that all points preceding the minimum correlate negatively with the frequency at the H% phrase boundary. When we subtract the estimated phrase curve drawn through the frequency at syllable start and the minimum, the pattern of correlations becomes quite clear. The points can be dichotomized into a pre-minimum and a postminimum group; all correlations between groups are negative, all correlations within groups are positive. This subtraction is justified when we assume that the local phrase curve varies independently of the superimposed accent curve.

One interpretation of this pattern of correlations is that continuation rises involve two quasi-independent gestures — one responsible for H* and the other for H%. Some evidence for this interpretation comes from an observation by Pierrehumbert (1980) that continuation rises can take on a multiplicity of shapes:

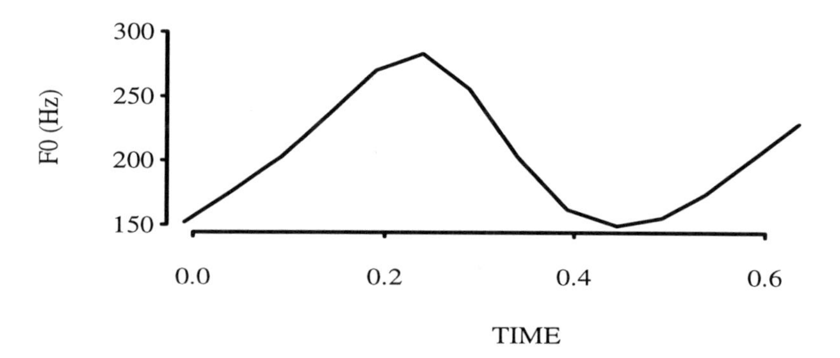

Figure 6.21: Average of 40 continuation rise contours.

some, like ours, with a clear peak, a valley, and a rise; others with an intermediate plateau instead of a well-defined valley.

Alignment parameters and parameterized time warps

We take a little conceptual detour here, by pointing out that the model given by Equation 6.8 can be interpreted as stating that accent curves are the result of a *parameterized time warp of an underlying template*; the precise shape of this time warp is dictated by the multi-dimensional temporal and segmental structure of the accent group. We define the template as consisting of n pairs $< i,\ P_i >, i = 1,\ \cdots,\ n$, where i is the index of the anchor point corresponding to P_i fraction of peak height. Thus, if the peak itself is the 7th anchor point in this template, then the peak corresponds to $i = 7$ and $P_i = 1.0$. When the modeled accent curves start and end at 0 Hz, $P_1 = 0.0$ and $P_n = 0.0$. The time warp is then given by:

$$\begin{aligned} Warp(i; D_{onset},\ D_{s-rhyme};\ C_o,\ v,\ C_c) &= \\ T_i(D_{onset},\ D_{s-rhyme};\ C_o,\ v,\ C_c) \end{aligned} \tag{6.9}$$

Warp maps template time (i) onto the time axis in recorded (or synthesized) speech, measured from accent group start; it is parameterized by the segmental and temporal structure of the accent group. It is important to realize that *even though the model in Equation 6.8 is linear, the time warp is not.*

6.3.3 A new alignment model: theory

The preceding section was largely model-free, and described detailed effects of the segmental composition and temporal structure of accent groups on empirically obtained curves which we called accent curves.

Obviously, the analysis model that we used to compute anchor points borrowed key concepts of the Öhman-Fujisaki model, described in Section 6.1. Here, we want to state our theory first in the broadest terms, and discuss how it generalizes or alters the Öhman-Fujisaki model. In the next section, we describe our implementation of this theory, which makes — of course — several assumptions that are far more specific.

We want to state the broad theory explicitly, because in future discussions and tests of the more specific model it is important to distinguish between assumptions that are part of the essence of the underlying theory and assumptions that are merely ad hoc assumptions made for convenience of the implementation.

Decomposition into curves with different time courses

We computed anchor points by estimating a local phrase curve, and subtracting it from the observed F_0 curve. In the Öhman-Fujisaki model, the observed F_0 curve is obtained by adding (in the logarithmic domain) a phrase curve, an accent curve, and a line representing the speaker's base pitch. We can generalize this assumption by stating that the F_0 curve is made up by adding various types of component curves:

$$F_0(t) = \bigoplus_{c \in C} \bigoplus_{k \in K_c} f_{c,k}(t) \tag{6.10}$$

Since addition in the logarithmic domain is the same as multiplication in the linear domain, in the Öhman-Fujisaki model, $\bigoplus$ is the multiplication operator. C is $\{Base, Phrase, Accent\}$, K_{Base} consists of just one element ($base$) whose corresponding curve is

$$f_{base,1}(t) = \text{base F}_0 \text{ value of speaker}$$

Likewise, K_{Phrase} also consists of just one element ($phrase$) whose corresponding curve is $f_{phrase,1}(t)$. Finally, each phrase contains multiple accent curves, where the i-th accent curve is denoted $f_{Accent,i}(t)$.

Obviously, Equation 6.10 is quite general, and we have to put some additional restrictions on it. First, we want the operator $\bigoplus$ to satisfy some of the usual properties of addition, such as *monotonicity* (if $a \geq b$ then $a \oplus x \geq b \oplus x$) and commutativity ($a \oplus b = b \oplus a$). The usual addition and multiplication operators have these properties, but there are also other operators that do, such as:

$$a \oplus b = \phi^{-1}\left[\frac{[\phi(a) + \phi(b)]}{[1 + \phi(a)\phi(b)]}\right]$$

For appropriately defined ϕ, this operation can have the interesting property that its value never exceeds some maximum (Krantz et al., 1971, page 93). This may be important to model F_0 ceilings (e.g., the top line in Pierrehumbert's model).

Now, an additional assumption we want to make is that each class of curves, c, corresponds to a *phonological entity with a distinct time course*. For example, the *Base* class has a larger temporal scope (all utterances of a given speaker) than the *Phrase* class, and the latter has a longer scope than the *Accent* class (which, in Möbius's work (Möbius, 1993; Möbius, Pätzold, and Hess, 1993) is tied to accent groups).

Accent group subinterval directional invariance

We make the following central assumption about alignment: when, holding all other parts of an accent group constant (in terms of segmental composition and duration), some part becomes longer, peak location shifts to the right. We refer to this assumption as the *accent group subinterval directional invariance assumption*.

The linear data analysis model posits that peak location is a *weighted sum* of the durations of these three subintervals. Thus, similar to the jump we make in duration modeling from directional invariance to sum-of-products models, we make the leap here from accent group subinterval directional invariance to weighted summation. Of course, this step is only partially justified. In fact, we now know that in the limit — when the unstressed remainder of the accent group becomes very long — linearity breaks down; we handle this simply by truncating durations at 400 ms. But in the range of durations observed in our data, linearity holds up remarkably well.

Generalized accent groups

We want to generalize the concept of accent group as defined by Möbius (Möbius, 1993; Möbius, Pätzold, and Hess, 1993). In Möbius's work, syllables are dichotomized into stressed and unstressed syllables, and an accent group is a sequence of successive syllables that starts with a stressed syllable and is terminated by a phrase boundary or a stressed syllable.

However, one has to consider non-dichotomous classifications of syllables. For example, the Bell Labs (American English) system classifies syllables in terms of lexical stress with three levels (primary, secondary, and unstressed) and in terms of the accent status attached to the syllable's word also with three levels (accented, deaccented, cliticized). This creates a large array of possibilities for different types of accent groups, different rules for their start and termination, and which accent groups can overlap.

For example, we trichotomize syllables into Strong, Medium, and Weak, and we could posit that there are two types of accent groups, Strong and Medium. Strong accent groups start with Strong syllables, are terminated only by Strong syllables and phrase boundaries, but not by Medium syllables. Medium accent groups start with Medium syllables, and are terminated by Strong and Medium syllables and by phrase boundaries. Thus, a sentence like *He lives in Massachusetts*, with Strong syllables *lives* and *chu* and Medium syllable *Ma* would have two Strong accent groups (*Lives in Massa* and *chusetts*) and one Medium accent group (*Massa*).

6.3.4 A new alignment model: implementation

We now describe the intonation component as currently implemented, emphasizing the fact that several aspects are expected to be modified as we encounter new languages.

Structure

The component uses as input the phone string with attached durations, syllables with attached lexical stress levels, words with attached prominence levels, minor phrases, and major phrases.

Based on the information in this input, the component first computes accent groups, and then three classes of curves: *minor phrase curves* (one for each minor phrase in a major phrase), *accent curves* (one for each accent group), and *segmental perturbation curves* (one for each obstruent-sonorant transition). These three classes of curves are added together in the log domain, and the result is passed to the synthesis component.

Accent groups

Defining an accented syllable in terms of the Bell Labs system as a syllable that has primary lexical stress and occurs in an accented word, we define an accent group as a sub-sequence of the sequence of syllables contained in a minor phrase, such that the first syllable is accented and the remaining syllables — if any — not accented. An example is given in Figure 6.22 for the phrase *The model contains hierarchical elements*, with accents on the words *model*, *hierarchical*, and *elements*.

This definition has several implications. First, accent groups cannot span minor phrase boundaries. Second, non-accented initial syllables in a minor phrase belong to no accent group. Third, accent groups do not overlap. Fourth, within a minor phrase successive accent groups are not separated by syllables.

Minor phrase curves

Minor phrase curves are obtained by interpolation between three points: first, a pre-nuclear part between the start of the minor phrase (initial boundary tone) and the start of the last accent group in the minor phrase; and, second, a nuclear part between this last point and the end of the minor phrase (final boundary tone; see below, however, on the treatment of Yes/No questions). The F_0 values at these points are computed by rule from the location of the minor phrase in

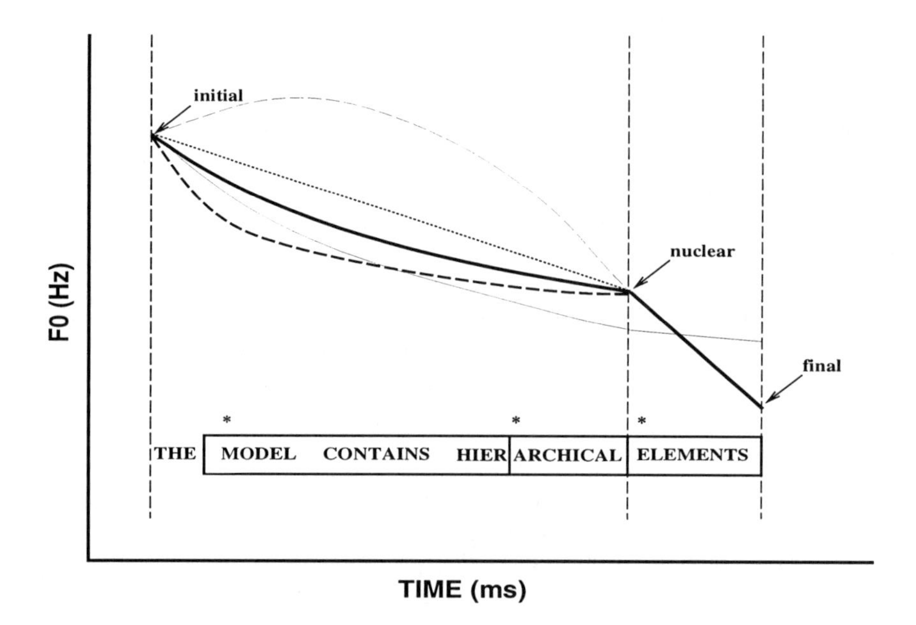

Figure 6.22: Minor phrase curve obtained by interpolation between pitch values located at "initial", "nuclear", and "final" time points, shown with vertical dashed lines. Accented syllables are shown with asterisks, accent groups with boxes. Several examples are given of the pre-nuclear part.

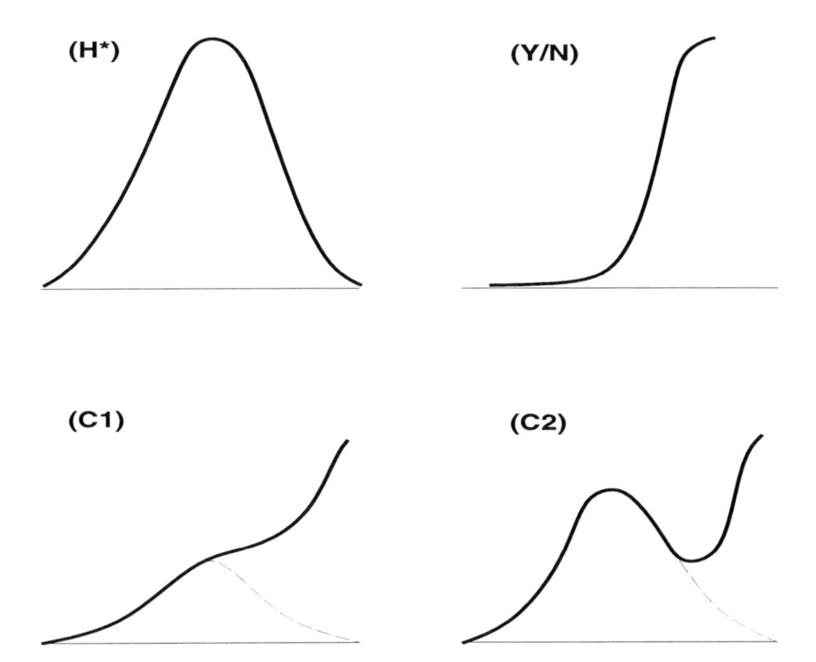

Figure 6.23: Templates for the H*, yes/no (Y/N) , and continuation rise (C1, C2) contours. The latter are compound templates produced by weighted addition of appropriately time-warped H* and yes/no contours. Horizontal axis: anchor point location. Vertical axis: frequency.

the major phrase. The pre-nuclear part of the curve is generated by a weighted sum of a linear component and a non-linear component. It includes the linear case, because the weight of the non-linear component can be set to zero. For now, the nuclear part is linear in the log domain.

Note that this class of curves includes the (linear) declination line, used in the Pierrehumbert and IPO models.

The essence of the phrase curve is that it captures a slowly changing local reference for accent-related pitch excursions. It is customary to think of the phrase curve as modeling elementary speech production processes (e.g., decrease in subglottal pressure as air in the lungs is used up). However, we prefer to be open to the possibility that the shape and height of phrase curves is under speaker control and exhibits considerable and meaningful variability. Therefore, we view the current parameterization of the phrase curve as quite preliminary, and anticipate results from production studies that will guide us to new parameterizations.

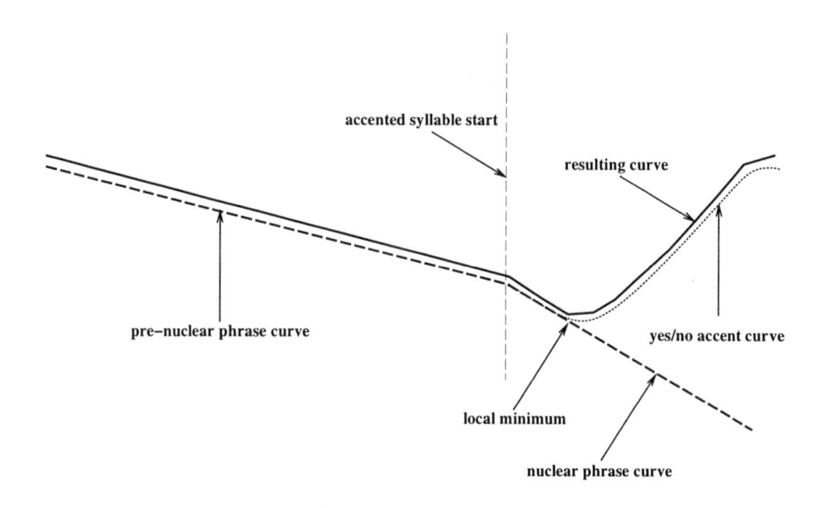

Figure 6.24: Generation of a Yes/No question contour, by adding a rise template to a standard phrase curve. The result is the creation of a local minimum in the accented syllable.

Accent curves

Computation of an accent curve for a given target accent group involves two ingredients: templates and alignment parameters. At run time, the temporal structure of an accent group is computed. This structure is characterized for now as the durations of the accented syllable onset and the sonorant rhyme (for monosyllabic accent groups), or the durations of the accented syllable onset, the accented syllable rhyme, and the unstressed remainder (for polysyllabic accent groups).

Also a *type* is computed for each accent group. Currently, these types are: (1) *H*, Phrase-medial*, (2) *H*, Phrase-final*, (3) *Yes/no question rise*, and (4) *Continuation rise*. Based on the type, the system retrieves the appropriate template and alignment parameters; after the template has been warped using the alignment parameters, the height or amplitude of the resulting curve is adjusted based on prominence and location in the phrase. We discuss this next.

Template for the target accent group type. Figure 6.23 shows templates for H* and yes/no contours. Note that continuation rises are produced by separately time-warping the H* and yes/no templates, and adding the result. Thus, the contours presented in 6.23, Panels C1 and C2, are not really templates, but show the result of the applying the warp + addition operation to the H* and yes/no templates. This treatment of continuation rises is suggested by the results reported in Section 6.3.2.

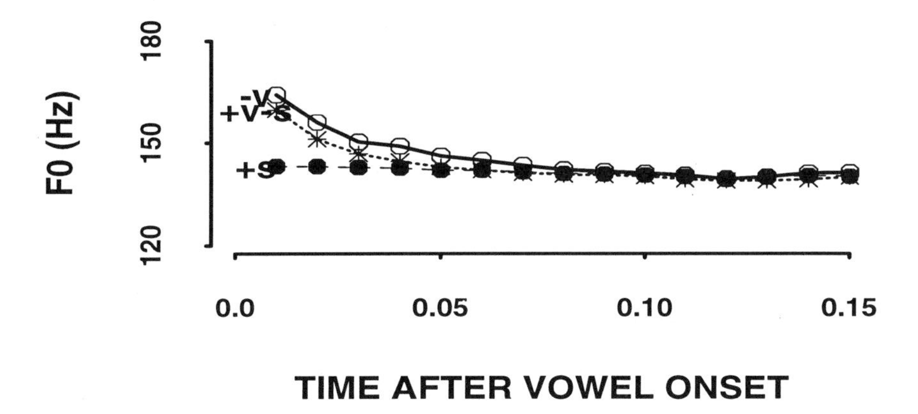

Figure 6.25: Effects of prevocalic consonant class (+S: sonorant, +V-S: obstruent voiced, vs. -V: voiceless) on F_0 during the initial 150 ms. of vowel regions in deaccented syllables.

Alignment parameters for the target accent group type. Equation 6.8 and Figure 6.19 explain the concept of alignment parameter. We have separate alignment parameter matrices for accent group types (1: H*, Phrase-medial), (2: H*, Phrase-final), and (3: Yes/no question rise). For type (4: Continuation rise), we use the matrices for (2: H*, Phrase-final) and (3: Yes/no question rise).

All these parameter matrices are estimated with the same procedure as was used to estimate the parameters in Figure 6.19. For the Yes/no question rise, we fitted the local phrase line to the points defined by accent syllable start and the local minimum (see Figure 6.24). The estimated accent curve was produced by subtracting the local phrase line from the observed F_0 curve, and dividing the resulting curve by its maximum value.

Accent curve amplitude. The amplitude of an accent curve is based on three classes of factors, prominence, locational, and temporal factors. Prominence factors are computed by previous TTS modules as noted in Section 6.2.

Locational factors have to do with whether the accent group is phrase-initial, phrase-final, or phrase-medial. In our current implementation, no distinction is made among the phrase-medial accent curves. In other words, although we may produce a "downstep" between the first and second accent group, no further

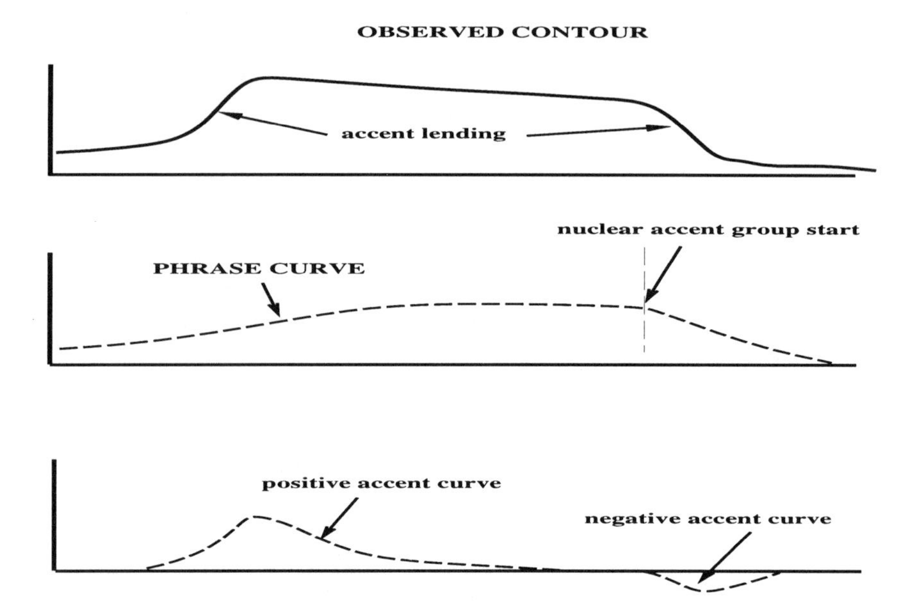

Figure 6.26: Stylized plateau-like contour, with accent-lending rise and fall flanking a straight-ine decline in the center, and terminated by a short flat region. This contour can be decomposed into a phrase curve with rising pre-nuclear portion, and two standard accent curves one of which has a negative amplitude.

reductions in accent group amplitude take place except when caused by prominence differences. In the absence of prominence differences, any reductions in absolute peak height between the second and third accent group are due to a phrase curve declination. We allow for amplitudes to be negative. In fact, for nuclear pitch accents in Italian declaratives, they always are.

Temporal factors refer to an overall effect of accent group duration on accent curve amplitude. We found a correlation in our data between these two factors; the relationship was sigmoid, meaning that for very short durations up to some lower bound amplitudes were quite low, while beyond some higher bound amplitudes did not further increase with duration.

Perturbation curves

Perturbation curves are associated with initial parts of sonorants following a transition from an obstruent. We have observed strong effects here, by contrasting vowels preceded by sonorants, voiced obstruents, and unvoiced obstruents in deaccented syllables where the local F_0 contour was flat in the post-sonorant condition (Figure 6.25). These effects are difficult to measure for accented syllables, because of non-flatness of the local F_0 due to the accent curve itself.

The curves shown in the figure are used directly, and are added to the local phrase curve.

6.4 Future Directions

The application of the current superposition based component to English, Spanish, German, French, Italian, Romanian, and Russian has been quite successful, despite the many simplifications. We view this new component as a first implementation of the general theory discussed in Section 6.3.3. We expect major modifications to follow.

There are some phenomena in Italian and Russian that may appear to be challenging for the superposition based component because they involve high plateau-like regions that are terminated by an accent-lending fall and are initiated by an accent-lending rise (See Figure 6.26). However, such curves can be generated by a *rising pre-nuclear phrase-curve*, a positive accent curve for the rise, and a negative accent curve for the fall. The flat region following the fall is thus the result of a locally decreasing nuclear phrase-curve and a locally increasing tail end of the negative accent curve. Whether this account is empirically tenable is currently under investigation, but it does show that these plateau regions do not pose a threat to the central tenets of the general theory.

Perhaps much more challenging to the general theory is the issue of whether we will be able to adapt this model to tone languages. Even more challenging is the fact that the definition of pitch accent implicit in the superposition based component is at variance with the definition as used by the accenting component (see Section 4.1). In general, more syllables are considered accented by the latter than can be comfortably handled by the former. Very good intonation can be generated, however, when some of these syllables are made deaccented using escape sequences (see Section 6.2) in the textual input; it is rarely the case that changing the status of a syllable from deaccented into accented improves speech quality. We are currently working on new accenting rules that will address this issue.

Chapter 7

Synthesis

Joseph Olive, Jan van Santen, Bernd Möbius, Chilin Shih

7.1 Introduction

The synthesis component takes a phone sequence and prosodic information as input, and produces digital speech as output. In this Introduction, we first give a brief — mostly historical — background on the generation of artificial speech, emphasizing the role of the *source-filter model* in synthesis. We next discuss key dimensions on which current approaches differ.

The remainder of this chapter has three main sections. Since, like the majority of current systems, the Bell Labs system produces speech by concatenating stored speech intervals, we examine in Section 7.2 the combinatorial situation for these intervals in unrestricted domains: *how many are needed to cover a domain*. Next, in Section 7.3, we discuss the process of generating these intervals. Finally, in Section 7.4, we discuss the run-time algorithms that generate speech from these intervals.

7.1.1 Basics of speech production

Brief history

Although the process of converting phonemes to speech is just one of the many modules in a TTS system, until recently it was the main concentration of research in synthesis.

In fact, as far back as the 18th century mechanical devices were constructed by C. Gottlieb Kratzenstein in 1779 (Kratzenstein, 1782), and by Wolfgang Ritter von Kempelen in 1791 (von Kempelen, 1791). These devices consisted of bellows (lungs), and a reed (vocal chords), which excited a resonance chamber (vocal tract). The shape of the resonance chamber could be altered manually to generate different speech sounds, in the same way as the positions of tongue, lips, and jaw alter the shape of the vocal tract. In other words, these machines were based on an understanding of some of the key features of the speech production apparatus.

A century later, studies of the relationship between the spectrum and the resultant sound led some to hypothesize (e.g., von Helmholtz (1954; 1870)) that different speech sounds could be produced by careful control of the relative loudness of different regions of the spectrum, and hence by electric means rather than mechanical replications of the vocal tract. The *voder* developed by Dudley at Bell Labs in 1939 was built on these principles (Dudley, Riesz, and Watkins, 1939). It consisted of two independent sound (or "excitation") generators, one for periodic sounds (vocal chords during voiced sounds) and the other for noise (turbulence caused by constrictions in the vocal tract). A manually operated filter mimicked the effects of the vocal tract.

Of course, these machines incorporated only a rudimentary understanding of the speech production apparatus. Nevertheless, the devices were important first steps towards constructing systems that are able to produce convincing synthetic speech — both because they produced intelligible speech sounds and because they were based on the critical concept of *independent control of a periodic (or noise-emitting) source and the contribution of a variable vocal tract*. This idea is the foundation of much of current speech synthesis.

Source-filter model of speech production

For digital speech synthesis, our understanding of speech production has to be given a computable form. An excellent presentation of mathematical details for acoustic models of speech production, and electrical analogs of the vocal tract, has been presented by Douglas O'Shaughnessy (1987, chapter 3.5). In this section, we will therefore only briefly discuss the acoustic theories of speech production (due to Fant) and vowel articulation (due to Ungeheuer), and their implications for speech synthesis.

Gunnar Fant's (1960) acoustic theory of speech production posits that speech production can be modeled by the following ingredients: an *excitation source* or voice source, an *acoustic filter* representing the frequency response of the vocal tract, and the *radiation characteristics* at the lips. The vocal tract transfer function is represented by a linear filter; the filter shapes the spectrum of the signal that is generated by the voice source, which is a quasi-periodic train of glottal pulses in the case of voiced sounds, and randomly varying noise for voiceless sounds. Fant's theory allows for mixed excitation and multiple simultaneous places of articulations; thus, it explains the production of voiced fricatives and can, in principle, also accommodate doubly articulated sounds such as the Igbo stops /kp/ and /gb/.

The main contribution of Gerold Ungeheuer's (1957; 1962) acoustic theory of vowel articulation is to show that the formant frequencies are determined only by the geometrical configuration of the vocal tract, i.e., by the local size of the cross-sectional area, varying along the z-axis between the vocal chords and the lips; in previous work, different resonance frequencies had always been assigned to different sections or sub-cavities (Helmholtz resonators) of the vocal tract. Ungeheuer uses a set of differential equations known as Webster's horn equations to mathematically describe the propagation of resonance frequencies in the vocal tract depending upon particular geometrical tract configurations. His theory is less general than Fant's approach because it only predicts formant frequencies for vowels without explaining consonant production, and even excludes the coupling of the nasal cavity with the vocal tract during nasalized vowels.

Both Fant's and Ungeheuer's theories make two simplifying assumptions about the configurations and processes involved in speech production. First, neither theory takes into account that the sub-glottal cavity is acoustically coupled with the supra-glottal cavities during voiced phonation. To make things more complicated, the degree of coupling varies during each cycle of vocal chord vibration; it ranges from an effective de-coupling at the moment of complete glottal closure, and increases and decreases gradually during the open glottis state. Second, both theories assume that the contributions of the vocal source and the vocal tract filter, respectively, can be analytically separated from each other, and that there is no back-coupling effect of the filter on the source.

Parametric synthesizers that are based on the source-filter model of speech production, most notably those applying *Linear Predictive Coding* (LPC) (Markel and Gray, 1976), make the same simplifying assumptions. For instance, while the source signal contains many spectral components, these components are assumed to be independent of the filter. Back-coupling feedback from the filter to the source, which is known to actually occur in human vocal tracts (Flanagan, 1972; Fujisaki and Ljungqvist, 1986), is assumed to cause effects of only minor importance and is therefore not modeled. Rabiner and Schafer (1978, chapter 3) discuss models based on simplified versions of the acoustic theory of speech production, such as lossless tube models, their discrete-time equivalents, and how they can be represented and implemented in the form of digital filters.

From the speech analysis point of view, it is often difficult to separate source from vocal tract characteristics, especially for female or child voices, which exhibit interactions between F_0 and F_1 because of their proximity. Moreover, the signal is less well-defined for higher F_0 values because of the wider distance between individual harmonics. An adequate model for the synthesis of female voices is therefore relatively hard to come by, and it is probably fair to say that the limited knowledge about the acoustic characteristics of female speech, compared to male speech, is the main reason that the parametric speech synthesis effort has largely focused on male voices, with a few notable recent exceptions (Karlsson, 1989; Klatt and Klatt, 1990).

7.1.2 Key differences between current approaches

With few exceptions, current speech synthesis approaches are based on the source-filter model. Yet, as we shall see, there exist fundamental differences between these approaches.

Rule-based vs. concatenative synthesis

One major distinction is between *rule-based* synthesis (e.g., DECtalk[1]), where mathematical rules are used to compute trajectories of parameters such as formants or articulatory parameters, and *concatenative* synthesis (e.g., (Küpfmüller and Warns, 1956; Peterson, Wang, and Sivertsen, 1958; Dixon and Maxey, 1968); for the earliest references to the Bell Labs system, see (Olive, 1977; Olive and Liberman, 1985)), where intervals of stored speech ("acoustic units") are retrieved, connected, and processed to impose the proper prosody.

Recently, there has been a trend away from rule-based synthesis towards concatenative synthesis. One cause of this trend is that while rule-based synthesis originally had the important advantage of requiring far less storage, this has become largely irrelevant because of vast increases in hard disk capacity over the last decade (the per-byte cost of storage has decreased by at least two orders of magnitude).

Another factor is that significant progress has been made in ensuring smoothness in concatenative synthesis. Where in the past concatenative synthesizers struggled with generating smooth speech from spectrally discrepant stored speech intervals and rule-based synthesis was always comparatively smooth, this smoothness gap has been drastically reduced.

A third circumstance is that it is simply quite hard to produce certain sounds by rule. For example, although convincing synthetic vowels can be generated easily, voiceless stop aspiration, bursts, and other turbulent sounds have proven to be very hard to generate. This reflects a profound limit of the rule-based approach: to produce such speech sounds it is necessary to model neuronal, physiological, mechanical, and/or fluid dynamical phenomena that are inherently too complicated for current analytic descriptions. On the other hand, by having a list of concatenative unit intervals that contain these potentially problematic regions in their entirety (e.g., a unit that starts towards the end the closure, continues through the burst and aspiration, and then terminates midway into the vowel), and performing relatively mild signal processing operations on them, concatenative synthesizers can generate convincing renditions of these sounds.

The fourth factor is pragmatic. As already mentioned in Chapter 5, and as will be discussed at length below, concatenative synthesis assumes that all perceptually significant coarticulatory phenomena can be captured by using units that span regions of heavy coarticulation. For example, to capture nasalization

[1] DECtalk is a trademark of Digital Equipment Corporation.

of a pre-nasal vowel one generates a word like "pin" with /p-ɪ/ and /ɪ-n/, but not with /p-ɪ/, /ɪ/, and /n/. There are practical limits to how many units one can use. These limits are dictated not so much by storage issues as by quality control issues and the cost and effort of recording a speaker. There is little doubt that the set of units required to capture *all known* coarticulatory phenomena (cf. (Coleman, 1992)) is simply too large. *Thus, although very good compromises between coverage and acoustic inventory size can be achieved, concatenative synthesis is inherently imperfect.* Unfortunately, there has been a move away from studying these phenomena in the context of synthesis, with notable exceptions (e.g., (Stevens and Bickley, 1991; Coleman, 1992; Bickley, Stevens, and Williams, 1997; Dirksen and Coleman, 1997)), because producing a concatenative synthesizer of acceptable quality is much easier and does not require the detailed knowledge of speech production processes needed for producing a rule-based synthesizer of similar quality. However, we maintain that in the future the inherent shortcomings of concatenative synthesis will once again become visible.

In the remainder of this chapter, we restrict ourselves to concatenative synthesis. Extensive references to rule-based synthesis can be found in (Allen, Hunnicutt, and Klatt, 1987; Klatt, 1987), and to coarticulatory phenomena in (Olive, Greenwood, and Coleman, 1993).

Speech analysis/synthesis techniques

The most obvious distinction among concatenative systems concerns speech analysis/synthesis technique. For an excellent review of these techniques, we refer to (Dutoit, 1997). The usual method is to use linear predictive analysis to encode stored speech into a sequence of filter coefficient vectors spaced several ms. apart (e.g., 5 ms., or pitch-synchronously) and to generate synthetic speech by applying the filter coefficients to an *excitation source* (or the speech *residual*) in the form of either pulse trains (for voiced segments) or noise (for voiceless segments). In the LPC model, the excitation source or residual corresponds to vocal chord activity. Concatenation consists of smoothing two frame sequences together, changing their timing, and adjusting the period of the pulse trains or changing the residuals.

During the last decade, a non-source-filter approach to synthesis known as *PSOLA* (for "*P*itch *S*ynchronous *O*ver*l*ap and *A*dd") (Charpentier and Moulines, 1989), has become popular. Here, pitch-synchronous single-peaked windows whose lengths equal roughly two pitch periods are multiplied with the speech signal, resulting in a set of localized signals that are each non-zero only in the intervals defined by the corresponding windows. Concatenation is performed by smoothly fading in and out of two such sets of localized signals. To modify pitch, these localized signals are shifted relative to each other so that their spacing reflects the desired pitch epoch durations, and then added. Provided that extreme pitch modification (say, outside of the 0.6–1.5 range) is avoided, this results in little spectral distortion. To modify duration, an integral number of pitch epochs is added or excised.

Many additional techniques exist. Some are variants of the LPC approach, e.g., replacing the pulse/noise excitation source by an explicit glottal waveform model (Oliveira, 1997), or a spectrally defined decomposition into periodic and aperiodic components (Richard and d'Alessandro, 1997); others are variants of PSOLA, including a hybrid of LPC and PSOLA (LP-PSOLA (Charpentier and Moulines, 1989)) where the overlap-add operation is applied to the LPC residual. At this point in time, it is unclear which coding scheme is preferable. PSOLA, by preserving the entire speech signal, can produce extremely high quality speech when the to-be-concatenated waveforms are highly similar. However, the technique appears to be more vulnerable to waveform differences than LPC methods, so that — given the difficulty of avoiding discontinuities — in practice PSOLA is often disappointing (with a notable exception being the French system produced by CNET (Bigorgne et al., 1993)). Relatively little systematic comparisons exist, except (Macchi et al., 1993; Dutoit, 1997).

Type of prosodic alteration

Acoustic units are usually altered during concatenation to obtain the required temporal, intonational, and perhaps additional acoustic-prosodic properties. The reason for these alterations is that for systems with no more than a few thousand units (and perhaps all systems), the prosodic context in which a unit is used may differ from the prosodic context in which the unit originally occurred in the speech corpus. Concatenative systems differ in which dimensions are altered.

In AcuVoice (AcuVoice, Inc.) and CHATR (ATR Interpreting Telecommunications Research Labs, (Campbell and Black, 1996)) no prosodic modification is applied in either the pitch domain or the temporal domain. This assumes, of course, that the combinatorics of speech can be captured by a set of units that does not pose undue stress on the recorded speaker. Much more about combinatorial issues will be said below in section 7.2.

The large majority of systems, however, apply various prosodic alterations. The usual variables are F_0 and timing, but many systems including the Bell Labs system also alter amplitude.[2] Note that alteration of F_0 and amplitude involves changing the entire F_0 and amplitude *profile* of a unit, not the average. This makes these forms of alteration more intricate than temporal modification. In the case of F_0, this is particularly hazardous, because here the success of the alteration depends on how well the speech parameters reflecting vocal tract shape (e.g., LPC coefficients) are uncontaminated by the source characteristics

[2] Specifically, the Bell Labs system computes an amplitude multiplier profile that spans the entire utterance. Its values (spaced at 5 ms intervals) are multiplied with the stored excitation or the modeled glottal waveform. The multiplier profile is computed by rule, and has a value of 1.0 (no change) in most locations; it has smaller values in unstressed syllables and very short syllables, and drops to a minimum value in the course of the utterance-final syllable.

(e.g., fundamental frequency, creaky voice) in the recordings. In the light of our remarks in section 7.1.1, there are reasons to be concerned about this.

In theory, prosodic alterations need not be confined to F_0, timing, and amplitude. Systems that use an explicit source model, such as the Bell Labs system, can alter source parameters, such as the per pitch epoch open quotient. This appears more difficult in the case of non-source-filter systems such as PSOLA, which seem largely confined to F_0, timing, and amplitude. Some consider this as a major drawback of this approach (e.g., (Kahn and Macchi, 1997)).

7.2 Combinatorics of Acoustic Units

In concatenative synthesis, a relatively small number of units is used in conjunction with prosodic alteration rules to cover the vast combinatorial space of all combinations of phoneme sequences and prosodic contexts that can occur in a language. How large is this space, and how does it limit acoustic inventory design choices?

Any scheme specifying which units are needed has to take into account the following basic facts. First, in American English, assuming a 43-phone alphabet, at least 70,000 of the theoretical maximum of 79,507 triphones actually occur in the language. Each of these triphones, in turn, can occur in thousands of contexts, many of which have strong coarticulatory effects. This includes triphones spanning word boundaries, with good reason: coarticulation does not respect word boundaries.

The second fact is that speech is highly variable, and that concatenative synthesis is particularly vulnerable to this variability. We have observed F_2 variations in excess of 50 Hz for repetitions of the exact same phrase, as uttered by a highly trained professional speaker. While the different renditions sounded identical to the human listener, when we excised two units and concatenated them, the 50 Hz discrepancy became audible. This means that in practice several renditions are needed for each phone sequence. It also means that speaking and recording conditions have to be tightly controlled, even for professional speakers. Spontaneous speech, let alone speech recorded from different persons or the same person in different environments, is not suitable for concatenative synthesis.

The third fact is that speakers cannot produce more than at most a few thousand successfully rendered and recorded utterances per day. Thus, schemes that propose 100,000 unit inventories are not realistic.

A seemingly convincing counterargument against these pessimistic statistics is that the frequency distributions of units are not only extremely uneven — with some units occurring orders of magnitude more often than others — but that we can in fact ignore a large fraction of the units because they simply occur too infrequently.

However, it appears that these frequency distributions have the somewhat perverse property that the number of rare types is so large that the combined probability mass of these types — even though the probability mass of each type individually is infinitesimal — approaches near certainty. We now describe some analyses leading to this conclusion. In one analysis, we look at the combinatorics of inventories in which different acoustic units are used depending on the prosodic context. In the other analysis, inventories consist of units that start and end with obstruents.

7.2.1 Prosodically annotated units

In our analysis, 250,000 sentences from the Associated Press Newswire were automatically transcribed by the text analysis components of our American English system. Next, we constructed a contextual vector for each diphone in the transcribed sentences, containing information about key factors such as accent status (accented vs. deaccented) and within-utterance position (initial, final, medial). The factors were kept deliberately coarse. The type count of diphone-context combinations was 222,678. Since most of these combinations are extremely rare, one cannot conclude that all these combinations must be covered in the training set (acoustic inventory) for adequate coverage of the input domain: if a system misses a unit once every thousand sentences, this should not be considered a problem. However, when we analyzed the data using a statistic we call the *coverage index* of a training set, the situation turned out to be more problematic than that.

The *coverage* index of a given training set with respect to a given domain is defined as the probability that all combinations occurring in a randomly selected test sentence are represented in the training set. Thus, 75% coverage means that the probability is 75% that all combinations in a randomly selected test sentence also occurred in the training set. We found that a training set of 25,000 combinations had an index of only 3%, and that an index of 75% required a training set of more than 150,000 combinations. Given that both training and test sentences were from the same text "genre" (Associated Press Newswire), the values of the coverage index are likely to be worse when there are large differences between the text genres used for training vs. test (see our second analysis, below).

7.2.2 Obstruent-terminated units

We performed the same analysis for units that were not prosodically annotated, but were required to have no internal obstruents. The idea here is that spectral discrepancies inside obstruent regions are relatively harmless because speech amplitude is low, so that, it is hoped, very smooth speech can be achieved. Of course, this is not quite true, because certain obstruents (e.g., /š/ in English) are quite loud and are subject to coarticulation (e.g., anticipatory lip rounding). However, the point here is more general: we want to show the combinatorial non-viability of any simple scheme for determining phone sequences that drastically limits in which phones cuts can be made.

Not surprisingly, the results were similar to those obtained in the analysis of prosodically annotated units. The number of units required for 75% coverage was 180,000, and the coverage provided by 25,000 units was only 25%.

7.2.3 Summary

These analyses make two useful points. First, the results show the necessity of prosodic modification. One cannot count on the database containing precisely the right phone sequences in the right contexts to match the requirements of an arbitrary input sentence. This shows the utmost importance of being able to have signal processing techniques that can produce these prosodic modifications with minimal audible distortions. In our experience, all currently known signal processing techniques cause some degree of distortion.

Second, "cuts" have to be made in regions other than obstruents. This makes acoustic unit inventory design (by which we mean constructing the list of unit types) a non-trivial exercise, in which one has to understand coarticulatory phenomena at a deep and detailed level.

There is an additional fundamental problem with very large inventories. When $p(n)$ denotes the probability that at least one out of n tokens of a given unit type is acoustically appropriate, the probability that each of N unit types has at least one such token is $p(n)^N$. When $p(n)$ is 0.99999 and N is 2000, then the probability is 0.018 that at least one unit type will have *no* appropriate tokens. This probability increases sharply with N (N = 10,000: 0.01, N = 25,000: 0.22). The only way to prevent this probability from becoming too large is by increasing n. But since the total number of to-be-recorded tokens is $n \times N$, this causes a sharp increase in the total amount of recordings.

7.3 Acoustic Unit Generation

Generating acoustic units is a multi-step process.

1. *Acoustic unit inventory design* involves constructing a list of phone sequences (unit types) that are to be recorded and excised.

2. *Text generation and recording.*

3. *Acoustic unit token selection.* Once appropriate text has been constructed and recordings are made, the next step is to select for each unit type the best unit token, with the goal of simultaneously minimizing the spectral differences between units and the distance of each sound to its "ideal". [3]

[3] An /X-i-Y/ unit may concatenate smoothly with a /Y-e-Z/ unit, but that is no guarantee that the /i/ and /e/ are good examples of these vowels. Iwahashi and Sagisaka (1992) refer to this type of discrepancy as *segmental distortion* and to the former type as *inter-segmental distortion*.

4. *Acoustic unit token excision.* The fourth step consists of excising the corresponding speech intervals from the speech files, and parameterizing the speech.

We are now convinced that each of these steps is critical. Superior parameterization is ineffective when the inventory design ignores certain major coarticulatory phenomena, while the best inventory design is rendered useless when not enough high-quality candidates are available. The latter can happen when the data base is too small or when there is unwanted variation in the candidates due to poor text selection. To illustrate, an older speech data base we used in the past was designed around an evolving inventory specification, so that some unit types which turned out to be required after the recordings were made had insufficient coverage (e.g., only one candidate, and, to make things worse, in utterance-final position).

After an inventory has been constructed, listening tests can be performed (Chapter 8). While useful, these listener tests are necessarily limited in that at most a small subset of all unit combinations can be covered.[4] Thus, for quality control based on listening experiments, it is essential that the number of unit types is kept to an absolute minimum.

Of course, objective measures of segmental and inter-segmental distortion that are used in unit selection (section 7.3.3) are meant to guarantee the quality of the inventory. However, these measures should be considered extremely preliminary as far as their capability to capture human speech perception is concerned. *Constructing perceptually valid objective measures for predicting the perception of segmental and inter-segmental distortion is a wide open research issue.* Hence, currently some degree of subjective listening is unavoidable, whether in the form of systematic experiments or of informal usage. In either case, the number of units should be as small as possible.

7.3.1 Acoustic unit inventory design

The acoustic inventory of a concatenative TTS system for a given language (*and* speaker, as we will argue below) is a set of stored speech segments that in its totality covers all legal phone sequences of that language, *including inter-word combinations.* Further requirements are that, first, all the necessary phonemic and allophonic distinctions are indeed reflected in the inventory structure and, second, the concatenation of two or more inventory units does not produce audible discontinuities in the resulting synthetic speech (see *acoustic unit selection*, section 7.3.3). Also, as we just stressed, the final inventory should have a manageable size.

As we pointed out in the preceding section, the challenge in inventory design is to capture key coarticulatory phenomena while at the same time keeping the

[4] With N phones and on the order of N^2 diphones, each diphone can be concatenated with N neighbors on the right, so that the number of concatenable diphone pairs is on the order of N^3 or 79,507 for $N = 43$ phones. For triphone-based systems, the number of combinations is 14,700,844.

language	total size	diphones	spec. diph.	polyphones
English	1162	1027	106	29
Mandarin	988	956	8	24
Spanish	1024	501	0	523
German	1177	1102	75	0
Russian	1682	1499	53	130
French	1123	1123	0	0
Romanian	629	628	1	0
Japanese	470	445	1	24

Table 7.1: Inventory structure for several languages implemented in the Bell Labs system: inventory size, number of diphones, specialized diphones, and polyphones.

number of units small.

Our approach to inventory design consists of two stages. In the first rule-based stage, we use general acoustic phonetics and phonological principles to (conservatively) specify where (in terms of phone identity, and phonemic-prosodic context) units can be cut. For example, voiceless stop closures in any phonemic-prosodic context are safe cut points. In the next data-based stage, we perform acoustic analyses of speech recorded by the target speaker to decide in which additional locations we might be able to safely cut units.

One approach to the rule-based stage has been described by Coleman (1993). He stated the following "cuttability" criteria. The overriding constraint for the cuttability of speech segments, according to Coleman's design, is the notion of *contaminators*. Essentially, any segment may be cut, if it is not spectrally contaminated by a neighboring segment. Nasal and rhotic phonemes, in particular, have strong coarticulatory effects on a preceding vowel. Short vowels, especially in unstressed position, may need to be embedded in triphonic units. Coleman also suggests that stress distinctions provided in the phonemic representation be kept in the acoustic unit description, thereby avoiding splitting a phoneme sequence of /X'aY/ into two units /X'a/ and /aY/, in which the stress information is omitted from the second unit. Relatively unproblematic classes of phonemes are long syllable nuclei, except when followed by nasal or rhotic contaminators, stops, voiceless fricatives (with the exception of /h/), voiceless affricates, and nasals and /l/ except before homorganic obstruents; all these phonemes can be split into unit-final and unit-initial subsegments, respectively.

The application of these very conservative criteria for cuttability results in large acoustic inventories. The total set of units that this method proposed for English was 26,387, and 17,479 for Spanish. However, even with a reduced set of criteria, such as dropping the distinction between primary and secondary stress in favor of a binary distinction between stressed and unstressed syllables

First Consonant	Second Consonant
stop	stop
stop	nasal
fricative	fricative
fricative	stop
fricative	nasal
nasal	stop
nasal	fricative
lateral	stop
lateral	fricative

Table 7.2: Consonant pairs with minimal coarticulation. These transition types do not have to be represented in acoustic units.

in English, and ignoring stress marks altogether in Spanish, the number of required units amounted to 10,553 for English and 3,240 for Spanish.

Table 7.1 shows the results of the second stage, in which we use acoustic and other analyses to further reduce inventory size. Clearly, drastic further reduction is possible. We now proceed to describe some of these analyses, which involve: (a) consonant pairs with minimal coarticulation; (b) phonotactic constraints; (c) cross-linguistic studies: trill and vowel space; (d) measurement of coarticulation (context clustering).

Consonant pairs with minimal coarticulation

In some cases, adjacent consonants have only slight coarticulatory effects on each other. Table 7.2 lists some of these cases. For example, we do not have to store units containing voiceless stop-to-voiceless stop transitions because unvoiced stops contain silent intervals that are not strongly affected by adjacent phones. Because of this minimal coarticulation, these phone sequences can be synthesized with units that have been cut at segment boundaries instead of inside segments. In other words, no units are required that contain these phone sequences; this significantly reduces the number of units required, regardless of whether one uses diphones or longer units.

The situation is not quite as obvious for other cases. Therefore, let us consider the transition types illustrated in Figures 7.1 through 7.3; the examples are segments of German speech.

Figure 7.1 shows the transition from the voiceless fricative [s] to the nasal consonant [n]. Due to the movements of the articulators involved in the production of this consonant pair, there is a short silence segment of approximately 10 ms. between the fricative and the nasal. This observation, together with

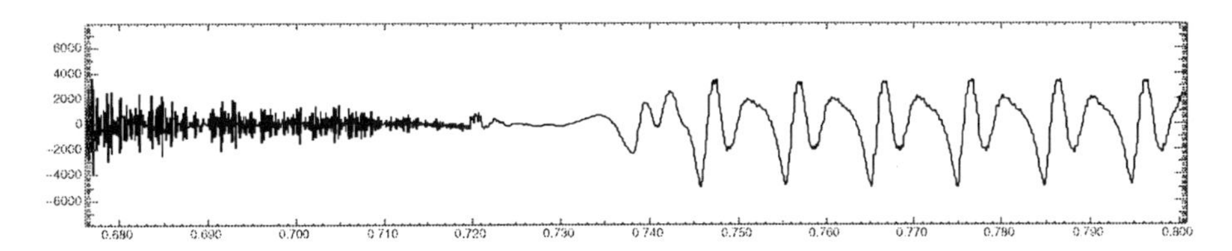

Figure 7.1: A silent interval with a duration of approximately 10 ms. is observed in the transition from a fricative to a nasal consonant ([s-n]).

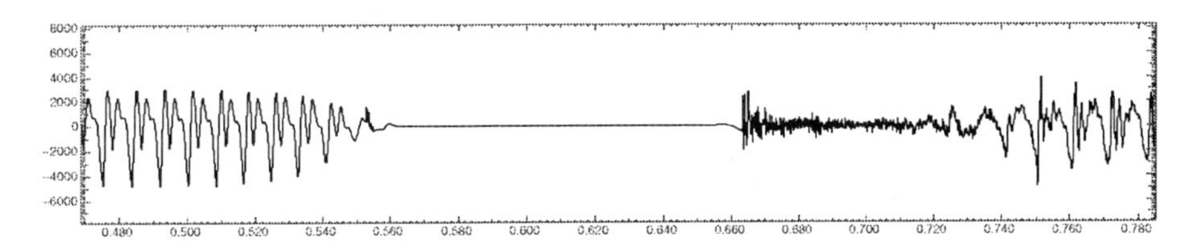

Figure 7.2: Silent interval in the transitional phase from a nasal consonant to a voiceless stop ([ŋ-t]).

the fact that the neighboring speech segments are relatively stable and do not affect each other, allows us to replace the connection with fricative-silence and silence-nasal units without losing crucial transitional information.

Figure 7.2 displays the transition from the velar nasal consonant [ŋ] to the voiceless stop [t]. Again, we observe a silent interval which allows the insertion of a silence unit, and we do not have to store inventory units containing this transition.

Similar observations can be made for the other cases given in Table 7.2. In the case of the transition from a nasal consonant to a homorganic voiced stop, where the closure of the vocal tract is sustained all the way from the beginning of the nasal to the release of the stop, we observe dampened oscillations in the transitional region, i.e., in the closure phase of the stop (Figure 7.3). These oscillations are caused by the vocal chords' continued vibration even after the uvula (soft palate) has been raised to de-couple the nasal cavity from the vocal tract in order to articulate the stop consonant. Yet, there is a short low-amplitude interval in the transitional phase (between 900 and 925 ms.) where a silence unit can be inserted during concatenation, and we do not have to store inventory units containing this consonant pair.

During synthesis, the concatenation algorithm inserts a silence unit with a duration of zero ms. between a pair of phone types represented by "First" and "Second" in Table 7.2. For consonant pairs that involve a stop as the second phone, the procedure is slightly different. Here, the type of unit to be inserted in the closure phase of the stop can be defined in a language specific phoneme

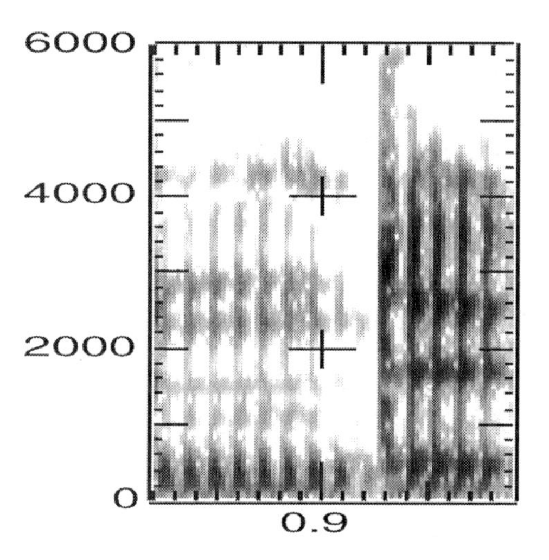

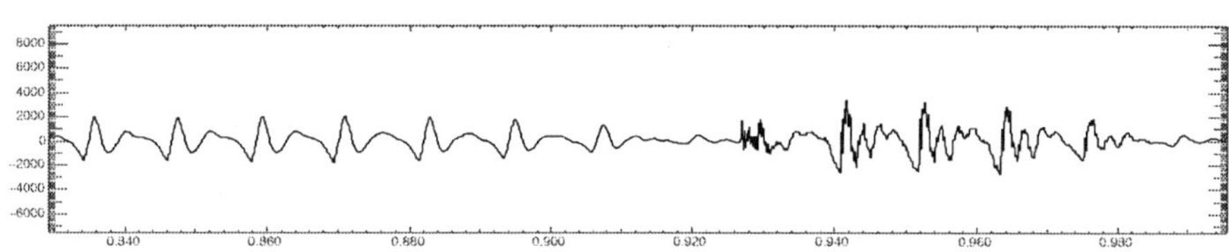

Figure 7.3: Transition from a nasal consonant into a homorganic voiced stop ([n-d]); spectrogram (top) and waveform.

definition file. All of our TTS systems use a silence segment for voiceless stops; for voiced stops, most systems (e.g., English, Spanish, French, and Japanese) use *filled* silence segments that contain some energy due to dampened oscillations (similar to what Figure 7.3 shows), while others use the same silence unit as for voiceless stops (e.g., German).

Considerations of this type allowed reducing the number of units for German by at least 15%.

Phonotactic constraints

The number of units can be further reduced by phonotactic constraints. Coleman (1993) showed that only 84 out of 1681 calculated diphones (Coleman counted 41 English phones, as opposed to 43 in the current version of our English TTS system) — or 5% — do not occur in English. There are strong phonotactic constraints only for the glides [j] and [w] as well as the fricative [h], all of which can only occur in pre-vocalic position. Most other diphones do occur in English.

There are significantly stronger phonotactic restrictions in German. The most effective constraint is the neutralized voicing opposition in morpheme- and word-final position, a phenomenon known as *Auslautverhärtung*. Phonologically voiced obstruents and clusters of obstruents in this position turn into their voiceless counterparts. This neutralization process also applies to foreign loan words and personal or place names. In addition, the distributional restrictions of the glide [j] and the fricative [h], as described for English above, are also valid for German. Further restrictions apply to vowel-to-vowel diphones, the majority of which can be replaced with vowel-to-glottal stop and glottal stop-to-vowel units, and to the occurrence of the palatal and velar fricatives whose distributions are complementary with respect to each other. In total, almost 30% of combinatorially possible diphones cannot occur under any conditions in German, not even across word boundaries or in foreign words or in names.

If we apply the principles based on acoustic properties of phone types as they were described in the previous section and combine them with phonotactic constraints, the total number of purely diphonic units needed for an acoustic inventory of German can be reduced from the maximum of 1849 (based on 43 phones) by 40.5% to slightly more than 1100. Although we do not have similar statistics for triphones and longer units, it is clear that the numbers of these longer units would also be significantly reduced.

The special treatment of the trill

A trill is a "vibration of one speech organ against another, driven by the aerodynamic conditions" (Ladefoged and Maddieson, 1996, page 217). Our experience in synthesizing trills is limited to apical (tongue tip) trills, so henceforth by "trill" we will always mean "apical trill". There is a wide range of variation in the pronunciation of the trill. It is clearly articulated in front of a stressed vowel, but less so in front of an unstressed vowel, and may be reduced to a

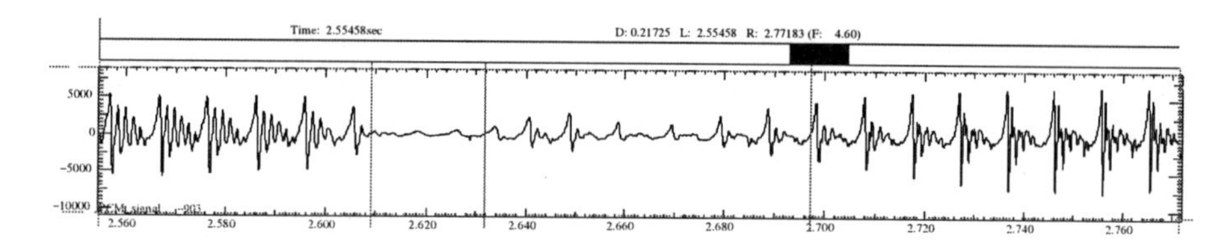

Figure 7.4: A syllable-initial trill.

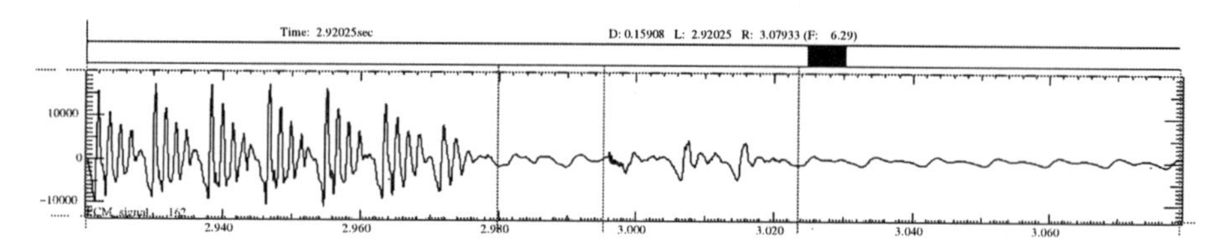

Figure 7.5: A syllable-final trill.

fricative in syllable-final position or in consonant clusters. The first question that needs to be addressed in designing the acoustic inventory involving the trill is whether these variations should be represented as the same sound, and whether triphone or context-sensitive units are needed. One of our systems — Spanish — uses triphones for the trill, while the other system uses diphones with carefully-selected cut points.

Most of the concatenative synthesis methods are designed primarily to handle steady state speech signals and perform poorly on trills, in particular if the timing of the trill needs to be adjusted. Trill vibration is certainly not a steady state and cannot be modeled by linear interpolation between edges of two concatenative units, uniform stretching, or uniform compression. Collecting triphones containing trills as the center unit avoids the problem of making connections in the middle of trills. But doing so has the unfortunate consequence of increasing the inventory size, furthermore, using triphones only bypasses the connection problem: it is still vulnerable when the timing of the trill needs adjustment.

We conducted acoustic studies of Italian trills and identified a region of low energy before the trill that is a good site for unit concatenation (Shih, 1996). Figure 7.4 shows a pre-vocalic, syllable-initial trill, while Figure 7.5 shows a post-vocalic, syllable-final trill. The first two lines in the figures mark the low-energy region, and the second and third line demarcate the section with the trill. The trill units can be cut and concatenated in the first region where the energy is low. In practice, all sound-to-trill units are cut before the trill starts, and the trills are reserved in the trill-to-sound units, thus eliminating the need for triphones.

When the timing of the trill needs adjustment, we only change the duration of the low-energy [r] region and leave the trill region intact. Although this is clearly not the ideal way to model a trill, a perception study shows that the synthesized results with different settings of speaking rate are all acceptable by native Italian speakers.[5]

The vowel space: some cross-linguistic examples

One basic issue that has to be solved before designing the acoustic unit set is to determine the proper phone set. Published reports on phonemic or phonetic inventories are a good starting point, but the optimal inventories for TTS systems are often different, partly due to idiosyncratic speech patterns of the chosen speakers, partly due to the practical concern of building a TTS system. We find the problems in the checklist below particularly susceptible to language variation and speaker variation, so that conducting exploratory acoustic studies of the chosen speaker will be worth the effort:

- Well-known phonological or phonetic processes, such as Russian "hard" and "soft" (palatalized or palatalizing) sounds.
- Stress variations: are there differences in the formant structure of stressed and unstressed vowels?
- Length variations: are there differences in the formant structure of long and short vowels?
- Are there differences in glides and corresponding high vowels?
- Should a diphthong be represented as a single unit or as combination of vowels?
- Pay attention to schwa, which is particularly susceptible to co-articulation, even in languages where this sound is stressed.

We need to juggle many factors in finalizing the phone inventory. Obviously, in dealing with text-to-speech synthesis, spectral quality has to be given higher priority than phonological economy or elegance. If consistent formant discrepancies are found between two populations of the realizations of a phoneme, one should seriously consider representing them as separate phones, provided that the distinction can be predicted from text, and that native listeners prefer having such distinctions. Our analysis of German long and short vowels and Russian soft and hard vowels, respectively, will illustrate this point.

Even if the formant structures of the two populations in question are similar, it is still necessary to check whether their transition patterns in context are similar. Mandarin glides are separated from corresponding high vowels,

[5] Of course, a correct model of trill needs to simulate the vibration of the trill. See (McGowan, 1992) for an articulatory model of this sound.

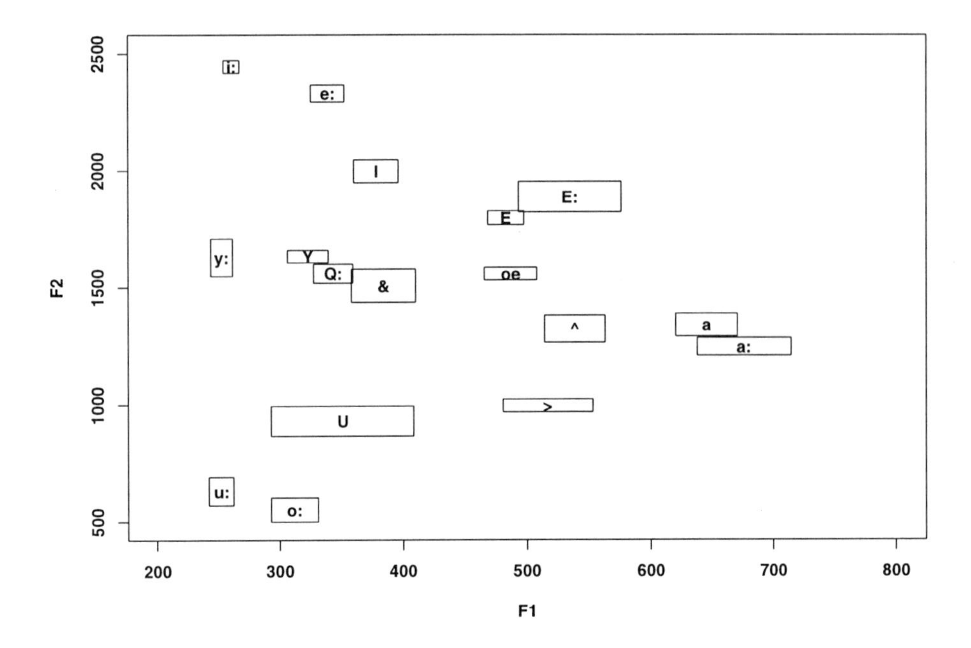

Figure 7.6: Vowel space of our German speaker: vowel symbols mark median values (in Hz) in F_1/F_2 space, boxes delimit standard deviations. Non-IPA symbols map on IPA symbols as follows: & = ə, ʌ = ɐ, E = ɛ, Q: = ø:, oe = œ, > = ɔ.

and diphthongs separated from monophthongs, because there are differences in transition patterns.

Finally, because our synthesis method can withstand rather heavy modification of duration and fundamental frequency, it is not necessary to split two sounds because of timing differences and pitch differences. This is the case for Japanese long and short vowels, which are combined in the inventory; there is no gain in synthesis quality by representing long and short vowels as separate phones in Japanese.

In the following we will present several scenarios from our multi-lingual system to discuss the decision-making process in more detail.

Let us first consider the rich and densely populated vowel space of German. The number of vowel phonemes in German depends on which kind of phonological analysis one prefers. The minimal pair *Miete* [mi:tə] vs. *Mitte* [mɪtə], for example, can be interpreted either by assuming a quantity opposition (long [i:] vs. short [ɪ]) or a quality opposition (tense [i] vs. lax [ɪ]). A widespread, since economic, solution is to assume the existence of a quantity phoneme /:/ representing phonological vowel length, thereby reducing the number of monophthong phonemes to eight.

From the point of view of speech synthesis, spectral quality is the key here. The vowel [ɪ] is not simply the short allophone of the phoneme /i/ ([i:] being the long variant); its typical formant values differ quite significantly from [i:], and its location in the F_1/F_2 space is rather distinct from [i:].

While this is a very clear-cut case where we have to include both vowel qualities in the inventory, the question of how many phonetic qualities there are of the German vowel phoneme /a/ has been addressed in the phonological and phonetic literature for many years and is largely undecided. Recently, Kohler concluded that the difference between phonologically long and short /a/ is clearly in the quantity domain (Kohler, 1995, page 170).

Figure 7.6 suggests, however, that for the definition of our TTS inventory a distinction between long [a:] and short [a] is preferable because the ranges delimited by the standard deviation boxes for the two vowels do not overlap. In fact, a *post hoc* analysis after running the acoustic unit selection procedure described in the following section, clearly indicates that distinguishing between two /a/ qualities is the right choice to make.

A similar situation is found in Russian soft, or palatalizing vowels, and hard, or non-palatalizing vowels, represented in upper case and lower case letters, respectively, in Figure 7.7. Only the stressed vowel population is included in the plot. It is clear that the soft vowels are consistently higher in the F_2 space. Furthermore, the unrounded vowels [I], [E] and [A] are lower in the F_1 space. The best way to capture the "soft" and "hard" distinction is to represent them as separate phones.

The schwa is another recurring problem, since it typically shows strong coarticulation effects with surrounding sounds and requires special treatment such as representing sub-populations separately, using context-sensitive units, or using multi-phone units. It is not surprising that the English schwa is a problematic case, since the schwa is always unstressed in English and it is subject to varying degree of reduction. However, we believe that at least part of the problem comes from the neutral, unmarked tongue position of this sound, since the fully stressed schwa in Mandarin also shows strong coarticulation effects.

Figure 7.8 compares the formant trajectories of the schwa [ə] in Mandarin in three different contexts: solid lines plot the average formant values of all the syllable final [ə]'s in our database; dotted lines plot the average formant values of all the [ə]'s followed by an alveolar nasal coda [n]; dashed lines plot the average formant values of all the [ə]'s followed by a velar nasal coda [ŋ]. The five points on the x-axis correspond to the 20%, 35%, 50%, 65%, and 80% points of the vowel duration. Averaged formant frequencies are plotted on the y-axis.

A following [n] raises the F_2 of [ə], while a following [ŋ] lowers it. The difference in F_2 ranges from 300 Hz at the 20% point to 600 Hz at the 80% point. Evidently, a unit excised from the [n] context will not connect well with

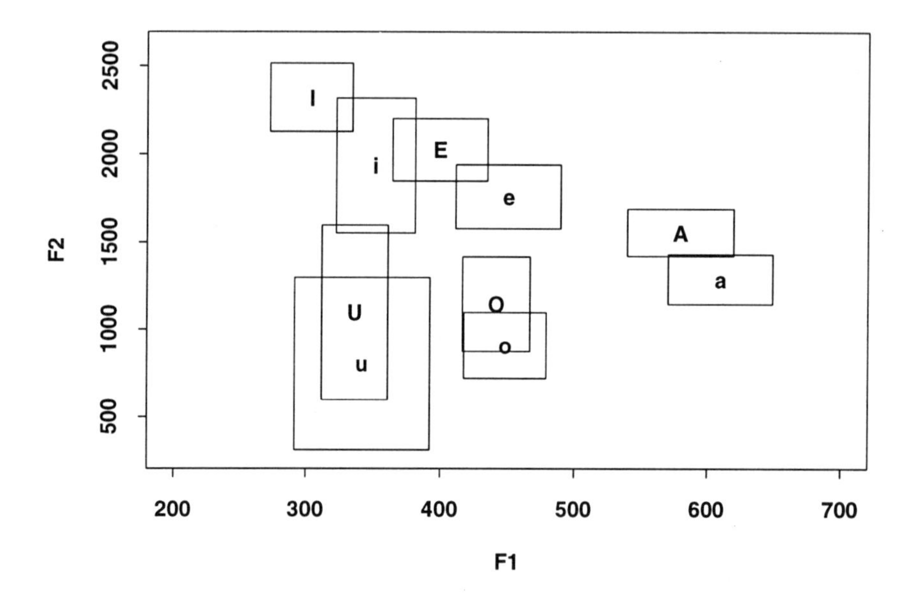

Figure 7.7: Partial vowel space of our Russian speaker: lower case symbols represent "hard" (non-palatalizing) vowels, upper case symbols are "soft" (palatalizing) vowels. Thus, i=ɨ, I=(j)i, e=ɛ, E=(j)e, A=ja O=jo, U=ju.

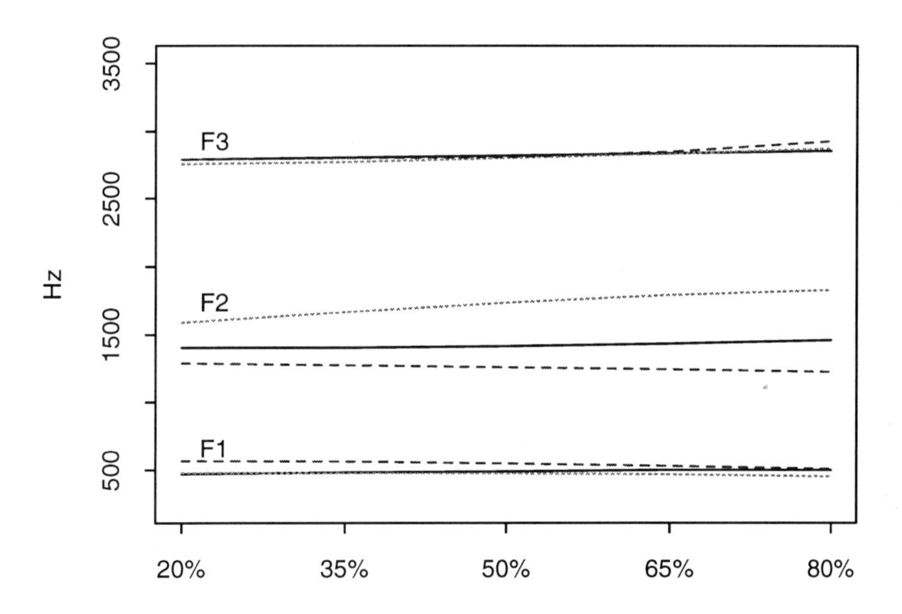

Figure 7.8: Coarticulation effects on Mandarin schwa. Formant trajectories (averages) at 20%, 35%, 50%, 65%, and 80% of [ə] duration, in three different contexts: syllable final (solid lines), [ə] followed by alveolar nasal coda [n] (dotted lines); [ə] followed by velar nasal coda [ŋ] (dashed lines).

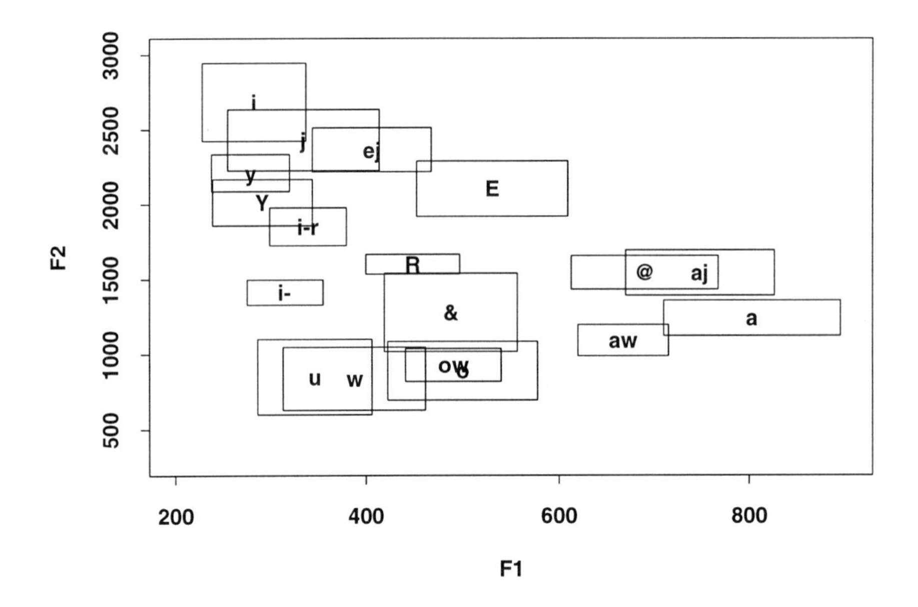

Figure 7.9: Mandarin vowels, glides, and diphthongs. Non-IPA symbols map onto IPA symbols as follows: & = ə, R = ɚ, E = ɛ, @ = aɹ, i- = ʐ̩, i-r = ʅ, aj = ai, aw = aʊ, ow = oʊ, ej = ei, Y = ɥ.

a unit taken from the [ŋ] context. To ensure smooth connection, we collect two sets of context sensitive units for [ə], one set being collected from the [n] context, the other one from either the syllable final [ə] or from the [ə] followed by [ŋ]. The unit selection component will check the context and select the appropriate unit during run time.

In certain cases, target formant values alone are not sufficient to decide whether or not a distinction has to be made in the inventory; they can be even deceptive at times. In our Mandarin inventory, the glides and the diphthongs are represented as separate phones from the monophthongs due to distinct but predictable transitional patterns characteristic of each class of sound (Shih, 1995).

Figure 7.9 gives the full vowel inventory of the Mandarin system. There are a few pairs of sounds with overlapping F_1/F_2 space, most notably, in two of the glide/vowel pairs [w]/[u] and [j]/[i], and two of the monophthong/diphthong pairs [o]/[oʊ] and [a]/[ai]. In particular, the vowel space of the diphthong [oʊ] is entirely contained within [o], and there is considerable overlap between the glide [w] and the vowel [u]. The principal motivation to separate these sounds is that a glide to vowel or vowel to off-glide (diphthong) transition is different from a vowel to vowel transition in Mandarin: the first two cases are syllable internal and the transition is smooth, while the last case spans a syllable boundary and often contains a creaky voice segment in the transition from the first vowel into the second. So, even though the formant values of [w] and [u] are similar, the two units [w]-[a] and [u]-[a] are quite distinct. For this reason, the glides, the diphthongs, and the vowels are represented as separate sounds in our Mandarin inventory.

Japanese has phonological vowel length opposition, but our plot in Figure 7.10 shows little difference in the F_1/F_2 space between long vowels (represented by ':'), and short vowels. This is particularly evident for the pairs [u:]/[u], [e:]/[e], and [o:]/[o]. Long [a:] has higher F_1 than short [a] on average, but the F_1 ranges overlap considerably. The only case that may cause trouble is the [i:]/[i] pair, which has similar F_1 but the long [i:] has higher F_2, which does not overlap with [i]. We conducted a listening test comparing the synthesis result of long vowels, using long vowel acoustic units as opposed to using short vowel acoustic units. Our listeners could not distinguish the two sets, and most importantly, did not prefer the long vowel synthesized with [i:] and [a:]. We therefore concluded that the formant distinction, if any, in long and short Japanese vowels, is not crucial for the acoustic inventory, so we only collect the short vowel acoustic units and lengthen their timing in the duration model, when they are phonologically long.

We should note, however, that the Japanese long and short vowels need to be modeled differently both in the timing component and in the intonation component, therefore they are represented as separate symbols in all previous components. The merging is done only in the inventory.

Unfortunately, there is no way around recording and segmenting a considerable amount of speech data that are intended to cover the vowels of the language in representative phone contexts, prior to exploratory studies of the type described above. As a consequence, recording speech samples for an acoustic inventory must be considered an iterative task. We will now briefly discuss the problem of context-sensitivity of phones.

Measurement of coarticulation: context clustering

We have already mentioned the fact that the inventories of most concatenative speech synthesis systems deviate from a strictly diphonic structure. The reason is that some phones, especially short vowels, fail to reach their articulatory (and therefore also acoustic) targets due to contextual effects coupled with durational constraints. In addition to this target "undershoot", we often observe an increased acoustic variability in reduced vowels. A classical, and rather straightforward, solution is to include triphonic consonant-vowel-consonant units in the inventory. Needless to say, this method drastically increases the size of the

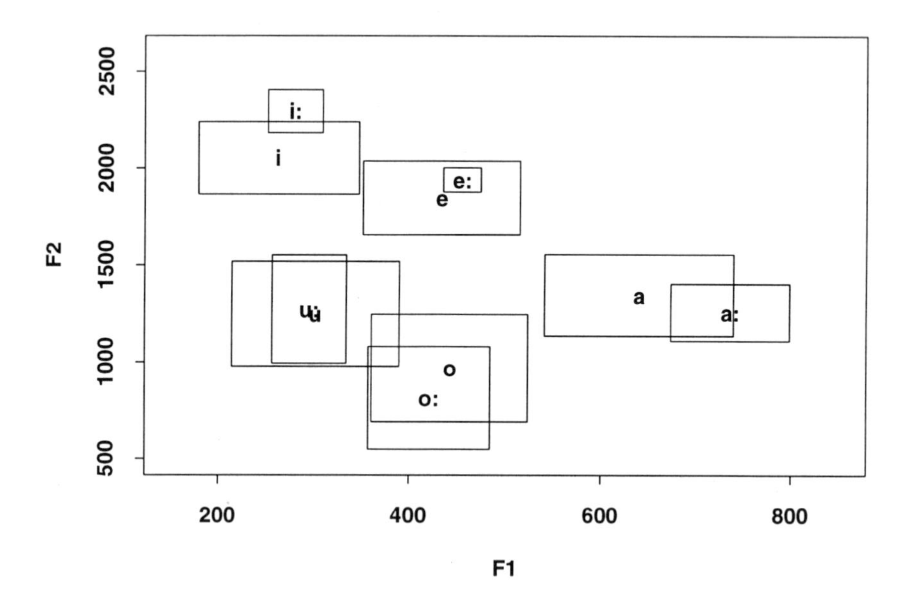

Figure 7.10: The vowel space of our Japanese speaker: long vs. short vowels.

inventory.

A more economical and at the same time elegant solution is the concept of *specialized units*. Let us consider two examples to explain this concept. In German, there is an enormous variability in the phonetic realization of the phoneme /r/, ranging from vocalization in the context of a preceding vowel and a following consonant or morpheme boundary, to a velar approximant in intervocalic position, to a velar voiced or unvoiced fricative when preceded by a voiceless obstruent. Now, the concatenation of, say, an /i-r/ unit that was excised from a pre-consonantal context to an /r-u/ unit cut from the context of a preceding voiceless obstruent, will result in a perceptually very disturbing discontinuity in the synthetic speech. Our solution is to store, for each unit involving /r/, two (or more) different tokens that are appropriately selected during synthesis according to the desired context (see section 7.4.1).

While *manner* of articulation of context phones is the decisive factor in the /r/ example, *place* of articulation is relevant for context-sensitive vocalic units. If we decide that we need specialized diphones for units that involve, say, a short [u], it is generally sufficient to store units for labial, dental and velar contexts.[6] Within one of these contexts, we do not usually have to distinguish between different manners of articulation, i.e., for the labial context between, e.g., [p], [f] and [m], because the impact on formant trajectories is assumed to be similar (but see the analysis below). For German, this method reduces the maximal number of context-sensitive units for [u] from 484 triphones (22 consonants x 1 vowel x 22 consonants) to 132 specialized diphones (2 x 22 consonants x 3 places of articulation).

We now present a method to measure these contextual effects, so that we do not have to rely on the coarse and possibly incorrect labial vs. dental vs. velar trichotomy. We illustrate this method with an example.

The analysis was confined to vowel instances from English with durations of at least 100 ms. (median duration of 136 ms.). For these instances, we measured formant values at 40 ms. from the start (*Early*), the center (*Mid*), and 40 ms. from the end (*Late*). We then fitted the following model to values for formant F_i, at time t (Early vs. Mid vs. Late), in vowel v, preceded by segment p_1, and followed by segment p_2:

$$\begin{aligned} F_i(v, p_1, p_2, t) \; = & \\ & \mathrm{PRE}_i(v,t) \times \mathrm{pre}_i(p_1,t) \; + \; \mathrm{Itc}_i(v,t) \; + \\ & \mathrm{POST}_i(v,t) \times \mathrm{post}_i(p_2,t) \end{aligned} \tag{7.1}$$

Here, $\mathrm{pre}_i(p_1,t)$ and $\mathrm{post}_i(p_2,t)$, for fixed t, are constrained to have zero sums over p_1 and p_2, respectively; $\mathrm{PRE}_i(v,t)$ and $\mathrm{POST}_i(v,t)$, for fixed t, are constrained to be positive and have mean values of 1.0 over vowels v. The constraints serve to make sure that the measures of coarticulatory effects can be directly interpreted in terms of Hz.

[6] More precisely, we use the terms labial, dental and velar to refer to consonants that decrease, leave unaltered, vs. increase the values on F_2 of adjacent vowels.

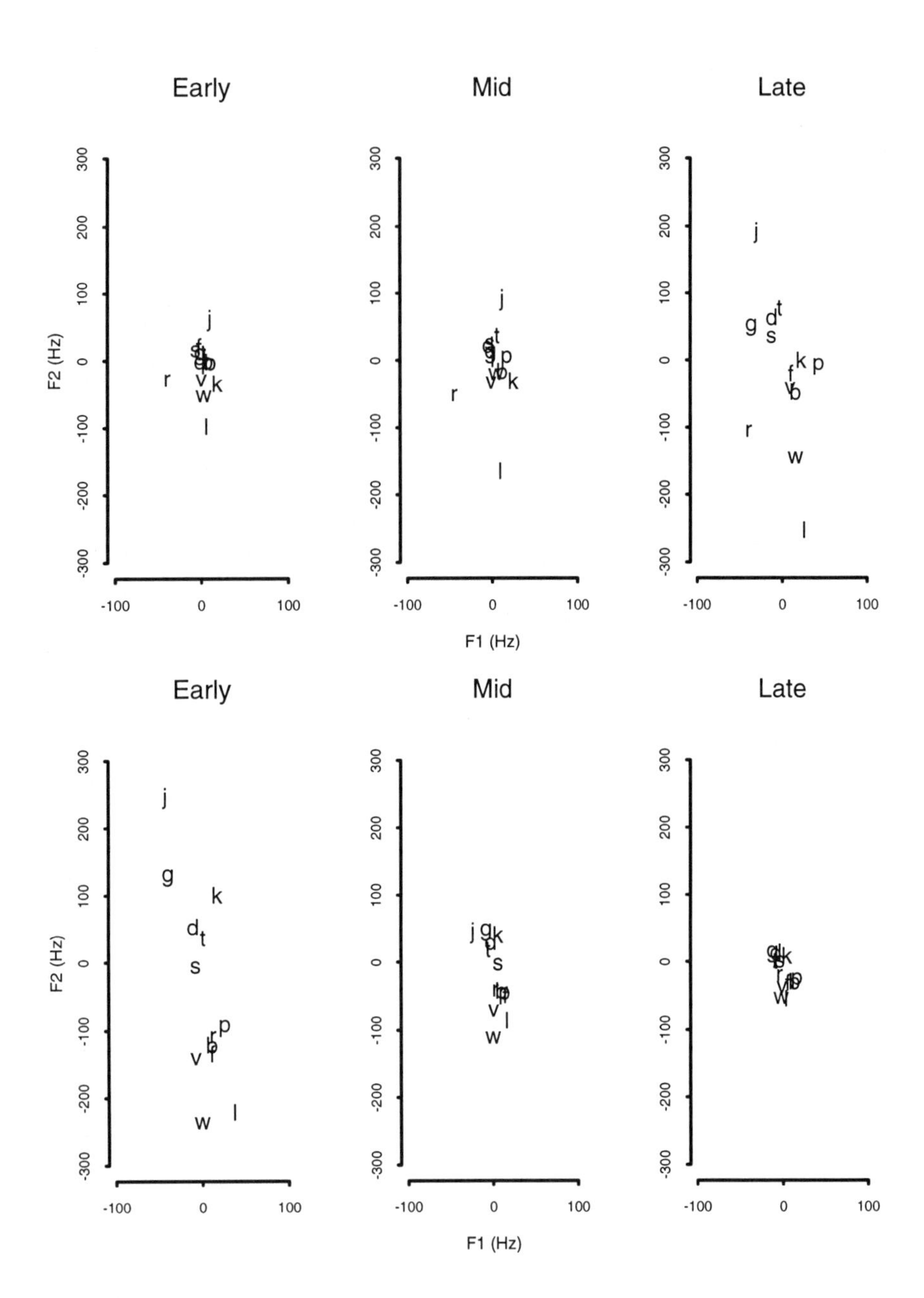

Figure 7.11: Effects (in Hz) of selected postvocalic consonants (top panels) and prevocalic consonants (bottom panels) on F_1 and F_2 of vowels (American English). Effects are normalized to have zero means over consonants.

This model makes the following substantive assumptions:

1. For a given vowel v and formant F_i, and for fixed t, the effects of the surrounding contexts are additive. This follows by holding $\mathrm{PRE}_i(v,t)$, $\mathrm{Itc}_i(v,t)$, and $\mathrm{POST}_i(v,t)$ constant, and realizing that Equation 7.1 then has the form $A(p_1) + B(p_2)$.

2. Differences between vowels are confined to:

   - Different intrinsic formant values [$\mathrm{Itc}_i(v)$].
   - Strength of the coarticulatory effects of the surrounding segments, as reflected by the weights $\mathrm{PRE}_i(v)$ and $\mathrm{POST}_i(v)$. However, because these weights are positive, it is assumed that the coarticulatory effects are in the same direction for each vowel.

3. It is *not* assumed that the coarticulatory effects have the same relative magnitudes across times. This would only be the case if we dropped the t argument in $\mathrm{pre}_i(p_1,t)$ and $\mathrm{post}_i(p_2,t)$.

Note that the total number of parameters for 15 vowels and 30 preceding/following consonants is 101 for each of the six combinations of time and formant (but obviously each vowel instance generates six simultaneous observations for each of these combinations). With judicious text selection, this means that only a few dozen sentences have to be read and segmented for this analysis to be carried out.

Figure 7.11 (top panels) shows the estimates of the contextual effects of the postvocalic consonant, $\mathrm{post}_i(p_2,t)$. The results are as expected. First, since we are looking at effects of the postvocalic consonant, the effects are strongest for the Late locations and weakest for the Early locations. Second, at the Late locations, and except for /k/, we see the usual separation into velar/alveolar consonants vs. labial sounds. However, at the early locations, many of the coarticulatory effects have vanished, leaving only /l/, /r/, /j/, and possibly /w/, as separate classes. At the Late location, the /d/-/b/ effect is as large as the /d/-/j/ effect, but at the Early location, the first effect has vanished while the second effect is still clearly there. Apparently, the leftward range of coarticulatory effects is not the same for stops as for glides (also see Assumption 3 above). A third important finding is that coarticulation is significantly stronger at the vowel centers (the customary cut-point in many diphone based synthesizers) than at the Late (for consonant-vowel units) or Early (for vowel-consonant units) time point. This can be seen by the larger spreads in the central panels in the Figure compared to those in the left-upper (effects of following consonant at Early time point) and right-bottom panels (effects of preceding consonant at Late time point). Thus, provided that vowels reach their targets at these non-central time points, much can be gained by cutting diphones here rather than in the vowel centers.

In the bottom panels of Figure 7.11, which plots the contextual effects of the preceding consonant, $\mathrm{pre}_i(p_1,t)$, we see that, again, the effects are in the predicted directions. When we compare the two figures, it is clear that the effects of the postvocalic consonant at Early formant values are stronger than the effects of the prevocalic consonant at Late formant values. This may mean that fewer context-sensitive diphones are needed for vowel-consonant than for consonant-vowel units.

This analysis is one of many that could be performed. For example, the model can also be applied to log area parameters and spectral tilt measures, as long as the two basic assumptions of the model seem reasonable. Second, variations of the model that incorporate either more remote coarticulatory effects or prosodic effects can be easily developed.

This subsection makes three important points. First, it shows the feasibility and necessity of systematic acoustic analyses. Second, these analyses precede the actual production of concatenative units, and are a much more efficient way of making inventory design decisions than producing a complete concatenative inventory and belatedly discovering ignored coarticulatory patterns by testing the entire inventory. Third, these results explain in part why the rule-based stage only get us so far: there are many acoustic details, such as the /d/-/b/ vs. /d/-/j/ effect discussed above, that are hard to predict from currently known first principles.

Conclusion

The procedures for acoustic unit inventory design described in this chapter can be applied to any language. They drastically reduce the complexity of inventory unit collection and the size of the speech database. A standardized and largely automatic procedure for the prediction of a list of units can be considered as part of the effort to produce inventories of higher acoustic quality for new languages or voices with significantly reduced time effort, even though a complete automation does not currently seem to be feasible.

7.3.2 Recordings

One crucial aspect in the construction of acoustic inventories is the size of the speech database to be recorded. As of now, the process of constructing acoustic unit inventories has not been completely automated. State-of-the-art automatic segmentors (Ljolje and Riley, 1993) do not provide the level of accuracy that is sufficient for unsupervised segmentation of speech. The situation is similar for automatic formant tracking. The majority of the Bell Labs TTS systems were built using hand-segmented speech data.

One suggestion for reducing the number of sentences for recordings is to apply a *greedy* algorithm that selects the minimal number of real words needed to cover all required inventory units; see Section 2.4.1. This dictionary-based greedy approach has a potentially serious drawback. While it does implicitly provide multiple examples for some inventory units due to redundancy inherent in natural language, it nevertheless results in only one token for the vast majority of units. The phonemic context in which each of these tokens is embedded is unpredictable. To give a hypothetical example, let us assume that the algorithm covers the target diphone /ɪ-m/ by the lexical entry *trim* simply because that word also covers two other targeted diphones, /t-r/ and /r-ɪ/. Other elements involving initial /ɪ/ will be covered by other words, and the phonemic context will vary almost randomly across all elements containing initial /ɪ/. As we have argued in the previous section on context-sensitivity, however, it is important to provide systematically varied phonemic contexts for candidate inventory units. The greedy algorithm's quest to cover the target set of units with the smallest number of words is incompatible with this goal.

An alternative approach has been used in the construction of the German TTS inventory. With the American English and Mandarin TTS systems as role models, the underlying working hypothesis was that the structure of the acoustic inventory would be basically diphonic. No immediate prediction about where specialized diphones or even triphones would be required was made before the recordings. The decision about context-sensitive units was later based on the results of the inventory unit selection procedure described in section 7.3.3..

The German database was constructed by systematically varying the contextual place of articulation for each diphone unit. In the case of the targeted /ɪ-m/ diphone, for instance, recordings were made for the labial /p-ɪ-m/, dental /t-ɪ-m/, velar /k-ɪ-m/, and rhotic /r-ɪ-m/ contexts, thus spanning the whole range of possible formant trajectories in the diphone /ɪ-m/. Stops were chosen as context phones because they have a minimal contaminating effect on the vowel spectrum.

The resulting triphonic segments were always embedded in the same carrier phrase, thereby keeping the prosodic context as constant as possible. The key words were placed in focal position in the carrier phrase. The syllable containing the target unit, however, did not carry primary stress; it carried secondary stress in order to avoid over-articulation (cf. (Portele et al., 1991)). An example sentence to be recorded for the diphone /ɪ-m/ in preceding dental context thus reads:

(58) *Er hatte Timmerei gesagt.*

with primary stress on *-rei*, secondary stress on the key syllable *Tim-*, and a dental context [t] for the target unit /ɪ-m/. Obviously, the carrier phrase needs to be varied slightly to accommodate utterance initial and final units but the construction principles are the same.

While using real words and sentences for the recordings as opposed to constructed carrier words and phrases may appear appealing at first glance, the latter approach has significant advantages: (a) controlled prosodic context and position in the phrase; (b) uniform stress level on all target inventory units; (c) systematically varied phonemic context; (d) no need for the syntactic constraints desirable for sentences using real words.

This approach, in combination with the principles outlined in the sections on phonetic properties of phones and on phonotactics, results in a far smaller number of sentences to be recorded — even though one sentence is needed for each target inventory unit.

7.3.3 Acoustic unit selection

Until recently, acoustic unit selection and cut point determination was performed manually. However, Iwahashi and Sagisaka (1992) at ATR and our group (Olive, 1994; van Santen, Möbius, and Tanenblatt, 1994) have introduced algorithms that significantly automate this process.

At first glance, the optimization problem (to select a candidate for each unit type such that the selected candidates jointly have minimal segmental and inter-segmental distortion) appears to be a "hard" problem, only resolvable by approximative methods (e.g., Iwahashi and Sagisaka (1992) used simulated annealing). However, two key ideas of our algorithm dramatically reduce the complexity of the problem.

The first idea concerns the concept of *ideal point*. Suppose that we want to select the optimal candidates and their cut points for diphones involving the vowel /i/. If there is a point *IP* in formant space (or some other spectrally-based vector space)[7] such that for each diphone type there is at least one candidate that contains at least one point in its formant trajectory within a distance of ϵ from *IP*, then we are guaranteed that if these candidates were cut at their points of smallest distance to *IP*, their spectral discrepancies would be at most 2ϵ. By making ϵ sufficiently small, these spectral discrepancies will become imperceptible.

The effect of this simple idea is that the optimization involves just three parameters (in the case of formants). In the approach by Iwahashi and Sagisaka the optimization has both combinatorial aspects (proposed sets of selected candidates) and continuous aspects (their cut points), adding up to a formidable parameter space.

The second idea is that *the selection procedure should take into account the concatenation process and any normalization that precedes this process*. After all, we are interested not in spectral distances per se but in spectral discontinuities in the output speech of the synthesizer. For example, if one normalizes (on a per-phone basis) the amplitudes of acoustic units, then amplitude discrepancies are irrelevant. As a consequence, consonants whose primary dimensions of variation are in the energy and temporal domains (but not in the formant domain) are assumed not to produce discrepancies provided that their energy and duration (and perhaps spectral tilt) are within rather wide tolerance bounds. Thus, selection of, say, a /p-i/ unit can be done exclusively on the basis of the /i/ part, assuming that any flawed /p/'s have been removed from the database by screening methods. This has obvious computational advantages over procedures that take into account the detailed acoustics of all segments in a unit

[7] The algorithm is not tied to formants. We use formants largely because, despite practical problems with formant tracking, they are the most concise and meaningful representation of vowels and semi-vowels.

(Iwahashi and Sagisaka, 1992).

We now describe the algorithm in more detail. For a given vowel or semi-vowel, we collect all speech intervals that contain that vowel and correspond to units having that vowel as a start or end phone. The task of the algorithm is to find an ideal point *IP* such that (1) for each relevant unit type there is at least one token trajectory that at some point gets within ϵ from *IP*, and (2) it has acceptably low segmental distortion.

We start by defining for a given collection of speech intervals a multidimensional interval (a cube in the case of F_1, F_2, F_3) of acceptable spectral values. For example, in the case of the vowel /ɪ/ we may select 300–500 Hz, 1700–2100 Hz, and 2200–3100 Hz. This multidimensional interval restricts the search space for *IP*.

Next, we quantize the multidimensional interval, and compute for each cell in this interval the following measure. Let $D(x, y)$ denote some distance measure in formant space, e.g. the weighted Euclidean distance. For a cell p and a trajectory Traj(t) (which is a mapping of the real-time interval occupied by the vowel into the multidimensional interval), let the distance Distance($p, Traj$) between p and Traj(t) be given by the minimum of p and Traj(t) over all values of time, t. Let *Type(Traj)* be the type of this trajectory (e.g., /p-ɪ/). Now we define the *type count of* p, Count(p), as the number of distinct types that each have at least one token, *Traj*, such that Distance(p, Traj) $< \epsilon$. An ideal point is a point p where the type count is maximal.

There are two possibilities. First, not all types are covered. In that case, either the inventory design was flawed, not enough tokens were recorded, or speaker variability was larger than expected. Each of these possibilities has to be considered. Second, all types are covered, but there are multiple ideal points, each providing full coverage. We select the point that is closest to standard formant values for that vowel. In the case of /ɪ/, these would be 400 Hz, 1900 Hz, and 2600 Hz.

A major issue to be resolved is computational. When one graphs the value of Count(p) as a function of the location of p in the multidimensional interval, it is seen that the resulting response surface is quite irregular. This is due to the fact that even with very many trajectories the multidimensional interval is still very sparsely populated, resulting in many quasi-randomly located empty areas where the value of Count(p) has local minima. As a consequence, we cannot use optimization routines that assume any form of smoothness.

We found, however, that for certain distance measures sufficiently fast algorithms can be found, so that even with quantization as fine as 50^3 cells the process takes only a few CPU hours on a standard workstation.

7.3.4 Acoustic unit token excision

Given the ideal points, making cuts in vowels is straightforward. For unit-final vowels, we take the earliest point in the vowel that yet is within distance ϵ from the ideal point, and for unit-initial vowels we take the latest such point.

This convention reflects the non-linear time-warping process (see below, in Section 7.4.2), where we lengthen vowels by inserting a linear path in the parameter space connecting the last frame of a unit with the first frame of the next unit. By keeping the stored vowel regions *as short as possible*, we can make the inserted path long, thereby reducing the spectral discrepancies between successive frames. An additional critical benefit of shortness is that the coarticulatory effects of context consonants is minimized, because vowels will be cut at a point distant from the possibly contaminating contextual consonant that is not part of the unit.

In the Bell Labs system, the excised and parameterized speech intervals are stored in a single indexed file. In other approaches, such as CHATR, pointers to locations in the speech corpus are stored, but the intervals are excised at run time. For inventories with many overlapping units as in CHATR, excision at run time is a necessity. For smaller inventories, pre-excision has the advantage of much more efficient search at run time.

7.3.5 Acoustic unit amplitude normalization

Due to a variety of factors, acoustic units vary in amplitude (as measured, e.g., by the root mean square of the speech wave over some appropriate averaging window). These factors include speaker instability and variations in the distance to the microphone. When left incorrected, these amplitude variations can cause serious problems during synthesis.

We correct these variations as follows. For a given unit, we measure the amplitude profile at either fixed or pitch synchronous intervals. For each phone, we compute the average of the amplitude in the temporal center of all occurrences of that phone (confined to those occurrences that are used in at least one selected acoustic unit). These average values are the target amplitudes that we would like the segments in an acoustic unit to attain. To accomplish this, we define for a given acoustic unit one or more *anchor points*, such as the starting frame, the final frame, and frames in the centers of unit-internal segments. We compute the ratios between the target and the actual values at these anchor points, and draw a smooth "multiplier" curve through the ratio/time point pairs. This curve is then multiplied with the amplitude profile to obtain a new amplitude profile that has the property that it attains the target values at the anchor points, but — due to the smoothness of the multiplier curve — preserves the local shape of the original amplitude profile.

During synthesis, the amplitude of the residuals or glottal waveform are adjusted so that the synthesized speech has an amplitude profile that matches (after temporal adjustments dictated by the timing component) the target amplitude profile.

7.4 The Synthesis Component

In the previous sections we discussed the design and construction of an acoustic inventory for concatenative speech synthesis. All the procedures involved in building the inventory are performed off-line. We will now continue with a description of the TTS system components that generate synthetic speech from symbolic input at run time. The input at this stage consists of a string of phonemes and their assigned durations, as well as a fundamental frequency (F_0) contour for the sentence.

It is convenient to distinguish the following main steps in the synthesis process: (a) selecting and retrieving the appropriate units from the acoustic inventory, (b) concatenating the retrieved units and adjusting durations and F_0, and (c) waveform synthesis (see also Section 2.6). This section will be organized accordingly. We will also discuss the issue of which coding method the Bell Labs TTS system applies, and which alternatives exist, and we will further describe the complex explicit source model that is used in our system.

7.4.1 From phonemes to acoustic units

Given a phonemic representation of the sentence to be synthesized, a collection of rewrite rules is applied that select a set of inventory units such that the phoneme string is covered in an optimal way (Olive, 1990). Obviously, in the case of a purely diphonic structure of the inventory, optimal coverage is straightforward because there are no alternative sets of units. However, while the majority of acoustic units in the inventories of our TTS systems are indeed diphones, the inventory structure deviates in specific ways from being purely diphonic, as we have explained in the previous section.

Because all inventories are built upon a baseline set of diphones that jointly cover all phoneme combinations in the language, the inclusion of any additional units creates ambiguities. There are at least four types of deviations from the diphonic structure.

- Not all pairs of phonemes are stored in the inventory; certain diphone types, for instance transitions between fricatives and stops, or between two fricatives, can be omitted based on the lack of spectral interaction between the two phones involved (see Section 7.3.1).
- For all fricatives, a steady-state unit is stored, which is labeled by a single phoneme symbol.
- As a consequence of the concept of context-sensitive or specialized units, there is often more than one entry that matches a desired unit type; ambiguities of this kind are usually resolved by the phonemic context.
- Coarticulatory effects may have required the storage of triphones or even longer units; in these cases, there are alternative ways of covering the target phoneme string that cannot be resolved by context specifications.

As a rule, the selection process attempts to match the longest possible phoneme substring to an existing inventory unit. The details of the procedure can best be explained by a concrete example; let us therefore consider the English sentence *I will tell you when you have new mail.* The input to the unit retrieval module contains the phoneme string /* ai w ɪ l t ɛ l j u w ɛ n j u h æ v n j u m ei l */; note that the string is expanded to include the silence element '*' at the beginning and end of the sentence.

Processing the phoneme string from left to right, the module will find the units /*-ai/ and /ai-w/ in the inventory table. For the next substring, the inventory contains the units /w-ɪ/ and /w-ɪ-l/; since the longer, triphonic unit matches the phoneme substring, it will be selected. Next, the algorithm searches for the longest string match beginning with the phoneme /ɪ/; the only matching unit, /ɪ-l/, is already covered by /w-ɪ-l/, so it is ignored and the search continues starting from the phoneme /l/. The next appropriate unit is /l-t/. Subsequently, the algorithm again finds two matching units, /t-ɛ/ and /t-ɛ-l/, and just as before, it selects the longer unit. Note that units starting with a stop require special treatment. Stops are made up of two acoustically distinct sub-segments, a closure phase and a burst. As we have explained in the section on acoustic inventory design, a silence element /*/ is inserted in the closure phase of the voiceless stop /t/, in our example right before the unit /t-ɛ-l/. The search now continues from the phoneme /ɛ/ through the end of the sentence, applying the same principles as described so far. The only phonemes in the remaining string that require any special treatment are /h/ and /v/. To produce high quality fricatives, we insert steady-state segments, /h/ and /v/, between the units /u-h/ and /h-æ/, and /æ-v/ and /v-n/, respectively. The phoneme /h/ is extremely context-sensitive, so we store specialized steady-state /h/ units in the inventory according to various following segmental contexts. Thus, the algorithm retrieves the /h/ version that most closely matches the following /æ/ context.

For the example sentence, the selection process will result in the following sequence of units, which are then retrieved from the acoustic inventory:

(59) /*-ai/ /ai-w/ /w-ɪ-l/ /l-t/ /t-ɛ-l/ /l-j/ /j-u/ /u-w/ /w-ɛ-n/ /n-j/ /j-u/ /u-h/ /h/ /h-æ-v/ /v/ /v-n/ /n-j/ /j-u/ /u-m/ /m-ei-l/ /l-*/

The subsequent steps in the synthesis process, namely the concatenation of the retrieved units and the adjustment of duration and F_0 contours, and the actual waveform synthesis, will be described in the following sections.

7.4.2 Concatenation

The concatenation operation used depends on whether the boundary phone (i.e., the phone that terminates the left unit and starts the next unit) is a vowel, semi-vowel, or a consonant, and on whether the target duration of the boundary phone exceeds the sum of the durations of the parts of the phone stored in the left and right units.

For vowels and semi-vowels where the target duration exceeds this sum, we generate new frames by linear interpolation between the final and initial frames of the two units.

This interpolation operation is obviously too simple. First, depending on the particular type of LPC parameterization used, the parameter frames may be required to lie on the unit circle, in which case linear interpolation will generate frames that violate this constraint, possibly leading to acoustic artifacts; of course, it is trivial to change the interpolation procedure to satisfy this constraint. This issue does not come up when log area coefficients are used. Second, the spectral trajectory produced by linear interpolation is not smooth, because the direction changes discontinuously at the terminal frames. This can be amended by smoothing the frames after interpolation. We are unsure whether this makes an audible difference, or is just a matter of mathematical esthetics.

Despite this simplicity, linear interpolation between terminal frames has the important property that long vowels are generated by lengthening the central portion. For most contexts, this is more appropriate than a uniform lengthening, where the transitional initial and final regions of the vowel are lengthened just as much as the central region. However, it is not correct for lengthening due to phrase boundaries or postvocalic consonant voicing, because, as we discussed in section 5, here it is primarily the final part of vowels that is stretched out.

In all other cases (i.e., consonants, or vowels or semi-vowels that must be shortened), the inter-frame interval is adjusted uniformly to generate the target phone duration. The same is true for phones that are internal to a unit, as the /ɪ/ in the /w-ɪ-l/ unit. Again, this manipulation does not capture known intra-segmental durational effects.

Once the entire sequence of LPC parameter frames has been computed by concatenation and temporal adjustment, the F_0 curve is imposed on the voiced regions by adjusting the number of samples per pitch epoch, and computing for each epoch the glottal waveform.

The end result of the concatenation process, then, is a sequence of frames, each consisting of an LPC vector and a vector of glottal waveform parameters; in voiced regions, the frames are pitch-synchronous, while in unvoiced regions they are spaced at 5 ms. intervals.

7.4.3 Glottal excitation modeling and waveform synthesis

Standard LPC synthesis uses a simple model for voiced excitation in the form of pulse-like waveforms repeated quasi-periodically at the rate of F_0. According to this model the vocal tract is excited once per fundamental period, and the spectral shaping of the source signal is uniform or stationary throughout the glottal cycle. Arguably, the most significant drawback of this simple approach is that it assigns identical (zero) phase information to all harmonics, a serious simplification given the considerable variability in the glottal waveforms of natural speech. This may cause a buzzy sound.

To overcome the lack of naturalness in the synthetic speech produced by the simple excitation model, explicit physiological models of sound generation in the vocal tract have been developed, ranging from the two-mass model of the vocal folds (Ishizaka and Flanagan, 1972) to the more recent theory of distinctive regions and modes (Mrayati, Carré, and Guérin, 1988). Fujisaki and Ljungqvist (1986) showed that explicit models of the voice source significantly reduce the LP prediction error, compared to single-pulse excitation models.

In the Bell Labs TTS system, we apply the glottal flow model proposed by Rosenberg (1971), which attempts to closely match the time-domain characteristics of a glottal pulse by using a third-order polynomial for the glottal opening and a second-order polynomial for the closing phase.

The Rosenberg model has recently been further modified by Luis Oliveira (1993), who provided control of, first, the spectral tilt of the voice source and, second, the level of aspiration noise. One partial motivation for this modification was the inability of previous models to synthesize convincing imitations of female voices. On average, female voices are breathier than male ones (Klatt and Klatt, 1990), and breathiness is known to be related to spectral tilt. Based on Fujisaki and Ljungqvist's findings, Oliveira extended the Rosenberg model by adding a decaying exponential, in the form of a first-order low-pass filter, during the closed glottis phase.

The new explicit source generator is also capable of modeling irregularities in the periodic component, such as vocal jitter, laryngealizations, and diplophonic double-pulsing. Thus, it opens up the possibility for the TTS system to dynamically vary the source parameters during an utterance and trigger voice quality changes according to the prosodic context. Listeners use voice quality cues, especially laryngealizations and creaky voice, as cues for prosodic phrase boundaries (Hedelin and Huber, 1990).

Figure 7.12 shows the block diagram of the waveform synthesizer used in the Bell Labs TTS system. Each source generator has its own amplitude control, similar to Klatt-style formant synthesizers (Klatt, 1980); simple source models typically use a switch between (quasi-periodic) voiced and (stochastic) unvoiced excitation. Oliveira's source model provides more complex voice source patterns and smooth transitions between voiced and unvoiced speech segments. The voice generator controls the modulation function of the aspiration noise as well as the damping of the filter memories, during the open glottis phase. In the standard implementation, the vocal tract parameters are the LPC reflection coefficients, and the vocal tract filter has a lattice structure.

7.5 Concluding Remarks

We have discussed how acoustic quality depends on many features of the synthesis component. Acoustic inventory design is the primary means by which

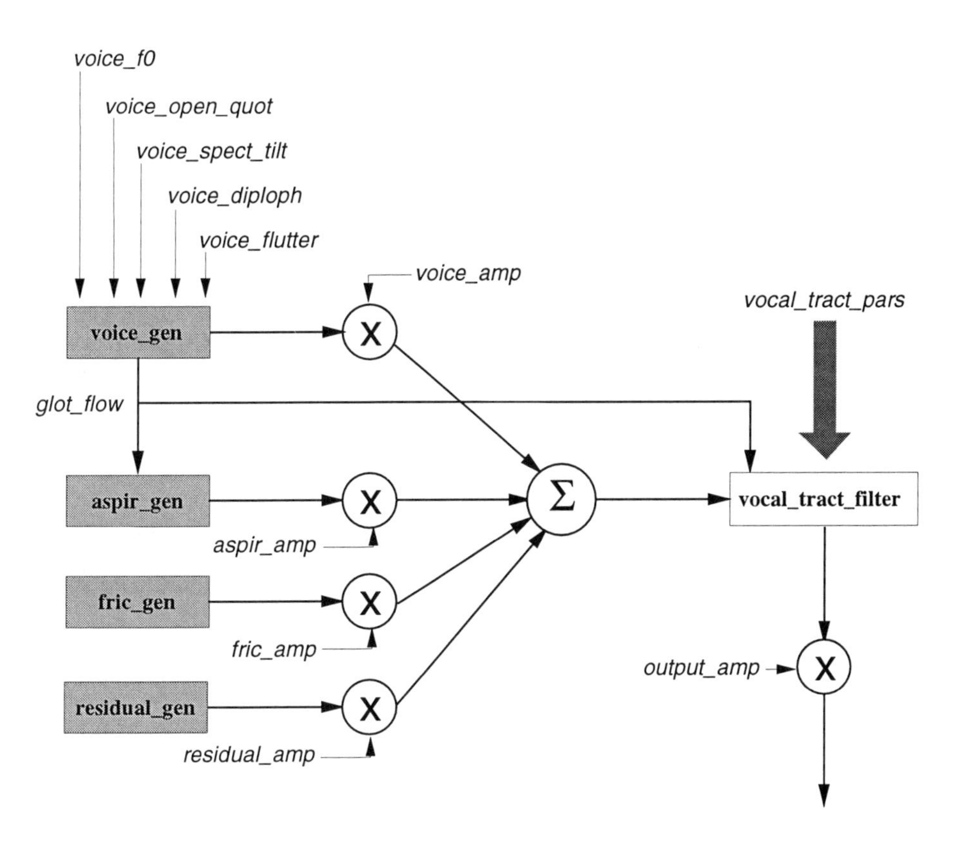

Figure 7.12: Block diagram of the Bell Labs waveform synthesizer using Oliveira's explicit voice source model.

coarticulatory phenomena are captured. Capturing coarticulation is critical because otherwise the system may produce the wrong sounds and audible discontinuities. Currently, only some systems (most notably, CHATR (Campbell and Black, 1996)) attempt to also capture prosody through inventory design. Based on statistical analyses of units in text corpora, we believe that such schemes can only work in extremely limited domains. In most systems, prosody is generated by altering acoustic features of the acoustic units, in particular timing, F_0, and amplitude. Conversely, no current concatenative systems perform rule-based coarticulation; they all use the inventory for this purpose.

We showed how inventory design is based both on general — phonological, phonotactic, and acoustic phonetics — principles and on acoustic analysis of the spectral effects of phonemic and prosodic context in a speech corpus.

Once an inventory has been designed, the capacity to find optimal unit tokens for each unit type is the next critical factor affecting speech quality. We discussed algorithms that make this process largely automatic. But mundane matters such as the constancy of the speaker can be important factors too.

There is little doubt that the speech parameterization plays an important role, but currently it is not clear which of several proposals is truly optimal. One reason for this is that very few systematic comparisons exist. Many systems have been produced, but quality differences are difficult to interpret because there typically are many differences in addition to those concerning speech parameterization.

Finally, the concatenation operation itself is of critical importance, both because it determines the smoothness of transitions between units, and because it affects the internal temporal structure of phonetic segments.

Chapter 8

Evaluation

Jan van Santen

8.1 Introduction

We use the term *system evaluation* here in the loose sense of any procedure that provides information about the performance either of the system as a whole or of specific components. Thus, system evaluation includes not only automatic checking of the pronunciation component against a pronunciation dictionary, formal assessment[1] with established listener tests (e.g., the Diagnostic Rhyme Test, or *DRT* (Voiers, Sharpley, and Hehmsoth, 1972)), and field tests in specific application settings, but also informal system demonstrations and interactive text-to-speech websites.

Evaluation serves two groups, text-to-speech users and text-to-speech developers. The primary interest of users is overall system performance in situations that are as similar as possible to the target use situation — field tests, if affordable, would be the ideal. Users typically have little interest in performance of the individual text-to-speech components. Developers, on the other hand, are mostly interested in *diagnostic* methods that help them debug or repair text-to-speech components.

This chapter focuses on formal assessment with listener tests. The emphasis in this discussion will be on broader issues (e.g., *domain coverage*, *perceptual multidimensionality*), and on tools for test construction. We discuss examples, including tests we developed internally. This chapter will not provide an overview of the details and merits of the many established tests, because these have been described at length in other publications, in particular by the Expert

[1] We often use the terms "assessment" and "test" interchangeably.

Advisory Group on Language Engineering Standards (*EAGLES*, (EAGLES, 1997)), and the SAM project (Pols and SAM-partners, 1992).

We emphasize that formal assessment with listener tests is only one — albeit prominent — evaluation method. At the end of this chapter, we will attempt to put this type of evaluation in a broader perspective, and discuss alternatives. To prepare the reader, the viewpoint developed in this chapter is that formal assessment is perhaps not the most effective method for system evaluation, and that more effective methods such as interactive text-to-speech websites should receive more attention. This chapter is more provocative than balanced, and is intended to generate a candid, realistic discussion about the concrete usefulness of assessment for the target groups.

8.2 General Issues in Text-to-Speech Assessment

In this section, we address general issues by way of theoretical background for the assessment methods that will be described in detail in subsequent sections. One of the main points in this chapter is that there are fundamental differences between assessment of text-to-speech systems and assessment of other speech technologies (in particular speech coding), and that these differences set limits on what can be expected in text-to-speech assessment.

8.2.1 Basic distinctions between assessment methods

Automatic vs. performance vs. subjective assessment

One can draw a distinction between three types of assessment: *automatic*, *performance*, and *subjective* assessment. Automatic assessment involves testing by automatic procedures, such as testing symbolic (i.e., non-speech) phonemic transcription component output against a pronunciation dictionary. While this may involve manual labor, it does not require a human to process speech output. Unfortunately, few aspects of synthesis output are amenable to automatic assessment. In performance assessment, a human listener performs a task requiring accurate perception of speech output, such as transcribing the output, or selecting between two vowels. Human performance can be measured objectively; the subject's answer is either correct or wrong. Finally, in subjective assessment, a human listener comments on (usually in the form of rating scales, such as the mean opinion score, *MOS*) speech output, or a specific aspect of speech output such as intonation.

Lexically open vs. closed text

In the more common lexically closed tests a fixed list of words is used, whereas in lexically open tests a *test text generation algorithm* is provided. Put bluntly, lexically open approaches have the critical advantage over lexically closed tests that they prevent systems developers from optimizing their systems in an ad hoc manner. For example, the DRT consists of a list of 96 monosyllabic word pairs. Clearly, adding the 192 test words *in toto* to the acoustic inventory is a small effort, given the current trend towards larger inventories. This form of ad hoc optimization would not be available for a lexically open test.

8.2.2 Domain coverage

One of the most central issues in evaluation is domain coverage, by which we mean the degree to which text used in an evaluation covers the input domain of the text-to-speech system.

Ad-hoc optimization of text-to-speech components

Since in unrestricted domain synthesis any finite amount of test text can only cover the domain partially, test text can at most be expected to be a *fair* sample of the domain. Equally important is that system performance on the test text be a fair sample of system performance in general. This brings us back to the distinction between lexically open vs. closed text: if a system has been tuned to the 192 DRT words, then performance on the DRT is not a fair sample of the system's performance on newly constructed DRT-like lists, let alone on unrestricted input.

To explain why we should be concerned about ad hoc tuning in text-to-speech system construction, we draw a contrast between text-to-speech systems and speech coders. Like text-to-speech systems, speech coders are complex software systems, often containing a dozen or more components. For example, a waveform interpolation coding system described by (Kleijn and Haagen, 1995) has ten analysis components and five synthesis components. Hence, as measured by the number of components, they are every bit as complex as text-to-speech systems. However, there is an important distinction. If a speech coder fails, the fault can be corrected only by making a *global* change in one or more components, i.e. a change that is likely to affect all or most speech input. By contrast, a text-to-speech system contains a myriad of tables, rules, and other *list-like* entities. For example, when a text-to-speech system mispronounces a word, this can be fixed either by entering the word and its correct pronunciation in an exceptions dictionary, or by changing one or two rules that affect few other words when they have a sufficiently narrow scope and do not interact with other rules. Or when we hear a click, it might be due to a single bad acoustic inventory element. Dictionaries, rule systems, and acoustic inventories are all examples of lists.[2] Of course, many components of a text-to-speech system are not list-like. For example, the duration component of the Bell Labs system

[2] Certain types of coders are based on code books, in which, for example, a list of vectors representing LPC residuals is stored. To our knowledge, however, no one has thought of ad hoc adjustment of codebook entries.

cannot be modified to produce different durations for a few special cases, and the synthesis component is even less amenable to such ad hoc modifications.

This list-like property has advantages and disadvantages. The advantage is that when changes are made in some list item, we may not have to worry about unintended side effects because the effects of the changes might be confined to a small subset of the input domain that can easily be tested exhaustively. For pronunciation of words and names that violate pronunciation rules, this property is critical for ultimately achieving very high pronunciation performance. Of course, many rule systems have the unfortunate property of rule interactions, where changes in one rule may interact in surprising ways with other rules, and may create serious problems.

The disadvantage is that we cannot generalize with any degree of confidence from performance on one input sample to performance in general. There is no way of checking the correctness of a pronunciation dictionary other than by checking every single entry, or at least checking very large samples. By the same token, this property also means that one has to be very careful when listening to a prepared demonstration of a system. The usually high quality of such demonstrations does not guarantee at all that novel input text will be processed equally successfully. In fact, it does not take much cynicism to be highly suspicious of non-interactive demonstrations.

The implication for assessment is that a component that contains list-like structures must be assessed with text materials that are *novel* in that during system development no special adjustments were made for these materials. This is a non-trivial request. To illustrate, a recent multilingual assessment of grapheme-phoneme converters (Pols, 1991) used as test materials newspaper texts of at least 100,000 words, the names of the 150 largest towns per country, names of 31 European capitals in the national spelling, and the 150 most frequent first names for each language. Although usage of multiple text types is to be applauded, and 100,000 words of newspaper text is certainly challenging, the remaining three text categories do not constitute serious tests of these systems' general capabilities to handle names of foreign capitals and towns, and first names. Most system builder perform at least some amount of testing, and the most frequent names are likely to occur in these tests. Since almost any pronunciation system uses a dictionary in addition to — or instead of — rules, it is all too likely that the developer has fixed pronunciation errors by adding or changing the appropriate dictionary items. The name parts of this test constitute at best a *minimal acceptability test* in the sense that a system that does not pass it should not be considered acceptable; however, a system may pass this test for the wrong reasons.

To summarize, the list-like structure of many text-to-speech components allows fixing problems by making ad hoc modifications. This makes it essential that tests use novel text materials. We also assert that good performance on published lexically closed tests is at most a necessary condition of system quality.

Domain coverage defined

How can we define coverage for unrestricted input domains? One way is to make use of the fact that the input domains of several text-to-speech components consist of discrete units. For example, the duration component has feature vectors as input, the accenting module tagged words, and the synthesis module phoneme sequences. Coverage can be defined with respect to a specific unit class.

Units need not correspond to a specific text-to-speech component; in fact, when evaluating a system developed elsewhere, these internal units may not be accessible or known. For example, diphone coverage may be useful even for concatenative synthesizers that use units larger than diphones. Likewise, one may define units in terms of combinations of general-purpose factors such as *consonant vs. vowel*, *onset vs. coda*, and *stressed vs. unstressed*. Even when no text-to-speech component uses units that consist of precisely these three factors, their coverage may still be considered useful.

Ideally, coverage would be measured with the coverage index (Section 7.2.1), which measures the probability that all units occurring in a sentence (or other appropriate text sampling unit) randomly selected from the domain is present in the evaluation text. This is difficult in practice, however, for it requires (1) the availability of text corpora that are sufficiently large that probabilities of occurrence of the units in various unit classes can be estimated reliably and (2) the ability to automatically compute these units in the first place. Elsewhere, we have shown that probabilities of occurrence of units cannot be estimated reliably when the unit type count is large, as is usually the case (van Santen, 1997). For example, we found that frequency distributions of word forms are quite different among text genres (e.g., Associated Press Newswire vs. the Bible vs. Department of Energy abstracts); similar results were obtained for triphone distributions.

Of course, text that has good coverage with respect to one unit class does not necessarily have good coverage with respect to a different class. For example, it is easy to generate text covering all possible diphones using the sentence frame *please say <noun> again*, but the resulting set of phrases does not remotely cover the input domain of the accenting module. In longitudinal studies comparing different versions of the Bell Labs system, we usually construct several lists of text items, each optimally covering the units of a specific system component.

8.2.3 Perceptual multidimensionality

The contrast between speech synthesis and speech coding reveals other unique aspects of synthesis evaluation as well. In coders, the range of distortions is complicated enough (Kroon, 1995), but does not include such dimensions as timing, intonation, let alone word pronunciation and stress assignment. The dimensions relevant for evaluation of speech synthesizers include all dimensions relevant for coders and many more.

This multidimensionality creates serious methodological problems, in particular for subjective assessment. In the extreme, multidimensionality can create self-contradictory intransitive behavior where in a pair-wise preference test a listener can systematically prefer A over B, B over C, and C over A.[3] It is plausibly assumed that the underlying psychological process (e.g., (Tversky, 1969)) involves listeners attaching "subjective weights" to the dimensions that vary as a function of context, *including the context provided by the set of systems being compared*. When a system is stronger than a second system on one dimension and weaker on a second dimension, weight variations can induce preference reversals, and this in turn can cause intransitive choices. The claim is not that intransitive behavior does not occur in coder evaluation, but that the problem is likely to be more prevalent the more perceptually distinct dimensions are present, and that more perceptual dimensions exist in synthetic speech than in coded speech.

Application specificity

This multidimensionality makes it quite likely that preferences are application specific, with each application creating its own constellation of subjective weights. As a result, for an assessment battery to be useful for text-to-speech users, it has to provide not only textual coverage but also coverage of the range of potential applications (*task coverage*). Obviously, this is a very tall order to satisfy. For this reason, field tests or other tests that are directly tied to the target application are eminently more informative than standardized assessment batteries administered in lab settings, regardless of how carefully the standardized tests have been constructed.

Usage of "reference" speech

An at first glance appealing way around multidimensionality is to have listeners compare synthetic speech to natural speech that is distorted in some unidimensional way, such as modulation by noise, also known as the modulated noise reference unit, or *MNRU*. The noise level necessary to reach indifference between noisy natural speech and synthetic speech can be considered as a measure of the quality of a synthetic speech sample.

However, the results of Bennett make it likely that a focus on noisy natural speech will affect the subjective weights. For example, if some systems have good prosody but poor voice quality, while other systems have poor prosody but good voice quality, then the introduction of a voice with perfect prosody and clearly imperfect voice quality will have an effect on the balance of the weights for these two dimensions. There appears to be no strong argument for why the weight constellation induced by noisy natural speech would be uniquely

[3] Ray Bennett, personal communication, 1986.

representative of weight constellations in target applications outside of the lab, so that it is unclear what this method has to offer.

8.2.4 Test text generation tools

A central role in our assessment work is played by on-line text corpora, text analysis tools, and greedy algorithms, which we use to generate test text with optimal coverage of either the input domain of specific text-to-speech components or of some other unit class not tied to a specific text-to-speech component. The availability of these facilities is not merely a matter of convenience, since it makes long hours of browsing through dictionaries unnecessary. Rather it signifies a critical leap in the efficacy of the produced test materials. Even more important is that tools allow us to construct lexically open tests.

Corpora

An ever increasing number of large text corpora has become available. Many of these are annotated with various descriptors, such as phonemic transcriptions, morphological structure, and parts-of-speech.

As the size and number of corpora increases, it will be more imperative that they can be annotated automatically. It is not clear what performance levels can be expected from linguistic analysis tools. Currently, part-of-speech tagging is still not very reliable, although it could be considered as being one of the easier annotation tasks.

To select text from corpora, we use variants of the greedy algorithm, applied to the units of interest. Example of units are words, n-phones, acoustic inventory units, and parts-of-speech-plus-punctuation sequences of length five.

Text generation tools

Test text can be selected from a corpus, but can also be generated by a rule system. One example is the Semantically Unpredictable Sentences test (SUS, (Benoit, 1990)), which consists of a set of syntactic structures (part-of-speech sequences) together with sets of words for each of the slots in these structures. Similar examples are given below for the Minimal Pairs Intelligibility test (Section 8.3.2) and our test used for detecting flawed acoustic units.

Words and quasi-words can be generated with phonotactic rules, such as in the Cluster Identification Test (*CLID*, (Jekosch, 1993)). In particular in languages like English, German or Russian, which have relatively complex syllable structure, this approach seems extremely useful. The obvious disadvantage of generated text is that it is meaningless, so that it confines us to strict segmental intelligibility tests while excluding comprehension tests. It is not impossible, however, that in the future algorithms will be developed for generating meaningful sentences from words.

8.3 Examples of Diagnostic Tests

We discuss here some of the tests we have developed internally. The main point is to illustrate actual usage of some of the tools described in the preceding section.

8.3.1 Detecting flawed acoustic units

Because of imperfections in the acoustic unit generation process, some percentage of the units are flawed. A subset of these units are poor enough that their problems are audible in any context, but others only show problems in certain prosodic contexts (requiring, e.g., very short or very long durations) or in combination with certain other units. Detecting the latter subgroup is quite difficult because of the combinatorial problems discussed in Section 7.3. But it should be feasible to construct text materials that allow listening to each acoustic unit at least once, because an average sentence contains at least a dozen units and a listener can process a few hundred sentences per hour; most current inventories have fewer than 3,000 units. However, finding sentences that cover all units is far from easy, as we shall see next; in fact, it is highly improbable that one will find those sentences by randomly drawing them from a corpus.

In (van Santen, 1993b), we describe a test for detecting flawed acoustic units. The text selection process proceeded as follows. First, we generated nonsense phrases. Using Church's part-of-speech tagger program (see (Church, 1988), and Section 3.8.1), words in the 1998 Associated Press Newswire corpus were tagged. Files were constructed containing adjectives, plural nouns, singular nouns, and plural verbs — present tense and past tense. By randomly combining these files, phrases were constructed with various structures, such as *the <adjective> <noun> was a <noun>*. We also added names, addresses, and numbers. Names were constructed by randomly combining a list of first names (from the Bell Laboratories online telephone directory) with a list of last names (from the *Donnelley* directory). Addresses were generated by randomly combining street numbers (up to 4 digits), street names, town names, and state names. Numbers were randomly selected in the range from 0 to 10 million. In total, we generated 67,440 *text items* — nonsense phrases, names, addresses, and numbers. These items contained 2533 distinct acoustic units.

Next, we used the capability of the Bell Labs system to print out the names of all acoustic units required for the pronunciation of a given piece of text input. Thus, each phrase could be represented as a set of acoustic unit labels. Finally, we applied the standard greedy algorithm, and found that only 650 text items were needed to cover all 2533 acoustic units. Random selection of text items from the list of 67,440 required 64,516 text items to achieve the same coverage, and random samples of 650 text items covered on average less than 1,000 acoustic units.

The experimental paradigm required naive listeners to highlight problem words displayed on a CRT screen using simple one-key commands, and to subsequently rate their seriousness on a three-point scale. On trials where they found a serious problem, they were asked to categorize this problem using the labels *outright mispronunciation*, *bad letters*, *missing letters or words*, or *choppiness*.

Statistical tests established that there was significant agreement between listeners on the ratings and the word pointing response (van Santen, 1993b). More important, we were able to compute a list of the 100 worst words, and to re-examine the acoustic units that occurred in these words. After replacing or re-excising[4] some of these acoustic unit tokens, we ran an experiment comparing the old with the new acoustic inventory, and found significant improvements. Thus, the test served the purpose of finding flawed acoustic units.

We now believe that it is unnecessary to use naive subjects for this task, unless it is more practical than having the system developers doing the listening. The most important aspect of this test is that the automated text generation and selection optimization, in conjunction with the automatically scored experimental paradigm, makes it possible to quickly test the units of a new inventory.

8.3.2 Diagnostic intelligibility tests

Perhaps the oldest and best-known intelligibility test is the DRT (Voiers, Sharpley, and Hehmsoth, 1972), which was originally developed for coder evaluation. In this test, the user hears a CVC word such as *den*, and has to choose between *den* and *ten*.

This test is obviously extremely limited. It only measures intelligibility of initial consonants in isolated monosyllabic CVC words. From results on this test, we cannot infer system performance on vowels, consonants in clusters, consonants in codas, unstressed syllables, polysyllabic words, and words in sentence contexts. In other words, this test covers at most a few percent of the space of possibilities. In addition, the minimal word pairs were chosen to differ on one phonetic feature only (voicing in the den/ten example). As pointed out by Pisoni (Pisoni, Nusbaum, and Greene, 1985) and by Greenspan (Greenspan, Bennett, and Syrdal, 1989), in open response tasks listeners make quite a few errors involving multiple features, at least with synthetic speech. Finally, the DRT is a lexically closed test, so that system developers can optimize their system for this specific list of words.

Tests developed subsequently improved on these serious limitations of the DRT. For example, the Modified Rhyme Test (*MRT*, (House et al., 1965)) also tests consonants in codas of CVC monosyllabic words, and uses a six- rather than 2-alternative forced choice paradigm, thereby allowing measurement of multi-feature errors. The Bellcore test (Spiegel et al., 1988) tests intelligibility of consonants in consonant clusters; it uses graphemic transcription by naive listeners, so that subsequent manual processing is needed to measure phonemic

[4] At the time these tests were conducted, no automatic unit generation methods were available.

error rates. Both the MRT and the Bellcore test are lexically closed tests. The Cluster Identification Test (*CLID*, (Jekosch, 1993)) is not a test, but an algorithm for generating monosyllabic phoneme strings satisfying pre-specified phonotactic constraints. These phoneme strings, which can contain complicated consonant clusters, can then be used as stimuli in a variety of (lexically open) experimental paradigms.

These and other tests did not quite meet our needs, which were for a test that

1. Can be scored automatically,

2. Is lexically open,

3. Has maximal coverage of the phoneme-context space, and

4. Yields not only an overall performance statistic, but also diagnostic information.

We now discuss two tests we developed internally, the *Minimal Pairs Intelligibility Test* and the *Automatically Scored Orthographic Name Transcription Task* (van Santen, 1992b; van Santen, 1993b).

Minimal Pairs Intelligibility Test

The *MPI* (minimal pairs intelligibility test) attempts to maximize coverage of the phoneme-context space, yet stays in the convenient minimal pairs framework of the DRT. This framework is convenient because of automatic scoring, and because the interpretation of results is straightforward.

The MPI is lexically open, because it presents an algorithm for generating a set of minimal word pairs, an example of which is published in the Appendix of (van Santen, 1993b). The system generates minimal pairs differing by one phonetic feature, but in contrast to the DRT it also generates pairs differing by two phonetic features, by deletion, or by insertion.

The key generalization from the DRT is that the MPI contains consonants and vowels; onsets, nuclei, and/or codas; consonant clusters; mono- and polysyllabic words; and stressed and unstressed syllables.

A typical pair is *intransigent* vs. *intransigence* in the sentence frame *the inferior perimeters glow an* ···, which tests a two-feature contrast (the acoustic features *sustention* and *sibilation* (Voiers, Sharpley, and Hehmsoth, 1972)) for word-final consonants in clusters in unstressed syllables in phrase-final nouns.

In an application of this test to two versions of the Bell Labs American English TTS system (1987 and 1991 versions) and natural speech, we found that no errors were made in either text-to-speech version or natural speech in the *word-initial/1-stress/–cluster* condition for 1-feature consonant contrasts. In this condition, minimal pairs differ by one consonant, and the contrasting consonants differ by one feature; moreover, the consonants are word-initial, head a stressed syllable, and are not part of consonant clusters (*–cluster*). In other words, this condition corresponds to *precisely the domain of the DRT*. This casts doubt on the usefulness of the DRT for high-end text-to-speech systems.

A second important result was that the 1991 version had more problems than the 1987 version with reduced consonants — consonants in codas and clusters; on the other hand progress was made in most other categories.

All speech versions — natural speech included — had serious problems with unstressed vowels. The error rates obtained for natural speech give a sense of the limits of the human speech signal. For segments in unstressed syllables, in a minimal pairs paradigm where it was completely impossible for the listener to make use of context, but where carefully pronounced speech was used under acoustically optimal conditions, error rates in excess of 15-25 percent were obtained. This suggests that under normal circumstances, which are acoustically considerably less optimal, many of the critical segmental features are simply not heard by the listener.

Finally, the effects of the number of features were mixed. Although the *consonant/1-feature* error rate was overall higher than the *consonant/2-feature* error rate, in the *word-initial/1-stress/–cluster* condition more errors were made with 2- than with 1-feature contrasts.

These and other results helped us focus our efforts in acoustic inventory construction, in particular in terms of rethinking our approach to reduced consonants.

Name transcription

The disadvantage of closed-response tests is that they test the intelligibility of only one phonetic segment, because it is difficult to construct balanced sets of alternatives that differ in multiple segment locations (e.g., *den*, *ten*, *Dan*, *tan*). Hence, although the listener hears an entire phrase containing twenty or more segments, only the intelligibility of at most 5% of the input is actually measured. Transcription tests such as the Bellcore test are more efficient in that more information is obtained from each utterance presentation. This does not necessarily translate into more information being obtained per time unit, because the listener has to write down or type the response, which takes significantly longer than a button push as in the MPI or DRT; in addition, the orthographic response has to be scored for phonemic errors, which is done manually. A drawback of transcription tests that use real words is that errors on some segments may be higher than on others not because of differences in acoustic intelligibility, but because of differences in predictability: the probability of a particular onset given the rhyme is usually lower than the probability of a particular coda given the onset and nucleus.

We present here an algorithm for automating transcription scoring (van Santen, 1993b). The basic idea is simple. For a given orthographic transcription, we generate a set of possible phonemic interpretations. These interpretations are compared to the correct phoneme sequence (we assume that the text-to-speech system is provided with correct phonemic input), and the best matching interpretation is found — thereby giving the listener the benefit of the doubt. We then score the discrepancies between this interpretation and the correct phoneme sequence.

For example, when the phonemic input to the text-to-speech system is /gúdrʊn wólkɑt/ (*Gudrun Wolcott*), and the orthographic response from the listener is *Gudro Welcot*, the best interpretation might be /gúdro wɛ́lkɑt/ (and not /gʌ́dro wɛ́lkɑt/). The discrepancies are /ʊ/ vs. /o/, /n/ vs. 0 (deletion), and /o/ vs. /ɛ/.

The scoring process has two phases. In the first, we generate the set of interpretations of a given orthographic response with an algorithm due to Michael Riley, based on classification and regression techniques (Breiman et al., 1984). The system was trained on 50,000 pairs of orthographic and phonemic transcriptions of names, which had been carefully checked manually. We then use a standard string alignment technique (Sankoff and Kruskal, 1983) to find the best matching interpretation. This alignment technique requires specification of a weight matrix, which indicates how much weight should be put on different error classes (e.g., vowel substitutions, onset consonant deletions, etc.).

We used this technique in an experiment where listeners transcribed names, such as *Gudrun Wolcott*. These names were generated by randomly combining last names selected from *Donnelley* with first names selected from the Bell Labs telephone directory. We selected names because there is less phonemic predictability than in regular words.

The technique proved to be more efficient than the MPI in the specific sense that statistically significant differences were found between various versions of the Bell Labs system using only one-quarter the number of listeners. Since similar diagnostic information can be obtained, this method seems preferable.

The key disadvantage of this technique is that it relies on Riley's algorithm and on the Bell Labs 50,000 name pronunciation dictionary, neither of which are in the public domain. However, as more transcribed text corpora are becoming available (e.g., through the Linguistic Data Consortium) and as decision-tree techniques are becoming commonplace, it should not be hard to produce algorithms that perform as well as Riley's. The string alignment algorithms are generally known.

In summary, we believe that the main drawback of transcription tests — the expense of phonemic scoring — can be overcome with procedures of the type described here. Of course, although phonemic scoring is a major obstacle for English, it is much less of an obstacle for languages where the orthographic-phonemic distance is smaller, such as in Romanian. The other main conclusion from these experiments is that names, because of their relative lack of phonotactic constraints (again, in particular in American English, with its enormous diversity of names), are more suitable for measuring segmental intelligibility than regular words, let alone meaningful sentences.

Concerning the diagnostic value for our system, these tests largely confirmed what we knew based on the MPI results.

8.3.3 Quality ratings with problem categorization

The tests discussed so far served primarily to diagnose problems with our acoustic inventory. We also needed a test that helped us with resource allocation: which text-to-speech components are responsible for the most prominent quality problems?

Towards this end, we constructed an experimental paradigm (van Santen, 1993b) based on earlier work by Wright and his coworkers (Wright, Altom, and Olive, 1986), in which a listener rates a word uttered by a text-to-speech system on a 1-6 scale (ranging from Excellent to Unrecognizable), and then answers various follow-up questions when the rating is 3 or worse. Wright's questions were replaced by broad problem categories specifically formulated to be meaningful to listeners and, at the same time, to have known relations to text-to-speech components.

The categories were the following: (1) *outright mispronunciation*, (2) *bad letters*, (3) *missing letters or words*, (4) *wrong stress*, (5) *wrong word emphasized*, (6) *overall voice quality*, (7) *choppiness*, (8) *bad rhythm*, and (9) *other*.

In an application of this experimental paradigm to several versions of the Bell Labs American English system and also to natural speech utterances, using a small sample of short sentences (not selected with any coverage optimization), we found that the problem areas were mostly confined to the synthesis component ((6) *overall voice quality*, (7) *choppiness*), while the text analysis components (1) (*outright mispronunciation*, (2) *bad letters*, (3) *missing letters or words*, (4) *wrong stress*, (5) *wrong word emphasized*) received the fewest complaints; the *bad rhythm* category (8) — presumably reflecting the duration component — was somewhere in between these two groups.

These results showed that, at that time, the text analysis components (at least for simple short sentences) did not require major investments of resources. Since then, we have produced new versions of the acoustic inventory and the duration component.

8.4 Conclusions

We described four tests that we used for diagnostics in our development efforts. One purpose of describing these tests was that some of them might be useful for other text-to-speech developers. But these examples also served to illustrate some of the general points made earlier, in particular concerning perceptual multidimensionality and coverage. Finally, the examples show how easy it is to construct tests that serve one's specific diagnostic needs, given a good set of test text generation tools and knowledge of the fundamentals of experimental design.

At the outset of this chapter, we posed the issue of how important assessment is for text-to-speech development and text-to-speech purchasing decisions. We now address this issue.

8.4.1 The role of assessment in text-to-speech development

The tests described above provided us with useful diagnostic information. Nevertheless, since resources were spent on developing and conducting these tests, it is reasonable to ask what information was not already known to us. The answer is that, based on (1) how many resources had been allocated in recent years to the various components of the Bell Labs system, (2) the intrinsic difficulty of the tasks faced by each component, and (3) our own informal impressions, many of the results were not surprising. The most useful results concerned the detection of individual problem acoustic units.

Overall, our assessment program has not helped us with either discovering major flaws, making new discoveries that had a real impact on speech quality, or deciding on drastic course changes in our research program. In other words, our experience does not confirm that "speech technology is typically evaluation-driven" (EAGLES, 1997). Instead, our speech technology largely has been driven by a mixture of *enabling technological developments* (e.g., availability of large text and speech corpora), progress in the underlying sciences, and priorities dictated by which components we decided were weak. It is our impression that much the same can be said about text-to-speech development efforts elsewhere. We agree, however, that development in other speech technologies such as automatic speech recognition, speech coding, and speaker verification, appears to have been significantly evaluation driven. Whether that is a good thing remains to be seen, of course.

8.4.2 The role of assessment in text-to-speech purchasing decisions

Our experience with buyers of text-to-speech products has been that their decisions are largely based on availability, software compatibility, size of memory footprint, price, potentially deceptive demonstrations by sales people and on WWW pages, and trade magazine articles (often based on those demonstrations). Speech quality plays a role, but is rarely assessed formally — and even if it is assessed formally, it may take only one manager who does not like the sound to brush aside the formal assessment results.

Why is the role of formal assessment in these decisions so limited? There would seem to be at least three possibilities. First, useful assessment methods or published results based on such methods exist, but purchasers are either unaware of their availability or are unwilling to use them for a variety of practical, sociological, or habit-based reasons. Second, although no useful methods are available yet, they will be available in the near future. Third, there are practical or fundamental reasons for why such methods will never be developed.

We may have to seriously entertain the third possibility. Our earlier remarks on some of the fundamental problems (perceptual multidimensionality and coverage) suggest that it may be very hard indeed to develop standardized assessment batteries that provide a picture of system performance that is sufficiently complete and relevant to allow a purchaser to confidently decide whether the system will perform well in the target application. There is at least one additional practical obstacle, which is that currently no impartial speech technology organizations exist that conduct large scale assessments of commercially available systems, so that it is the purchaser's task to conduct these tests. In the latter case, it might be more efficient for the purchaser to conduct a field study.

With some exceptions, most purchasers are not in a position to conduct extensive comparative tests — whether field studies or laboratory studies using established tests.

8.4.3 Some proposals

These largely negative remarks on the value of formal assessment should not be taken to imply that we deny that developers and purchasers need evaluative input. Purchasers should not be left at the mercy of canned demonstrations, and developers can benefit from quick, accurate, relevant diagnostics, because their own judgments can be clouded by over-exposure to their own system. In fact, we believe that there is an important role to play by formal assessment tools in the form of *minimal acceptability tests* (see below). Here are a few suggestions.

Assessment tool kit

As far as text-to-speech developers is concerned, we feel that given the specificity of their diagnostic needs and hence the low probability of a match between these needs and existing assessment instruments, it is more useful for them to have an *assessment tool kit* for quickly and efficiently assembling a diagnostic assessment instrument. We discussed some of the test text generation tools; adding tools for experimental design and statistical analysis should not be too difficult. Tools that are hard to install locally could be provided on interactive Web sites (see below).

Minimal acceptability tests

Currently available text-to-speech systems span a large range in terms of price and performance. While it may be difficult to choose between high-end text-to-speech systems, some differences are sufficiently large and across-the-board, that several of the currently available tests should be able to measure them reliably. We suggest that a battery of standard tests should be agreed upon that specifies minimal acceptability standards; impartial organizations would apply these tests to available text-to-speech systems, and the results would be public knowledge. We understand from our customers that this type of information would be genuinely useful.

As an example, few text-to-speech purchasers would be interested in a system that cannot even pronounce the 150 most frequent personal names correctly. Because, as we mentioned earlier, the ability to pronounce these names correctly is no guarantee for general pronunciation quality, minimal acceptability tests can only serve the purpose of excluding text-to-speech systems from consideration, and do not help selecting among the remaining majority of systems.

By aiming at these relatively modest assessment goals, reaching a consensus on which tests should be included in this minimal acceptability test battery may not be impossible.

Web-based text and text-to-speech servers

For text-to-speech purchasers, we propose that in addition to published minimal acceptability statistics, the most helpful evaluation aid would consist of Web sites that enable the user to obtain *honest demonstrations*. This proposal involves two components. One is the development of *interactive text-to-speech web sites*, where the user types in text and can listen to text-to-speech output as fast as the net allows. Many interactive text-to-speech web sites already exist, and are pointed to by an increasing number of master web sites.

The second component consists of *text servers*, which are sites that generate text (either by rule, or selected from text corpora) to be used in the interactive web sites. These servers would list various *domain categories* such as personal names, newspaper text, telephone numbers, minimal word pairs, nonsense sentences (e.g., based on SUS (Benoit, 1990)), and non-words (e.g., generated by CLID (Jekosch, 1993)). In addition, these servers would allow the user to choose from a variety of *text selection methods* (e.g., random selection, greedy selection with respect to some unit type such as bigrams or words, greedy selection with vs. without frequency weighting). These servers would also compute some *measure of coverage* of the selected text, such as the coverage index. For obvious reasons, these text servers should be localized at non-commercial organizations such as the European Speech Communication Association, the Acoustical Society of America, and the Acoustical Society of Japan. Finally, a useful addition would be the capability to automatically generate stimulus lists having specific experimental design properties (e.g., counterbalanced lists of minimal pair items).

Such a setup would allow anyone seriously considering purchasing a text-to-speech system to obtain arbitrary quantities of quality-controlled, categorized text from the server, and compare text-to-speech systems using the interactive web sites. It is not difficult to design text servers that pass selected text directly to the interactive text-to-speech servers. This all would be possible without the usually prohibitive burden of obtaining text corpora, typing in large quantities of text, and installing text-to-speech systems.

Chapter 9

Further Issues

Richard Sproat, Jan van Santen, Joseph Olive

One of the difficulties in putting together this description of our work on multilingual text-to-speech synthesis has been deciding when to consider the work "complete" enough to present in an archival format. Naturally, and fortunately, one of the properties of any such actively ongoing enterprise is that there is never a stage at which all the loose ends are tied up. The purpose of this chapter is to sketch some of these loose ends, and to give a sense of some of the directions that we believe are going to be important for multilingual TTS research in the years to come.

We organized this chapter along the lines of the goals that can be served by future research. These are, first, improving intelligibility and naturalness; second, facilitating, and when possible automating, the process of generating systems for new languages, dialects, or speakers; and, third, adding new capabilities.

9.1 Improving Intelligibility and Naturalness

The most common complaint about even the best synthesizer is that it "still does not sound natural". Our answer to this complaint is twofold. On the one hand, we do not think that for widespread applicability of synthesis it is necessary that users think they are interacting with a human speaker. Indeed, in some instances, it may be inappropriate to deceive people into thinking they are dealing with another human, since their expectations of a system's performance will undoubtedly be raised; and this will lead subsequently to frustration when they realize that the system is not as intelligent as it sounds. As a corollary, we believe that research on synthesis of emotion, although fascinating, is misguided. But this point aside, the main point is that there are limits to how natural we need a synthesizer to sound to be acceptable for broad areas of application.

On the other hand, there is no doubt that synthesizers could sound significantly more natural, and that also intelligibility — and in particular comprehensibility — should improve.

The causes for the unnatural sound of synthesized speech are complex and as yet not fully understood. We know that the unnaturalness of synthetic speech is due both to the speech's sound quality and to the imperfect prosody, which results in a mechanical accent. The use of better analysis/synthesis techniques should improve the sound quality of the synthesized speech. The mechanical accent is caused by flaws in many of the system's modules. The synthesized speech will sound degraded if either F_0 or segmental durations are not appropriate, even if all other prosodic features are computed correctly. It is often difficult to assess which module is responsible for a perceived flaw. Using segmental duration and F_0 as an example, a bad syllable duration can force the F_0 to take a strange shape, and this F_0 peculiarity may be perceptually more salient than the underlying temporal problem. Likewise, poor accent assignment can produce too many pitch peaks, again leading listeners to conclude that the F_0 component is broken. As a result, we expect improvements in naturalness to be slow, not only because so many components may be involved, but also because the issues to be resolved may in fact be quite hard.

9.1.1 Synthesis

We noted a trend in synthesis to take advantage of increased computer storage and processing by maximizing acoustic inventory size, and hence being able to minimize the degree of alteration applied to units during synthesis. We are sceptical about this trend for combinatorial reasons, at least as far as unrestricted domain synthesis is concerned. There may be special types of domain restrictions, however, where this approach might work. For example, if a task allows one to restrict the vocabulary to a (small) finite list, then such approaches are certainly viable. Examples of such tasks include reading information about bank accounts (excluding customers' names!), travel timetables, and so forth. We are not convinced, however, that large-inventory approaches hold the answer to achieving high quality speech for domains that are significantly less restricted than these.

Nevertheless, we agree with proponents of this trend that current speech parameterizations and signal processing operations performed during concatenation leave much to be desired.

We can think of at least two areas of research here. One is to stay within the current framework, the other is to investigate a hybrid between rule-based and concatenative synthesis.

Improvement within the current framework

An obvious area for improvement is the acoustic inventory construction process. For example, our current discrepancy measure (distance in formant space) is not based on systematic perceptual research. In fact, very little is known about how spectral discontinuities are perceived.

Another area for improvement is speech parameterization. As we pointed out earlier, there may be a direct relation between the amount of information about the original speech signal that is preserved in the speech parameterization and the likelihood of spectral discrepancies between units. This direct relation means that the acoustic unit selection process has to be more selective as the speech parameterization preserves more original acoustic detail. In the limit, increased levels of preserved detail may require either enormous numbers of tokens to be recorded for each acoustic unit type or speakers with an uncanny spectral constancy.

But there are other aspects of speech parameterization than the amount of spectral detail preserved. In particular, we need parameterizations that are robust with respect to the alterations performed during concatenation.[1] Since with few exceptions (e.g., PSOLA, MBROLA) no current speech parameterizations have been specifically developed for speech synthesis, we feel that there may be room for improvement here.

Rule-based concatenative synthesis

The other approach to synthesis improvement would depart from the current framework, and, by moving closer to rule based synthesis, lead in a direction opposite to that taken by approaches based on very large acoustic inventories. We propose to develop speech parameterizations that allow independent manipulation of parameters to mimic coarticulatory and prosodic spectral effects. Thus, instead of limiting prosodic alterations to F_0 and timing (and possibly source parameters), we would also change such parameters as spectral tilt (which may be done at the source level, however), formant bandwidth, and the actual formant values (or the correlates of these variables in the new parameterization scheme). In addition, we would also change these parameters to mimic long-range or subtle coarticulatory effects that are currently out of reach for inventories of reasonable sizes. Better source/tract separation and more accurate source modeling may play an important role here.

This approach would require the same type of knowledge base as current rule based synthesizers, but it does not face the primary obstacle encountered by rule based synthesis — that of requiring a complete analytic description of extremely complicated speech sounds. Instead, only certain aspects of these sounds would be altered.

[1] For example, imperfect source/tract separation in speech analysis and inaccurate source modeling cause a degradation of speech when it is altered during the concatenation process.

9.1.2 Acoustic prosody

Timing

Currently, speech timing is controlled by manipulating segmental duration. At the sub-segmental level, the concatenation operation uses a non-linear time warping approach for lengthening vowels, but uses uniform warping for all other cases. We know that in natural speech vowels are not always lengthened in the center; for example, phrase-final vowels or vowels followed by voiced sounds are lengthened primarily by stretching their final parts; see Section 5.2.1. And sounds other than vowels are unlikely to be warped uniformly. In other words, the current system does not perform the correct sub-segmental temporal operations.

These shortcomings can be addressed by developing models for sub-segmental timing. One way to do this is by acoustically defining segment-internal boundaries (delineating, for example, the early, middle, and final part of a segment), and developing the same type of models for these subsegmental intervals as we applied to segmental duration. More continuous time warp based approaches may be based on the type of time warp studies that were discussed in (van Santen, Coleman, and Randolph, 1992).

Hybrid synthesis opens the possibility of independently manipulating the trajectories of different speech parameters. This is complicated, because it would require highly accurate speech analysis procedures for estimating these parameters in natural speech recordings.

Needless to say, there are are many other issues in timing that will need to be addressed. For example, effects of speaking rate on various segmental classes, possibly interacting with prosodic context, remain largely unexamined.

Intonation

Besides testing the validity of our general superpositional theory, our immediate need is to test the application of this theory to tone languages. In addition, analyses of natural speech data have to be conducted to measure the variety of shapes of phrase curves, and which factors determine their shapes. Finally, the concept of accent groups may need re-addressing, in particular to provide a more convincing account of secondary stress (in the current system, secondary stress is treated the same way as zero stress).

9.1.3 Prosodic text analysis

An obvious area for improvement is that the current accenting module assigns pitch accents to far too many syllables, often resulting in a bumpy intonation contour; see Section 4.1. The prosodic labeling scheme applied to the data on which this accenting module is based uses a different implicit definition of what constitutes a pitch accent than the current intonation module. This issue needs addressing.

Much more formidable is the task of proper prominence and phrase boundary assignment for paragraph-length text. We feel that performing this task at high levels of accuracy will not be possible without major advances in several areas of computational linguistics and artificial intelligence, in particular discourse analysis and semantics. Many of the problems here may very well have no general solution, and can only be approximately solved in highly restricted contexts.

9.2 Automation

Currently, acoustic inventory construction, duration modeling, and F_0 modeling (alignment parameters) are largely automated. One obstacle is segmentation, in particular for prosodically rich and varied speech. A second obstacle is the incompleteness of our knowledge of objective criteria for measuring the "goodness" or intelligibility of specific segments.

Much more difficult is automatization of text analysis component construction for new languages. Yet, because this phase is the most time consuming in the process of dealing with a new language, its automatization is the most urgent. We have already discussed recent work on automatizing the process of building word-pronunciation modules (Sections 3.7 and 3.9.3), and we have argued that while such approaches may ultimately prove useful, the results (to date) have not been stunningly successful, in our opinion. As we also noted, word pronunciation is but one small part of the problem, and in many languages it is not even the most difficult part. Computing the information that one would need in order to predict pronunciation — given that the orthography by itself is often not sufficient — can be a much more onerous task.

Failing general solutions to the fully automatic construction of text analysis components, one might for the interim concentrate on *semi-automatic* or *machine-assisted* techniques. For example, one of the major problems faced by TTS systems is lexical disambiguation. As we discussed in Section 4.2, there are good methods for performing lexical disambiguation, but these methods typically presume that one already knows the set of cases to be disambiguated. But it can be surprisingly difficult a priori to come up with a list of all the elements of a language that might possibly be read in more than one way: dictionaries and thesauruses are usually only partially helpful here, and for many languages such electronic resources are in any event lacking. One area worth exploring, therefore, is corpus-based methods for constructing lists of potentially polysemous forms. The problem can be simply stated: given a database of text for a language, produce a list of all words in the language, listing first those words which are likely to be polysemous, and which therefore at least have the potential for having multiple sense-related pronunciations.

```
Hi Brad:

Here is the final flight schedule that you requested:

Date     Departs          Arrives          Airline
4/3/97   EWR, 10am        SFO, 12:30pm     United
4/6/97   SFO, 12pm        DFW, 4pm         American

Let me know if you will need a rental car in either San Francisco
or Dallas.

Best wishes,

Tom

------------------------------------------------------------------

Tom Hogan                               /  ___   ___
Dick Campbell Enterprises           ____/  /      /__/
hogan@campbell.com                 /___/  /___  /___

------------------------------------------------------------------
```

Figure 9.1: An typical example of an email message.

9.3 New Capabilities

Up to this point, we have discussed how text-to-speech can do things better, or how we can construct improved text-to-speech systems more efficiently. Here, we discuss new things that text-to-speech systems might do.

9.3.1 Encoding and analysis of document structure

In the model of text analysis presented in Chapters 3 and 4, text is viewed simply as a string of characters from some writing system. While this view is one that is shared with most TTS systems, there is an important respect in which it is deficient, namely that human readers are clearly aware of more structure in documents than a simple linear string representation would suggest.

Consider for example the (constructed) example of an email message shown in Figure 9.1. While the example is artificial, it illustrates a number of features typical of electronic mail (as well as other types of text). In addition to "plain" text, we also find tabular material (the flight schedule); and a signature field where, unlike plain text, line breaks typically correspond to a phrasing break; the signature also contains some non-text material in the form of a graphic.

A person asked to read aloud the text of this email message would certainly make use of this structure in deciding how to render the message in speech. The plain text portion would presumably be read as normal prose, with phrasing breaks being cued partially off punctuation. The signature field, however would be treated differently, since any experienced reader would know that line breaks in such regions are significant in a way that they are not in plain text. The treatment of the table will be different still. Not only are line (row) breaks significant in tables, but so is the column structure. Furthermore, it is unlikely that the reader will present the table by simply reading the elements of the rows and columns from left to right and top to bottom as this would not in general be a very useful way of presenting the information in the table. For instance if one reads the column headers (the first row of the table in the example in Figure 9.1) only once, the listener is likely to forget which column represents which information, and will therefore have difficulty following the subsequent presentation of the rows.[2]

Clearly none of this rich structure is adequately represented in a framework where the input text is merely a string of characters: some richer *document* representation is required. A reasonable framework for representing such structure is the *Standard Generalized Markup Language* (SGML) (Goldfarb, 1990). Despite its name, SGML is actually a *metalanguage* for defining markup languages. Some well-known markup languages based on SGML are the markup guidelines developed under the auspices of the Text Encoding Initiative (TEI) (Sperberg-McQueen and Burnard, 1994; Véronis and Ide, 1996); and the Hypertext Markup Language — HTML (see, e.g., (Graham, 1996)), used on the Worldwide Web.

An SGML-type markup of the email message in Figure 9.1 is given in Figure 9.2. In this version of the text, the document structure is clearly indicated, and a speech synthesis system that is sensitive to such markup would be able to mimic what a human reader would normally do with such a document. For example since the system would have access to information about the structure of the table, it would be possible for it to render it in some fashion other than in the somewhat less than useful left-to-right, top-to-bottom fashion. Indeed, having such a rich representation of the input text opens up the possibility of having fine listener control of the behavior of the TTS system, changing it from a text reading device into a much more useful *audio rendering* device (Raman, 1994).[3]

The need for such a markup scheme, plus the desire to have such a scheme be conventional and widely adopted, has motivated the development of the

---

[2] On tabular information and its interpretation see (Hurst and Douglas, 1997).

[3] The question of how one would derive this kind of markup for text that does not already come marked up according to some formally defined scheme is a separate issue which we will not address here.

```
<PLAIN>
Hi Brad:

<P>

Here is the final flight schedule that you requested:

<TABLE>
<TR> <TD> Date    <TD> Departs    <TD> Arrives      <TD>  Airline  </TR>
<TR> <TD> 4/3/97 <TD> EWR, 10am <TD> SFO, 12:30pm <TD>  United   </TR>
<TR> <TD> 4/6/97 <TD> SFO, 12pm <TD> DFW, 4pm     <TD>  American </TR>
</TABLE>

Let me know if you will need a rental car in either San Francisco or
Dallas.

<P>

Best wishes,

Tom

</PLAIN>

<SIGNATURE>
<SEPARATOR>
--------------------------------------------------------------------
</SEPARATOR>

Tom Hogan <DIV>
Dick Campbell Enterprises <DIV>
hogan@campbell.com <DIV>

<GRAPHIC>
      /  ___   ___
 ____/  /     /__/
/___/  /___  /___
</GRAPHIC>

<SEPARATOR>
--------------------------------------------------------------------
</SEPARATOR>
</SIGNATURE>
```

Figure 9.2: A marked up version of the email message in Figure 9.1. For the sake of simplicity, we have followed HTML conventions where possible, namely for the representation of paragraph breaks (<P>) and the markup of tables: <TR> ... </TR>, delimits a table row, and <TD> marks the start of a table column. We somewhat arbitrarily use the element <DIV> to represent a significant division in the text of the signature field.

Spoken Text Markup Language (STML) (Taylor and Isard, 1996; Sproat et al., 1997). Since STML is in the early stages of development it is not possible at this stage to give many details of the standard (see (Sproat et al., 1997) for an initial proposal): in particular the markup in Figure 9.2 should *not* be taken as an instance of STML. The following general details are agreed upon, however: in addition to containing markup for *text description*, such as that exemplified in Figure 9.2, STML will also contain *speaker directives*, which indicate properties of the speaker reading the text — e.g. what voice to use, what the pitch range of the voice should be, and in a system with a "talking head", what kind of face to use. Other obvious markup would include the language of the text.

Of course, extant TTS systems typically allow for some limited markup of the input text. The Bell Labs TTS systems described in this book, for example, have a fairly rich set of escape sequences for controlling various aspects of the output speech (see Section 6.2). Similarly the current industry standard for TTS markup, Microsoft's SAPI (Microsoft, 1996), includes functionality for controlling the properties of the speech, plus a limited amount of further functionality, such as the ability to set the language of the document.

Schemes such as SAPI or the Bell Labs escape sequences are undesirable from two points of view. First of all, the syntax of the tags is quite ad hoc and unconventional. In contrast STML will adopt SGML, the international standard for text markup. Second, such markup schemes focus almost exclusively on what we have termed speaker directives: almost no attention is given to document structure.[4] Yet in order for TTS systems to allow for maximally flexible capabilities in rendering a document, a (standardized) method for marking up salient aspects of that structure is required. The purpose of STML is to provide such a scheme. TTS systems will thus no longer view text as merely a string of characters, but rather as the richly structured two-dimensional object that it is.

9.3.2 "Style" based text analysis and prosody

An important element of document structure analysis is that a document is partitioned into text regions that have to be made as distinctive in the output speech as they are separate in the document input. This requires control over speaking *style*.

The concept of style introduces a novel element in text analysis. Not only might we be able to use typographic cues such as formats and STML symbols to define stylistically distinct text regions, we also may be able to recognize such regions by other means. For example, a sentence such as *He bought blueberries, strawberries, bananas, and pears* is not typographically distinct, yet has

[4] In the case of SAPI the lack of attention to document structure appears to be by design, since it is stated that "[a]n application should use text-to-speech only for short phrases or notifications, not for reading long passages of text" (Microsoft, 1996, page 6). Given that TTS systems are routinely used in many applications to read long passages of text, this motivation seems at best bizarre.

a specific semantic or lexical form (that of a *list*) and should be uttered with a distinctive F_0 contour. An issue here is whether a significant percentage of arbitrary textual input can be assigned to distinctive style regions, each with its own prosodic rules.

As for the acoustic-prosodic aspects of style, we currently know little about how to model stylistic differences. Merely altering F_0, speaking rate, and loudness is not likely to be sufficient. A more interesting proposal would be to change not these overt acoustic speech parameters, but the parameters of the prosodic modules themselves. For example, one could increase the range of the stress parameters in the duration module and decrease the range of the boundary-related parameters, so that stress has a larger effect on output speech timing while less phrase-final lengthening occurs. What is needed is to make systematic comparisons of natural speaking styles in terms of these underlying parameters.

9.3.3 Voice mimic

The issue of how to mimic a specific person's voice is interesting both scientifically and commercially — many prospective customers have expressed interest in this capability. Currently, one would have to construct an entirely new acoustic inventory for the to-be-mimicked speaker, and estimate new acoustic-prosodic parameters as well. Until we can map acoustic units onto units that mimic a specific speaker, new acoustic inventory construction is unavoidable.

There are at least three aspects of a person's speech that would need to be copied in order to produce a successful mimic. These are: vocal tract characteristics; prosody, including both intonation and segmental timing; and patterns of allophonic variation. The first two of these are relatively easy using currently available techniques. The third is, of course, much more difficult. However, for all of these problems, the amount of training data necessary for a new speaker could in principle be minimized using sophisticated text-selection techniques; see Section 2.4.

9.3.4 Summary

Text-to-speech synthesis has more open issues than can be addressed within the next five or ten years. Which of these issues will be resolved depends on their difficulty and on how priorities are set. Many of the issues are too hard and open-ended to be appropriate for development organizations, while at the same time the TTS community is not large enough to make quick progress with these problems. Moreover, these questions, while quite exciting for researchers directly involved with TTS, may not appear that interesting from a wider phonetics and linguistics standpoint because they seem specific, model-dependent, and overly concerned with quantitative detail. Generating broader interest in academe in these types of questions is an important challenge for the TTS community.

Appendix A

Character Set Encodings

Concomitant with the use of different scripts for different languages is the use of different electronic coding schemes for those scripts. There exist various international standards for coding of different character sets. Among these are the iso8859 series for single-byte codes; various interconvertible two-byte Japanese standards (EUC-JP/Shift JIS) for Japanese text; and two separate two-byte standards, Big5 and GB for Chinese. While these standards are widespread, they are by no means universal. European languages which could be coded using the iso8859-1 standard are often instead written in plain ascii, with the various diacritics being translated in sometimes standard and sometimes nonstandard ways (see below). There are several standards for coding Cyrillic besides iso8859-5. The universal adoption of the uniform two-byte standard UNICODE (The Unicode Consortium, 1996) will hopefully eliminate this confusion, but this to-be-hoped-for state is probably a while off. So the current state of character coding is messy, and any multilingual TTS system must be flexible enough to deal with this fact.

By default, the Bell Labs TTS systems are set up to work with the following character sets for the following languages:

Character Set	**(Set Name)**	**Languages**
iso8859-1	(Western European)	Spanish, French, German, Italian, English
iso8859-2	(Central European)	Romanian
iso8859-5	(Cyrillic)	Russian
Big5	(Traditional Chinese)	Mandarin
GB	(Simplified Chinese)	Mandarin
EUC		Japanese

Input that deviates from these standards must be handled specially, something which the gentex architecture makes it relatively easy to do. For the purposes of the present discussion we can divide the types of “recodings” that one finds into three basic classes:[1]

- *Recoding*, or the use of a different coding scheme. An example would be the use of Apple Standard Cyrillic rather than iso8859-5.

- *Respelling*, or the spelling of a symbol with a diacritic with some sequence of characters. Examples include spelling accented vowels <é> with a sequence such as <'e>, or German umlauted vowels <ü> with a vowel-e sequence <ue>.

- *Elimination*. In this most extreme case, a diacritic is simply eliminated in favor of the diacriticless symbol. So Spanish <Méjico> would simply be written as <Mejico>.

The last category, *elimination*, is clearly the most tricky for TTS systems since it frequently creates homographic sets that would not be homographic sets in the fully specified orthographic representation. Thus French <côté> 'side' and <côte> 'coast', which have different pronunciations, are indistinguishable when accents are removed: <cote>. One solution to this problem, of course, is to apply the same kinds of homograph disambiguation techniques as were discussed in Chapter 4. Yarowsky (see, e.g., (Yarowsky, 1996)) has presented some very promising results in this area. Nonetheless, we want to stress that while the approach that should be taken to this problem is clear enough, we have not to date addressed it in any general way in our TTS systems.

The *respelling* category is (mostly) much simpler since it lends itself fairly straightforwardly to a simple WFST-based solution. For example, for umlauted vowels in German one can simply have a set of rules that rewrites them as their respelled equivalents:

$$
\begin{array}{ll}
\text{ü} & \rightarrow \text{ue} \\
\text{ö} & \rightarrow \text{oe} \\
\text{ä} & \rightarrow \text{ae}
\end{array}
$$

The transducer derived from this set of rules, when inverted, will *optionally* transduce sequences like <ue> into their umlauted equivalents and it will be up to the lexical analysis component to decide whether or not to treat them as an umlauted vowel or as a sequence of vowels. Where homographs are created by such respellings, one would again have to resort to homograph-disambiguation techniques.

The *recoding* possibility is in principle the easiest since it merely involves replacing a single table. Labels on arcs in transducers are represented as integers, and a separate label file maps those internal labels to strings in a human-readable alphabet. Our current version of Russian for example has the following correspondences as part of this table:

[1] The true situation, as we shall see in the discussion of Big5 and GB below, is considerably less clear-cut than this classification would suggest, but this is at least a good starting point for discussion.

iso8859-5 character	internal code
а	108
б	109
в	110
г	111
д	112
е	113
ж	114

Adapting the system for another coding scheme is as simple as replacing the iso8859-5 characters in this table with their equivalents in the new scheme.

Of course, things are not always so simple, and a good instance of this is the two Chinese coding schemes Big5 and GB. Big5 is a two-byte code where the first byte always has the high bit set, and the second byte may be set or not. In other words, a Big5 character must match the following bit pattern, where *1* means that the bit is set and *?* means that it may be either set or unset:

1	?	?	?	?	?	?	?	?	?	?	?	?	?	?	?

In contrast GB requires the high bit to be set in both bytes:

1	?	?	?	?	?	?	?	1	?	?	?	?	?	?	?

The Big5 character set thus has much more space than the GB set, and indeed there are approximately 13,000 Chinese characters in the Big5 set, but only about 6,700 in the GB set. Necessarily, then, some information will be lost when one converts from Big5 into GB, and replacing our original Big5 codes in the table with GB codes will result in a lot of gaps in the table. This would be straightforwardly acceptable for GB-coded text were it not for an additional complication, namely that while Big5 is used to code traditional Chinese characters (as used in Taiwan), GB is (mostly) used to code simplified characters (as used in Mainland China). Part of the process of simplification involved eliminating some distinctions that are still made in the traditional character set (as well as, in some cases, introducing some distinctions that were previously not made). Thus the characters 干 *gān* 'intervene', and 幹 *gàn* 'manage' are both written as 干 in simplified characters. The table entries:

Big5 character	internal code
干	595
幹	4198

are replaced by the following in GB:

GB character	internal code
干	595
干	4198

Fortunately, given the architecture of gentex this is not a problem, at least for characters that can be disambiguated using lexical information. For a GB-coded text, the FSA representing the input would be a lattice containing all internal codes corresponding to the GB character 干. Lexical information from the lexical analysis transducers will frequently be sufficient to disambiguate the intended reading. Consider, for example, the 干 of (Big5) 干擾 *gānrǎo* 'interfere', and the 幹 of 幹部 *gànbù* 'cadre' (note the difference in tone between the two *gan* syllables), which as we have noted are both written 干 in GB.

The information already present in the lexical transducers is sufficient to disambiguate the 干 in GB 干扰 *gānrǎo* 'interfere', and the 干 in GB 干部 *gànbù* 'cadre', so no further work is necessary other than providing an appropriate GB version of the correspondence table.

The gentex model is thus flexible enough to handle a variety of character coding schemes, requiring relatively minimal effort to port the system to a new scheme. Often it is simply a matter of replacing a table with an appropriately table recoded with the new scheme. For respellings, it is necessary to provide a special transducer for the new respelling. In the worst case (elimination) it will be necessary to provide additional homograph disambiguation capabilities. But none of these options require any fundamental change to the way gentex operates with the language in question.

Appendix B

Glossary of Grammatical Labels

3sg: third singular (verb).

Abbr: abbreviation.

adj: adjective.

Adv: adverb.

clit: clitic.

conj: conjunction.

cor: coronal.

csub: subordinating conjunction.

D: determiner.

fem: feminine.

gen: genitive (case).

instr: instrumental (case).

masc: masculine.

N: noun.

nom: nominative (case).

Num, num: number.

P: preposition.

Pl, pl: plural.

PPos: possessive pronoun.

pref: prefix.

prep: prepositional (case).

Sg, sg: singular.

sp: simple past (French verbs).

imp: imperfect.

ind: indicative.

strid: strident.

suff: suffix.

V: verb.

REFERENCES

Adamson, Martin and Robert Damper. 1996. A recurrent network that learns to pronounce English text. In *Proceedings of the Fourth International Conference on Spoken Language Processing*, volume 3, pages 1704–1707, Philadelphia, PA. ICSLP.

Adriaens, Léon. 1991. *Ein Modell deutscher Intonation.* Ph.D. thesis, Technical University Eindhoven.

Allen, Jonathan, M. Sharon Hunnicutt, and Dennis Klatt. 1987. *From Text to Speech: the MITalk System.* Cambridge University Press, Cambridge.

Andersen, Ove, Roland Kuhn, Ariane Lazaridès, Paul Dalsgaard, Jürgen Haas, and Elmar Nöth. 1996. Comparison of two tree-structured approaches for grapheme-to-phoneme conversion. In *Proceedings of the Fourth International Conference on Spoken Language Processing*, volume 3, pages 1700–1703, Philadelphia, PA. ICSLP.

Anderson, Mark, Janet Pierrehumbert, and Mark Liberman. 1984. Synthesis by rule of English intonation patterns. In *Proceedings of IEEE*, pages 2.8.1–2.8.4.

Baayen, Harald. 1989. *A Corpus-Based Approach to Morphological Productivity: Statistical Analysis and Psycholinguistic Interpretation.* Ph.D. thesis, Free University, Amsterdam.

Bachenko, Joan and Eileen Fitzpatrick. 1990. A computational grammar of discourse-neutral prosodic phrasing in English. *Computational Linguistics*, 16:155–170.

Balestri, Marcello, Stefano Lazzaretto, Pier Luigi Salza, and Stefano Sandri. 1993. The CSELT system for Italian text-to-speech synthesis. In *Proceedings of the 3rd European Conference on Speech Communication and Technology*, volume 3, pages 2091–2094, Berlin. ESCA.

Barbosa, Plínio and Gérard Bailly. 1994. Characterisation of rhythmic patterns for text-to-speech synthesis. *Speech Communication*, 15:127–137.

Barlow, R., Bartholomew D., J. Bremner, and H. Brunk. 1972. *Statistical Inference under Order Restrictions.* Wiley, New York, NY.

Belhoula, Karim. 1993. Rule-based grapheme-to-phoneme conversion of names. In *Proceedings of the Third European Conference on Speech Communication and Technology*, volume 2, pages 881–884, Berlin. ESCA.

Benoit, Christian. 1990. An intelligibility test using semantically unpredictable sentences: Towards the quantification of linguistic complexity. *Speech Communication*, 9:293–304.

Bickley, Corine, Kenneth Stevens, and David Williams. 1997. A framework for synthesis of segments based on pseudoarticulatory parameters. In Jan van Santen, Richard Sproat, Joseph Olive, and Julia Hirschberg, editors, *Progress in Speech Synthesis*. Springer, New York, NY, pages 211–220.

Bigorgne, D., O. Boëffard, B. Cherbonnel, Françoise Emerard, D. Larreur, J. Le Saint-Milon, I. Métayer, Christel Sorin, and S. White. 1993. Multilingual PSOLA text-to-speech system. In *Proceedings of the 1993 International Conference on Acoustics, Speech and Signal Processing*, volume 2, pages 187–190. IEEE.

Bird, Steven and Ewan Klein. 1994. Phonological analysis in typed feature structures. *Computational Linguistics*, 20:455–491.

Bolinger, Dwight. 1951. Intonation—levels vs. configurations. *Word*, 7:199–210.

Brandt Corstius, Hugo. 1968. *Grammars for Number Names*. D. Reidel, Dordrecht.

Breiman, Leo, Jerome Friedman, Richard Olshen, and Charles Stone. 1984. *Classification and Regression Trees*. Wadsworth and Brooks, Pacific Grove, CA.

Bril, G., R. Dykstra, C. Piller, and T. Robertson. 1984. Algorithm AA 206. Isotonic regression in two independent variables. *Applied Statistics*, 35:352–357.

Brill, Eric. 1992. A simple rule-based part of speech tagger. In *Proceedings of the Third Conference on Applied Natural Language Processing*, Trento. ACL.

Bruce, Gösta. 1977. *Swedish Word Accents in Sentence Perspective*. Number XII in Travaux de l'Institut de Phonétique. Gleerup, Lund.

Bruce, Gösta. 1990. Alignment and composition of tonal accents. In John Kingston and Mary Beckman, editors, *Papers in Laboratory Phonology I: Between the Grammar and Physics of Speech*. Cambridge University Press, Cambridge, pages 105–114.

Buchsbaum, Adam and Jan van Santen. 1996. Selecting training text via greedy rank covering. In *Proceedings 7th ACM-SIAM Symposium on Discrete Algorithms*, Atlanta, GA. ACM.

Campbell, W. Nick. 1992. Syllable-based segmental duration. In Gérard Bailly and Christian Benoit, editors, *Talking Machines: Theories, Models, and Designs*. Elsevier, Amsterdam, pages 211–224.

Campbell, W. Nick and Alan Black. 1996. CHATR: a multi-lingual speech re-sequencing synthesis system. In *Proceedings of Institute of Electronic, Information and Communication Engineers-89*, Tokyo. IEICE.

Carlson, Rolf, Björn Granström, and M. Sharon Hunnicutt. 1989. Multi-language text-to-speech development and applications. In William Ainsworth, editor, *Advances in Speech, Hearing, and Language Processing*. JAI Press, London.

Chao, Yuen Ren. 1968. *A Grammar of Spoken Chinese*. University of California Press, Berkeley, CA.

Charpentier, Francis and Eric Moulines. 1989. Pitch-synchronous waveform processing techniques for text-to-speech synthesis using diphones. In *Proceedings of the First European Conference on Speech Communication and Technology*, volume 2, pages 13–19, Paris. ESCA.

Chou, Fu-Chiang, Chiu-yu Tseng, Keh-jiann Chen, and Lin-shan Lee. 1997. A Chinese text-to-speech system based on part-of-speech analysis, prosodic modeling and non-uniform units. In *Proceedings of the 1997 International Conference on Acoustics, Speech and Signal Processing*, volume 2, pages 923–926, Munich. IEEE.

Church, Kenneth. 1988. A stochastic parts program and noun phrase parser for unrestricted text. In *Proceedings of the Second Conference on Applied Natural Language Processing*, pages 136–143, Austin, TX. ACL.

Church, Kenneth Ward and William Gale. 1991. A comparison of the enhanced Good-Turing and deleted estimation methods for estimating probabilities of English bigrams. *Computer Speech and Language*, 5:19–54.

Coker, Cecil, Kenneth Church, and Mark Liberman. 1990. Morphology and rhyming: Two powerful alternatives to letter-to-sound rules for speech synthesis. In Gérard Bailly and Christian Benoit, editors, *Proceedings of the ESCA Workshop on Speech Synthesis*, pages 83–86, Autrans, France. ESCA.

Coleman, John. 1992. "Synthesis-by-rule" without segments or rewrite-rules. In Gérard Bailly and Christian Benoit, editors, *Talking Machines: Theories, Models, and Designs*. Elsevier, Amsterdam, pages 43–60.

Coleman, John. 1993. Computation of candidate synthesis units. Technical report, AT&T Bell Laboratories. 11222-930719-07TM.

Cooper, Franklin. 1983. Some reflections on speech research. In Peter MacNeilage, editor, *The Production of Speech*. Springer, New York, NY, pages 275–290.

Cormen, Thomas, Charles Leiserson, and Ronald Rivest. 1990. *Introduction to Algorithms*. MIT Press, Cambridge, MA.

Crystal, Thomas and Arthur House. 1982. Segmental durations in connected speech signals: Preliminary results. *Journal of the Acoustical Society of America*, 72:705–716.

Crystal, Thomas and Arthur House. 1988a. The duration of American-English stop consonants: an overview. *Journal of Phonetics*, 16:285–294.

Crystal, Thomas and Arthur House. 1988b. Segmental durations in connected-speech signals: Current results. *Journal of the Acoustical Society of America*, 83:1553–1573.

Crystal, Thomas and Arthur House. 1988c. Segmental durations in connected-speech signals: Syllabic stress. *Journal of the Acoustical Society of America*, 83:1574–1585.

Crystal, Thomas and Arthur House. 1990. Articulation rate and the duration of syllables and stress groups in connected speech. *Journal of the Acoustical Society of America*, 88:101–112.

Daelemans, Walter and Antal van den Bosch. 1997. Language-independent data-oriented grapheme-to-phoneme conversion. In Jan van Santen, Richard Sproat, Joseph Olive, and Julia Hirschberg, editors, *Progress in Speech Synthesis.* Springer, New York, NY, pages 77–89.

Daniels, Peter and William Bright. 1996. *The World's Writing Systems.* Oxford University Press, New York, NY.

de Pijper, Jan Roelof. 1983. *Modelling British English intonation.* Foris, Dordrecht.

Dedina, Michael and Howard Nusbaum. 1996. PRONOUNCE: a program for pronunciation by analogy. *Computer Speech and Language*, 5:55–64.

Dirksen, Arthur and John Coleman. 1997. All-Prosodic Speech Synthesis. In Jan van Santen, Richard Sproat, Joseph Olive, and Julia Hirschberg, editors, *Progress in Speech Synthesis.* Springer, New York, NY, pages 91–108.

Dixon, N. and H. Maxey. 1968. Terminal analog synthesis of continuous speech using the diphone method of segment assembly. *IEEE Transactions on Audio and Electroacoustics*, 16:40–50.

Dodge, Yadolah. 1981. *Analysis of Experiments with Missing Data.* Wiley, New York, NY.

Duden. 1984. *Duden Grammatik der deutschen Gegenwartssprache.* Dudenverlag, Mannheim.

Dudley, Homer, R. Riesz, and S. Watkins. 1939. A synthetic speaker. *Journal of the Franklin Institute*, 227:739–764.

Dutoit, Thierry. 1997. *An Introduction to Text-to-Speech Synthesis.* Kluwer, Dordrecht.

Dykstra, R. and R. Robertson. 1982. An algorithm for isotonic regression for two or more independent variables. *Annals of Statistics*, 10:708–716.

EAGLES. 1997. *Spoken Language Systems*, volume III. Mouton de Gruyter, Berlin, 2nd edition.

Edwards, Jan and Mary Beckman. 1988. Articulatory timing and the prosodic interpretation of syllable duration. *Phonetica*, 45:156–174.

Fant, Gunnar. 1960. *Acoustic Theory of Speech Production.* Mouton, The Hague.

Ferri, Giuliano, Piero Pierucci, and Donatella Sanzone. 1997. A complete linguistic analysis for an Italian text-to-speech system. In Jan van Santen, Richard Sproat, Joseph Olive, and Julia Hirschberg, editors, *Progress in Speech Synthesis.* Springer, New York, NY, pages 123–138.

Fischer-Jørgensen, Eli. 1987. The phonetic manifestation of the 'stød' in Standard Danish. *Annual Report of the Institute of Phonetics (University of Copenhagen)*, 21:55–282.

Flanagan, James. 1972. *Speech Analysis, Synthesis and Perception*, volume 3 of *Communications and Cybernetics.* Springer, Berlin.

Franks, Steven. 1994. Parametric properties of numeral phrases in Slavic. *Natural Language and Linguistic Theory*, 12(4):597–674.

Fudge, Eric. 1984. *English Word-Stress.* Allen and Unwin, London.

Fujio, Shigeru, Yoshinori Sagisaka, and Norio Higuchi. 1995. Stochastic modeling of pause insertion using context-free grammars. In *Proceedings of the 1995 International Conference on Acoustics, Speech and Signal Processing*, volume 1, pages 604–607, Detroit, MI. IEEE.

Fujisaki, Hiroya. 1983. Dynamic characteristics of voice fundamental frequency in speech and singing. In Peter MacNeilage, editor, *The Production of Speech.* Springer, New York, NY, pages 39–55.

Fujisaki, Hiroya. 1988. A note on the physiological and physical basis for the phrase and accent components in the voice fundamental frequency contour. In Osamu Fujimura, editor, *Vocal Fold Physiology: Voice Production, Mechanisms and Functions.* Raven, New York, NY, pages 347–355.

Fujisaki, Hiroya and Mats Ljungqvist. 1986. Proposal and evaluation of models for the glottal source waveform. In *Proceedings of the 1986 International Conference on Acoustics, Speech and Signal Processing*, volume 3, pages 1605–1608, Tokyo. IEEE.

Gårding, Eva. 1983. A generative model of intonation. In Anne Cutler and D. Robert Ladd, editors, *Prosody: Models and Measurements.* Springer, Berlin, pages 11–25.

Gårding, Eva. 1987. Speech act and tonal pattern in standard Chinese: Constancy and variation. *Phonetica*, 44:13–29.

Gaved, Maggie. 1993. Pronunciation and text normalisation in applied text-to-speech systems. In *Proceedings of the Third European Conference on Speech Communication and Technology*, volume 2, pages 897–900, Berlin. ESCA.

Goldfarb, Charles. 1990. *The SGML Handbook.* Clarendon Press, Oxford.

Golding, Andrew. 1991. *Pronouncing Names by a Combination of Rule-Based and Case-Based Reasoning.* Ph.D. thesis, Stanford University.

Golding, Andrew. 1995. A Bayesian hybrid method for context-sensitive spelling correction. In *Proceedings of the Third Workshop on Very Large Corpora*, pages 39–53, Cambridge, MA. ACL.

Goldsmith, John. 1990. *Autosegmental and Metrical Phonology.* Blackwell, Oxford.

Graham, Ian. 1996. *HTML Sourcebook.* Wiley, New York, NY, 2nd edition.

Greenspan, Steven, Ray Bennett, and Ann Syrdal. 1989. A study of two standard speech intelligibility measures. *Journal of the Acoustical Society of America*, 85(S43):2444, October.

Grosz, Barbara and Candace Sidner. 1986. Attention, intentions, and the structure of discourse. *Computational Linguistics*, 12:175–204.

Grønnum, Nina. 1992. *The Groundworks of Danish Intonation—An Introduction.* Museum Tusculanum Press, Copenhagen.

Halliday, Michael. 1967. *Intonation and Grammar in British English.* Mouton, The Hague.

Harris, James. 1983. *Syllable Structure and Stress in Spanish.* MIT Press, Cambridge, MA.

Harrison, Michael. 1978. *Introduction to Formal Language Theory.* Addison Wesley, Reading, MA.

Hays, William. 1981. *Statistics.* Holt, Rinehart and Winston, New York, NY, 3rd edition.

Hedelin, Per and Dieter Huber. 1990. Pitch period determination of aperiodic speech signals. In *Proceedings of the 1990 International Conference on Acoustics, Speech and Signal Processing*, volume 1, pages 361–364, Albuquerque, NM. IEEE.

Heemskerk, Josée and Vincent van Heuven. 1993. MORPA: A morpheme lexicon based morphological parser. In Vincent van Heuven and Louis Pols, editors, *Analysis and Synthesis of Speech: Strategic Research towards High-Quality Text-to-Speech Generation.* Mouton de Gruyter, Berlin, pages 67–85.

Heggtveit, Per Olav. 1996. A generalized LR parser for text-to-speech synthesis. In *Proceedings of the Fourth International Conference on Spoken Language Processing,* volume 3, pages 1429–1432, Philadelphia, PA. ICSLP.

Hemphill, C., John Godfrey, and George Doddington. 1990. The ATIS spoken language systems pilot corpus. In *Proceedings of the DARPA Speech and Natural Language Workshop,* Cape Cod MA, June. DARPA, Morgan Kaufmann.

Hirschberg, Julia. 1993. Pitch accent in context: Predicting intonational prominence from text. *Artificial Intelligence,* 63:305–340.

Hirschberg, Julia and Janet Pierrehumbert. 1986. Intonational structuring of discourse. In *24th Annual Meeting of the Association for Computational Linguistics,* pages 136–144, New York, NY. ACL.

Hirschberg, Julia and Pilar Prieto. 1994. Training intonational phrasing rules automatically for English and Spanish text-to-speech. In *Proceedings of The Second ESCA/IEEE Workshop on Speech Synthesis,* pages 159–162, New Paltz, NY. AAAI/IEEE/ESCA.

Hoaglin, D., F. Mosteller, and John Tukey. 1983. *Understanding Robust and Exploratory Data Analysis.* Wiley, New York, NY.

Hopcroft, John and Jeffrey Ullman. 1979. *Introduction to Automata Theory, Languages and Computation.* Addison-Wesley, Reading, MA.

Horne, Merle and Marcus Filipsson. 1994. Generating prosodic structure for Swedish text-to-speech. In *Proceedings of the Third International Conference on Spoken Language Processing,* volume 2, pages 711–714, Yokohama. ICSLP.

House, Arthur, C. Williams, M. Hecker, and K. Kryter. 1965. Articulatory-testing methods: Consonantal differentiation with a closed-response set. *Journal of the Acoustical Society of America,* 37:158–166.

Hurford, James. 1975. *The Linguistic Theory of Numerals.* Cambridge University Press, Cambridge.

Hurst, Matthew and Shona Douglas. 1997. Layout and language: Preliminary experiments in assigning logical structure to table cells. In *Proceedings of the Fifth Conference on Applied Natural Language Processing,* Washington, DC. ACL.

Institut für Phonetik und digitale Sprachverarbeitung, 1994. *The Kiel Corpus of Read Speech, Volume 1.* Institut für Phonetik und digitale Sprachverarbeitung, Universität Kiel. (CD-ROM).

Ishizaka, K. and James Flanagan. 1972. Synthesis of voiced sounds from a two-mass model of the vocal cords. *Bell System Technical Journal*, 50(6):1233–1268.

Iwahashi, Naoto and Yoshinori Sagisaka. 1992. Speech segment network approach for an optimal synthesis unit set. In *Proceedings of the Second International Conference on Spoken Language Processing*, volume 1, pages 479–482, Banff, Alberta. ESCA.

Jannedy, Stefanie and Bernd Möbius. 1997. Name pronunciation in German text-to-speech synthesis. In *Proceedings of the Fifth Conference on Applied Natural Language Processing*, pages 49–56, Washington, DC. ACL.

Jekosch, Ute. 1993. Speech quality assessment and evaluation. In *Proceedings of the Third European Conference on Speech Communication and Technology*, volume 2, pages 1387–1394, Berlin. ESCA.

Kahn, Dan and Marianne Macchi. 1997. Recent approaches to modeling the glottal source for TTS. In Jan van Santen, Richard Sproat, Joseph Olive, and Julia Hirschberg, editors, *Progress in Speech Synthesis*. Springer, New York, NY, pages 3–8.

Kaplan, Ronald and Martin Kay. 1994. Regular models of phonological rule systems. *Computational Linguistics*, 20:331–378.

Karlsson, Inger. 1989. A female voice for a text-to-speech system. In *Proceedings of the First European Conference on Speech Communication and Technology*, volume 1, pages 349–352, Paris.

Karn, Helen. 1996. Design and evaluation of a phonological phrase parser for Spanish text-to-speech. In *Proceedings of the Fourth International Conference on Spoken Language Processing*, volume 3, pages 1696–1699, Philadelphia, PA. ICSLP.

Karttunen, Lauri. 1993. Finite-state lexicon compiler. Technical Report P93–00077, Xerox Palo Alto Research Center.

Karttunen, Lauri. 1995. The replace operator. In *33rd Annual Meeting of the Association for Computational Linguistics*, pages 16–23, Cambridge, MA. ACL.

Karttunen, Lauri. 1996. Directed replacement. In *34th Annual Meeting of the Association for Computational Linguistics*, pages 108–115, Santa Cruz, CA. ACL.

Karttunen, Lauri and Kenneth Beesley. 1992. Two-level rule compiler. Technical Report P92–00149, Xerox Palo Alto Research Center.

Karttunen, Lauri, Ronald Kaplan, and Annie Zaenen. 1992. Two-level morphology with composition. In *COLING-92*, pages 141–148, Nantes. COLING.

Kenstowicz, Michael. 1994. *Phonology in Generative Grammar.* Blackwell Publishers, Cambridge, USA.

Kiraz, George. 1996. *Computational Approach to Non-Linear Morphology.* Ph.D. thesis, University of Cambridge. (Also forthcoming, Cambridge University Press).

Klatt, Dennis. 1973. Interaction between two factors that influence vowel duration. *Journal of the Acoustical Society of America*, 54:1102–1104.

Klatt, Dennis. 1976. Linguistic uses of segmental duration in English: Acoustic and perceptual evidence. *Journal of the Acoustical Society of America*, 59:1209–1221.

Klatt, Dennis. 1980. Software for a cascade/parallel formant synthesizer. *Journal of the Acoustical Society of America*, 67:971–980.

Klatt, Dennis. 1987. Review of text-to-speech conversion for English. *Journal of the Acoustical Society of America*, 82:737–793.

Klatt, Dennis and L. Klatt. 1990. Analysis, synthesis and perception of voice quality variations among female and male talkers. *Journal of the Acoustical Society of America*, 87:820–857.

Kleijn, W. Bastiaan and Jesper Haagen. 1995. Waveform interpolation for coding and synthesis. In W. Bastiaan Kleijn and Kuldip Paliwal, editors, *Speech Coding and Synthesis.* Elsevier, Amsterdam, pages 175–207.

Kohler, Klaus. 1977. *Einführung in die Phonetik des Deutschen.* Schmidt, Berlin.

Kohler, Klaus. 1987. The linguistic functions of F0 peaks. In *Proceedings of the Eleventh International Congress of Phonetic Sciences*, volume 3, pages 149–152, Tallinn.

Kohler, Klaus. 1990. Macro and micro F0 in the synthesis of intonation. In John Kingston and Mary Beckman, editors, *Papers in Laboratory Phonology I: Between the Grammar and Physics of Speech.* Cambridge University Press, Cambridge, pages 115–138.

Kohler, Klaus. 1991. Prosody in speech synthesis: the interplay between basic research and TTS application. *Journal of Phonetics*, 19:121–138.

Kohler, Klaus. 1995. *Einführung in die Phonetik des Deutschen.* Erich Schmidt Verlag, Berlin, 2nd edition.

Konst, Emmy and Louis Boves. 1994. Automatic grapheme-to-phoneme conversion of Dutch names. In *Proceedings of the Third International Conference on Spoken Language Processing*, volume 2, pages 735–738, Yokohama. ICSLP.

Koskenniemi, Kimmo. 1983. *Two-Level Morphology: a General Computational Model for Word-Form Recognition and Production.* Ph.D. thesis, University of Helsinki, Helsinki.

Krantz, D., R. Luce, P. Suppes, and A. Tversky. 1971. *Foundations of Measurement, Volume I.* Wiley, New York, NY.

Kratzenstein, C. Gottlieb. 1782. Sur la naissance de la formation des voyelles. *Journal de Physique*, 21:358–380. French translation of: Tentamen coronatum de voce, *Acta Acad. Petrop.*, 1780.

Kroon, Peter. 1995. Evaluation of speech coders. In W. Bastiaan Kleijn and Kuldip Paliwal, editors, *Speech Coding and Synthesis.* Elsevier, Amsterdam, pages 467–494.

Kubozono, Haruo. 1988. *The Organization of Japanese Prosody.* Ph.D. thesis, University of Edinburgh. Published by Kurosio, Tokyo (1993).

Küpfmüller, K. and O. Warns. 1956. Sprachsynthese aus Lauten. *Nachrichtentechnische Fachberichte*, 3:28–31.

Kutik, E., W. Cooper, and S. Boyce. 1983. Declination of fundamental frequency in speaker's production of parenthetical and main clauses. *Journal of the Acoustical Society of America*, 73:1731–1738.

Ladd, D. Robert. 1983a. Peak features and overall slope. In Anne Cutler and D. Robert Ladd, editors, *Prosody: Models and Measurements.* Springer, Berlin, pages 39–52.

Ladd, D. Robert. 1983b. Phonological features of intonational peaks. *Language*, 59:721–759.

Ladd, D. Robert. 1988. Declination 'reset' and the hierarchical organization of utterances. *Journal of the Acoustical Society of America*, 84:530–544.

Ladd, D. Robert. 1990. Metrical representation of pitch register. In John Kingston and Mary Beckman, editors, *Papers in Laboratory Phonology I: Between the Grammar and Physics of Speech.* Cambridge University Press, Cambridge, pages 35–57.

Ladd, D. Robert and C. Johnson. 1987. 'metrical' factors in the scaling of sentence-initial accent peaks. *Phonetica*, 44:238–245.

Ladefoged, Peter and Ian Maddieson. 1996. *The Sounds of the World's Languages.* Blackwell, Oxford.

Leben, William. 1976. The tones in English intonation. *Linguistic Analysis*, 2:69–107.

Lee, Sangho and Yung-Hwan Oh. 1996. A text analyzer for Korean text-to-speech systems. In *Proceedings of the Fourth International Conference on Spoken Language Processing*, volume 3, pages 1692–1695, Philadelphia, PA. ICSLP.

Lehiste, Ilse. 1977. Isochrony reconsidered. *Journal of Phonetics*, 5:253–263.

Lewis, Harry and Christos Papadimitriou. 1981. *Elements of the Theory of Computation.* Prentice-Hall, Englewood Cliffs, NJ.

Liberman, Mark and Janet Pierrehumbert. 1984. Intonational invariance under changes in pitch range and length. In Mark Aronoff and Richard Oehrle, editors, *Language Sound Structure.* MIT Press, Cambridge.

Liberman, Mark and Alan Prince. 1977. On stress and linguistic rhythm. *Linguistic Inquiry*, 8:249–336.

Liberman, Mark and Richard Sproat. 1992. The stress and structure of modified noun phrases in English. In Anna Szabolcsi and Ivan Sag, editors, *Lexical Matters.* CSLI (University of Chicago Press), Chicago, IL.

Litman, Diane and Julia Hirschberg. 1990. Disambiguating cue phrases in text and speech. In *COLING-90*, Helsinki, August. COLING.

Ljolje, Andrej and Michael Riley. 1993. Automatic segmentation of speech for TTS. In *Proceedings of the Third European Conference on Speech Communication and Technology*, volume 2, pages 1445–1448, Berlin. ESCA.

Ljungqvist, Mats, Anders Lindström, and Kjell Gustafson. 1994. A new system for text-to-speech conversion and its application to Swedish. In *Proceedings of the Third International Conference on Spoken Language Processing*, volume 4, pages 1779–1782, Yokohama. ICSLP.

Local, John and Richard Ogden. 1997. A model of timing for nonsegmental phonological structure. In Jan van Santen, Richard Sproat, Joseph Olive, and Julia Hirschberg, editors, *Progress in Speech Synthesis.* Springer, New York, NY, pages 109–122.

Luk, Robert and Robert Damper. 1993. Experiments with silent-e and affix correspondences in stochastic phonographic transductions. In *Proceedings of the Third European Conference on Speech Communication and Technology*, volume 2, pages 917–920, Berlin. ESCA.

Luk, Robert and Robert Damper. 1996. Stochastic phonographic transduction for English. *Computer Speech and Language*, 10:133–153.

Macchi, Marianne, Mary Jo Altom, Dan Kahn, S. Singhal, and Murray Spiegel. 1993. Intelligibility as a function of speech coding method for template-based speech synthesis. In *Proceedings of the Third European Conference on Speech Communication and Technology*, volume 2, pages 893–897, Berlin. ESCA.

Magata, Ken-ichi, Tomoki Hamagami, and Mitsuo Komura. 1996. A method for estimating prosodic symbol from text for Japanese text-to-speech synthesis. In *Proceedings of the Fourth International Conference on Spoken Language Processing*, volume 3, pages 1373–1376, Philadelphia, PA. ICSLP.

Maghbouleh, Arman. 1996. An empirical comparison of automatic decision tree and hand-configured linear models for vowel durations. In *Computational Phonology in Speech Technology: Proceedings of the Second Meeting of ACL SIGPHON*, pages 1–7, Santa Cruz, CA. ACL.

Markel, J. and A. Gray. 1976. *Linear Prediction of Speech.* Number 12 in Communication and Cybernetics. Springer, Berlin.

McCulloch, Neil, Mark Bedworth, and John Bridle. 1987. NETspeak – a re-implementation of NETtalk. *Computer Speech and Language*, 2:289–301.

McGowan, Richard. 1992. Tongue-tip trills and vocal tract wall compliance. *Journal of the Acoustical Society of America*, 91:2903–2910.

Meyer, Peter, Hans-Wilhelm Rühl, Regina Krüger, Marianne Kugler, Leo Vogten, Arthur Dirksen, and Karim Belhoula. 1993. PHRITTS - a text-to-speech synthesizer for the German language. In *Proceedings of the Third European Conference on Speech Communication and Technology*, volume 2, pages 877–880, Berlin. ESCA.

Microsoft, 1996. *Microsoft Speech Software Development Kit Developer's Guide.* Microsoft, Redmond, WA. Version 2.0.

Möbius, Bernd. 1993. *Ein quantitatives Modell der deutschen Intonation—Analyse und Synthese von Grundfrequenzverläufen.* Number 305 in Linguistische Arbeiten. Max Niemeyer Verlag, Tübingen.

Möbius, Bernd. 1995. Components of a quantitative model of German intonation. In *Proceedings of the Thirteenth International Congress of Phonetic Sciences*, volume 2, pages 108–115, Stockholm.

Möbius, Bernd, Matthias Pätzold, and Wolfgang Hess. 1993. Analysis and synthesis of German F0 contours by means of Fujisaki's model. *Speech Communication*, 13:53–61.

Möbius, Bernd and Jan van Santen. 1996. Modeling segmental duration in German text-to-speech synthesis. In *Proceedings of the Fourth International Conference on Spoken Language Processing*, volume 4, pages 2395–2399, Philadelphia, PA. ICSLP.

Mohri, Mehryar. 1994. Syntactic analysis by local grammars automata: an efficient algorithm. In *Papers in Computational Lexicography: COMPLEX '94*, pages 179–191, Budapest. Research Institute for Linguistics, Hungarian Academy of Sciences.

Mohri, Mehryar. 1997. Finite-state transducers in language and speech processing. *Computational Linguistics*, forthcoming.

Mohri, Mehryar and Richard Sproat. 1996. An efficient compiler for weighted rewrite rules. In *34th Annual Meeting of the Association for Computational Linguistics*, pages 231–238, Santa Cruz, CA. ACL.

Monaghan, Alex. 1990. Rhythm and stress-shift in speech synthesis. *Computer Speech and Language*, 4:71–78.

Morrison, Donald. 1967. *Multivariate Statistical Methods*. McGraw-Hill, New York, NY.

Mrayati, M., René Carré, and B. Guérin. 1988. Distinctive regions and modes: a new theory of speech production. *Speech Communication*, 7:257–286.

Needham, John. 1959. *Science and Civilisation in China: Volume 3 — Mathematics and the Sciences of the Heavens and the Earth*. Cambridge University Press, Cambridge. (With the collaboration of Wang Ling).

Nunberg, Geoffrey. 1995. *The Linguistics of Punctuation*. CSLI (University of Chicago Press), Chicago, IL.

Nunn, Anneke and Vincent van Heuven. 1993. MORPHON: Lexicon-based text-to-phoneme conversion and phonological rules. In Vincent van Heuven and Louis Pols, editors, *Analysis and Synthesis of Speech: Strategic Research towards High-Quality Text-to-Speech Generation*. Mouton de Gruyter, Berlin, pages 87–99.

Odé, Cecilia. 1989. *Russian Intonation: a Perceptual Description*. Number 13 in Studies in Slavic and General Linguistics. Rodopi, Amsterdam.

Öhman, Sven and J. Lindqvist. 1966. Analysis-by-synthesis of prosodic pitch contours. *Speech Transmission Laboratory—Quarterly Progress and Status Report*, 4:1–6.

Olive, Joseph. 1977. Rule synthesis of speech from dyadic units. In *Proceedings of the 1977 International Conference on Acoustics, Speech and Signal Processing*, volume 1, pages 568–570, Hartford, CT. IEEE.

Olive, Joseph. 1990. A new algorithm for a concatenative speech synthesis system using an augmented acoustic inventory of speech sounds. In *Proceedings of the ESCA Workshop on speech synthesis*, pages 25–30, Autrans. ESCA.

Olive, Joseph. 1994. Objective methods for selecting concatenative acoustic inventory elements. Technical Report 11222-940523-02TM, AT&T Bell Laboratories.

Olive, Joseph, Alice Greenwood, and John Coleman. 1993. *Acoustics of American English Speech: a Dynamic Approach.* Springer, New York, NY.

Olive, Joseph and Mark Liberman. 1985. Text to speech – an overview. *Journal of the Acoustic Society of America, Suppl. 1*, 78:s6.

Oliveira, Luis. 1993. Estimation of source parameters by frequency analysis. In *Proceedings of the Third European Conference on Speech Communication and Technology*, volume 1, pages 99–102, Berlin. ESCA.

Oliveira, Luis. 1997. Text-to-speech synthesis with dynamic control of source parameters. In Jan van Santen, Richard Sproat, Joseph Olive, and Julia Hirschberg, editors, *Progress in Speech Synthesis.* Springer, New York, NY, pages 27–40.

Onomastica. 1995. Multi-language pronunciation dictionary of proper names and place names. Technical report, European Community, Linguistic Research and Engineering Programme. Project No. LRE-61004, Final Report, 30 May 1995.

Ooyama, Yoshifumi, Hisako Asano, and Koji Matsuoka. 1996. Spoken-style explanation generator for Japanese kanji using a text-to-speech system. In *Proceedings of the Fourth International Conference on Spoken Language Processing*, volume 3, pages 1369–1372, Philadelphia, PA. ICSLP.

O'Shaughnessy, Douglas. 1987. *Speech Communication—human and machine.* Addison-Wesley, Reading, UK.

O'Shaughnessy, Douglas. 1989. Parsing with a small dictionary for applications such as text to speech. *Computational Linguistics*, 15:97–108.

Pereira, Fernando and Michael Riley. 1996. Speech recognition by composition of weighted finite automata. `http://xxx.lanl.gov/ps/cmp-lg/9608018`.

Pereira, Fernando, Michael Riley, and Richard Sproat. 1994. Weighted rational transductions and their application to human language processing. In *Proceedings of the Human Language Technology Workshop*, pages 249–254, Plainsboro, NJ. ARPA.

Pereira, Fernando and Rebecca Wright. 1997. Finite-state approximation of phrase-structure grammars. In Emmanuel Roche and Yves Schabes, editors, *Finite-State Language Processing.* MIT Press, Cambridge, MA.

Peterson, G., William Wang, and E. Sivertsen. 1958. Segmentation techniques in speech synthesis. *Journal of the Acoustical Society of America*, 30:793–742.

Pierrehumbert, Janet. 1980. *The Phonology and Phonetics of English Intonation.* Ph.D. thesis, Massachusetts Institute of Technology, Cambridge, MA. Distributed by the Indiana University Linguistics Club.

Pierrehumbert, Janet. 1981. Synthesizing intonation. *Journal of the Acoustical Society of America*, 70:985–995.

Pierrehumbert, Janet and Mary Beckman. 1988. *Japanese Tone Structure.* MIT Press, Cambridge, MA.

Pike, Kenneth. 1958. *The Intonation of American English.* University of Michigan Press, Ann Arbor, MI, 7th edition. (1st edition 1945).

Pisoni, David, Howard Nusbaum, and B. Greene. 1985. Perception of synthetic speech generated by rule. *Proceedings of the IEEE*, 73:1665–1676.

Pitrelli, John and Victor Zue. 1989. A hierarchical model for phoneme duration in American English. In *Proceedings of the First European Conference on Speech Communication and Technology*, volume 2, pages 324–327, Paris, September. ESCA.

Pols, Louis. 1991. Quality assessment of text-to-speech synthesis-by-rule. In S. Furui and M. Mohan Sondhi, editors, *Advances in Speech Signal Processing.* Marcel Dekker, New York, NY, pages 387–416.

Pols, Louis and SAM-partners. 1992. Multi-lingual synthesis evaluation methods. In *Proceedings of the Second International Conference on Spoken Language Processing*, volume 1, pages 181–184, Banff, Alberta. ICSLP.

Port, Robert. 1981. Linguistic timing factors in combination. *Journal of the Acoustical Society of America*, 69:262–273.

Portele, Thomas, Birgit Steffan, Rainer Preuss, and Wolfgang Hess. 1991. German speech synthesis by concatenation of non-parametric units. In *Proceedings of the Second European Conference on Speech Communication and Technology*, volume 1, pages 317–320, Genoa. ESCA.

Poser, William. 1984. *The Phonetics and Phonology of Tone and Intonation in Japanese.* Ph.D. thesis, Massachusetts Institute of Technology.

Quené, Hugo and René Kager. 1992. The derivation of prosody for text-to-speech from prosodic sentence structure. *Computer Speech and Language*, 6:77–98.

Quené, Hugo and René Kager. 1993. Prosodic sentence analysis without parsing. In Vincent van Heuven and Louis Pols, editors, *Analysis and Synthesis of Speech: Strategic Research towards High-Quality Text-to-Speech Generation*. Mouton de Gruyter, Berlin, pages 115–130.

Rabiner, Lawrence and Ronald Schafer. 1978. *Digital Processing of Speech Signals*. Prentice-Hall, London.

Radzinski, Daniel. 1991. Chinese number-names, tree adjoining languages, and mild context-sensitivity. *Computational Linguistics*, 17:277–299.

Rajemisa-Raolison, Régis. 1971. *Grammaire Malgache*. Centre de Formation Pédagogique, Fianarantsoa, Madagascar. 7th edition.

Raman, T.V. 1994. *Audio System for Technical Readings*. Ph.D. thesis, Cornell University.

Rao, C. Radhakrishna. 1965. *Linear Statistical Inference and its Applications*. Wiley, New York, NY.

Richard, Gael and Christophe d'Alessandro. 1997. Modification of the aperiodic component of speech signals for synthesis. In Jan van Santen, Richard Sproat, Joseph Olive, and Julia Hirschberg, editors, *Progress in Speech Synthesis*. Springer, New York, NY, pages 41–56.

Riedi, Marcel. 1995. A neural-network-based model of segmental duration for speech synthesis. In *Proceedings of the Fourth European Conference on Speech Communication and Technology*, volume 1, pages 599–602, Madrid. ESCA.

Riley, Michael. 1989. Some applications of tree-based modelling to speech and language. In *Proceedings of the DARPA Speech and Natural Language Workshop*, pages 339–352, Cape Cod, MA, October. DARPA, Morgan Kaufmann.

Riley, Michael. 1992. Tree-based modeling for speech synthesis. In Gérard Bailly and Christian Benoit, editors, *Talking Machines: Theories, Models, and Designs*. Elsevier, Amsterdam, pages 265–273.

Roche, Emmanuel. 1997. Finite-state transducers: Parsing free and frozen sentences. *Journal of Natural Language Engineering*. To appear.

Roget, Peter. 1977. *Roget's International Thesaurus*. Harper and Row, New York, NY. Revised by Robert Chapman.

Rosenberg, Aaron. 1971. Effect of glottal pulse shape on the quality of natural vowels. *Journal of the Acoustical Society of America*, 49:583–590.

Sanders, Eric and Paul Taylor. 1995. Using statistical models to predict phrase boundaries for speech synthesis. In *Proceedings of the Fourth European Conference on Speech Communication and Technology*, volume 3, pages 1811–1814, Madrid. ESCA.

Sankoff, David and Joseph Kruskal. 1983. *Time Warps, String Edits, and Macromolecules: the Theory and Practice of Sequence Comparison*. Addison-Wesley, London.

Schane, Sanford. 1968. *French Phonology and Morphology*. MIT Press, Cambridge, MA.

Sejnowski, Terence and C. Rosenberg. 1987. Parallel networks that learn to pronounce English text. *Complex Systems*, 1:145–168.

Shen, Jiong. 1985. Beijinghua shengdiao de yinyu he yudiao (Pitch range and intonation of the tones of Beijing Mandarin). In *Beijing Yuyin Shiyan Lu (Acoustic Studies of Beijing Mandarin)*. Beijing University Press, Beijing, pages 73–130.

Shen, Xiao-Nan Susan. 1990. Tonal coarticulation in Mandarin. *Journal of Phonetics*, 18:281–295.

Shih, Chilin. 1988. Tone and intonation in Mandarin. In *Working Papers of the Cornell Phonetics Laboratory, Number 3: Stress, Tone and Intonation*. Cornell University, pages 83–109.

Shih, Chilin. 1995. Study of vowel variations for a Mandarin speech synthesizer. In *Proceedings of the Fourth European Conference on Speech Communication and Technology*, volume 3, pages 1807–1810. ESCA.

Shih, Chilin. 1996. Synthesis of trill. In *Proceedings of the Fourth International Conference on Spoken Language Processing*, volume 4, pages 2223–2226, Philadelphia. ICSLP.

Shih, Chilin and Benamin Ao. 1997. Duration study for the Bell Laboratories Mandarin text-to-speech System. In Jan van Santen, Richard Sproat, Joseph Olive, and Julia Hirschberg, editors, *Progress in Speech Synthesis*. Springer, New York, NY, pages 383–399.

Shih, Chilin and Richard Sproat. 1996. Issues in text-to-speech conversion for Mandarin. *Computational Linguistics and Chinese Language Processing*, 1:37–86.

Silverman, Kim. 1987. *The Structure and Processing of Fundamental Frequency Contours*. Ph.D. thesis, Cambridge University, Cambridge.

Silverman, Kim. 1988. Utterance-internal prosodic boundaries. In *Proceedings of the Second Australian International Conference on Speech Science and*

Technology, pages 86–91, Sydney. Australian Speech Science and Technology Association.

Sperberg-McQueen, C. Michael and Lou Burnard. 1994. *Guidelines for Electronic Text Encoding and Interchange (TEI P3)*. Chicago/Oxford, Text Encoding Initiative. Available as `http://www.uic.edu/orgs/tei/p3/elect.html`.

Spiegel, Murray, Mary Jo Altom, Marianne Macchi, and K. Wallace. 1988. Using a monosyllabic test corpus to evaluate the intelligibility of synthesized and natural speech. In *Proceedings of the American Voice I/O Systems Conference*. AVIOS.

Sproat, Richard. 1992. *Morphology and Computation*. MIT Press, Cambridge, MA.

Sproat, Richard. 1994. English noun-phrase accent prediction for text-to-speech. *Computer Speech and Language*, 8:79–94.

Sproat, Richard. 1997a. Computational morphology. In Robert Dale, Hermann Moisl, and Harold Somers, editors, *Handbook of Natural Language Processing*. Marcel Dekker, New York, NY. Forthcoming.

Sproat, Richard. 1997b. Multilingual text analysis for text-to-speech synthesis. *Journal of Natural Language Engineering*. To appear.

Sproat, Richard, Julia Hirschberg, and David Yarowsky. 1992. A corpus-based synthesizer. In *Proceedings of the Second International Conference on Spoken Language Processing*, volume 1, pages 563–566, Banff, Alberta. ICSLP.

Sproat, Richard and Joseph Olive. 1997. A modular architecture for multilingual text-to-speech. In Jan van Santen, Richard Sproat, Joseph Olive, and Julia Hirschberg, editors, *Progress in Speech Synthesis*. Springer, New York, NY, pages 565–573.

Sproat, Richard and Michael Riley. 1996. Compilation of weighted finite-state transducers from decision trees. In *34th Annual Meeting of the Association for Computational Linguistics*, pages 215–222, Santa Cruz, CA. ACL.

Sproat, Richard, Chilin Shih, William Gale, and Nancy Chang. 1996. A stochastic finite-state word-segmentation algorithm for Chinese. *Computational Linguistics*, 22:377–404.

Sproat, Richard, Paul Taylor, Michael Tanenblatt, and Amy Isard. 1997. A markup language for text-to-speech synthesis. In *Proceedings of the Fifth European Conference on Speech Communication and Technology*, Rhodes. ESCA.

Stevens, Kenneth and Corine Bickley. 1991. Constraints among parameters simplify control of Klatt formant synthesizer. *Journal of Phonetics*, 19:161–174.

't Hart, Johan. 1984. A phonetic approach to intonation: from pitch contours to intonation patterns. In Dafydd Gibbon and Helmut Richter, editors, *Intonation, Accent and Rhythm—Studies in Discourse Phonology*. Mouton de Gruyter, Berlin, pages 193–202.

't Hart, Johan, René Collier, and Antonie Cohen. 1990. *A Perceptual Study of Intonation—An Experimental-Phonetic Approach to Speech Melody*. Cambridge University Press, Cambridge.

Talkin, David and James Rowley. 1990. Pitch-synchronous analysis and synthesis for TTS systems. In *Proceedings of the ESCA Workshop on Speech Synthesis*, pages 55–59, Autrans. ESCA.

Taylor, Paul and Amy Isard. 1996. SSML: A speech synthesis markup language. *Speech Communication*, 21.

Terken, Jacques. 1993. Synthesizing natural-sounding intonation for Dutch: rules and perceptual evaluation. *Computer Speech and Language*, 7:27–48.

The Unicode Consortium. 1996. *The Unicode Standard, Version 2.0*. Addison-Wesley, Reading, MA.

Thielen, Christine. 1995. An approach to proper name tagging in German. In *From Texts to Tags: Issues in Multilingual Language Analysis: Proceedings of the ACL SIGDAT Workshop*, pages 35–40, Dublin. ACL.

Thorsen, Nina. 1985. Intonation and text in Standard Danish. *Journal of the Acoustical Society of America*, 77:1205–1216.

Thorsen, Nina. 1988. Standard Danish intonation. *Annual Report of the Institute of Phonetics (University of Copenhagen)*, 22:1–23.

Traber, Christof. 1993. Syntactic processing and prosody control in the SVOX TTS system for German. In *Proceedings of the Third European Conference on Speech Communication and Technology*, volume 3, pages 2099–2102, Berlin. ESCA.

Traber, Christof. 1995. SVOX: The implementation of a text-to-speech system for German. Technical Report 7, Swiss Federal Institute of Technology, Zurich.

Trager, G. and H. Smith. 1951. *An Outline of English Structure*. Battenburg Press, Norman, OK.

Tranel, Bernard. 1981. The treatment of French liaison: Descriptive, methodological, and theoretical implications. In Helen Contreras and J. Klausenberger, editors, *Proceedings of the Tenth Anniversary Symposium on Romance Linguistics*, volume 3 (supplement 2), pages 261–281, Seattle, WA.

Tranel, Bernard. 1986. French liaison and extrasyllabicity. *Studies in Romance Linguistics*, pages 283–395.

Tranel, Bernard. 1990. On suppletion and french liaison. *Probus*, 2:169–208.

Tversky, A. 1969. Intransitivity of preferences. *Psychological Review*, 76:1–12.

Tzoukermann, Evelyne. 1994. Text-to-speech for French. In *Proceedings of The Second ESCA/IEEE Workshop on Speech Synthesis*, pages 179–182, New Paltz, NY. AAAI/IEEE/ESCA.

Tzoukermann, Evelyne and Mark Liberman. 1990. A finite-state morphological processor for Spanish. In *COLING-90*, volume 3, pages 277–286. COLING.

Tzoukermann, Evelyne and Olivier Soumoy. 1995. Segmental duration in French text-to-speech synthesis. In *Proceedings of the Fourth European Conference on Speech Communication and Technology*, volume 1, pages 607–611, Madrid.

Umeda, Noriko. 1975. Vowel duration in American English. *Journal of the Acoustical Society of America*, 58:434–445.

Umeda, Noriko. 1977. Consonant duration in American English. *Journal of the Acoustical Society of America*, 61:846–858.

Ungeheuer, Gerold. 1957. *Untersuchungen zur Vokalartikulation.* Ph.D. thesis, University of Bonn, Bonn.

Ungeheuer, Gerold. 1962. *Elemente einer akustischen Theorie der Vokalartikulation.* Springer, Berlin.

van Hemert, J.., Ute Adriaens-Porzig, and Léon Adriaens. 1987. Speech synthesis in the SPICOS project. In Hans Tillmann and Gerd Willée, editors, *Analyse und Synthese gesprochener Sprache.* Olms, Hildesheim, pages 34–39.

van Heuven, Vincent and Louis Pols. 1993. *Analysis and Synthesis of Speech: Strategic Research towards High-Quality Text-to-Speech Generation.* Mouton de Gruyter, Berlin.

van Holsteijn, Yvonne. 1993. TextScan: A preprocessing module for automatic text-to-speech conversion. In Vincent van Heuven and Louis Pols, editors, *Analysis and Synthesis of Speech: Strategic Research towards High-Quality Text-to-Speech Generation.* Mouton de Gruyter, Berlin, pages 27–41.

van Leeuwen, Hugo. 1989a. A development tool for linguistic rules. *Computer Speech and Language*, 3:83–104.

van Leeuwen, Hugo. 1989b. *Too$_L$iP: A Development Tool for Linguistic Rules.* Ph.D. thesis, Technical University Eindhoven.

van Santen, Jan. 1992a. Contextual effects on vowel duration. *Speech Communication*, 11:513–546.

van Santen, Jan. 1992b. Diagnostic perceptual experiments for text-to-speech system evaluation. In *Proceedings of the Second International Conference on Spoken Language Processing*, volume 1, pages 555–558, Banff, Alberta. ICSLP.

van Santen, Jan. 1993a. Analyzing N-way tables with sums-of-products models. *Journal of Mathematical Psychology*, 37:327–371.

van Santen, Jan. 1993b. Perceptual experiments for diagnostic testing of text-to-speech systems. *Computer Speech and Language*, 7:49–100.

van Santen, Jan. 1993c. Quantitative modeling of segmental duration. In *Proceedings of the Human Language Technology Workshop*, pages 323–328, Princeton, NJ, March. ARPA.

van Santen, Jan. 1994. Assignment of segmental duration in text-to-speech synthesis. *Computer Speech and Language*, 8:95–128.

van Santen, Jan. 1997. Combinatorial issues in text-to-speech synthesis. In *Proceedings of the Fifth European Conference on Speech Communication and Technology*, Rhodes. ESCA.

van Santen, Jan, John Coleman, and Mark Randolph. 1992. Effects of postvocalic voicing on the time course of vowels and diphthongs. *Journal of the Acoustical Society of America*, 92(4, Part 2):2444.

van Santen, Jan and Julia Hirschberg. 1994. Segmental effects on timing and height of pitch contours. In *Proceedings of the Third International Conference on Spoken Language Processing*, volume 2, pages 719–722. ICSLP.

van Santen, Jan, Bernd Möbius, and Michael Tanenblatt. 1994. New procedures for constructing acoustic inventories. Technical Report 11222-941012-08TM, AT&T Bell Laboratories.

van Santen, Jan and Chilin Shih. 1997. Syllabic and segmental timing in Mandarin Chinese and American English. Submitted for publication.

Véronis, Jean and Nancy Ide. 1996. *The Text Encoding Initiative.* Addison-Wesley, Reading, MA.

Vitale, Anthony. 1991. An algorithm for high accuracy name pronunciation by a parametric speech synthesizer. *Computational Linguistics*, 17:257–276.

Voiers, W., A. Sharpley, and C. Hehmsoth. 1972. Research on diagnostic evaluation of speech intelligibility. Research Report AFCRL-72-0694, Cambridge Research Laboratories.

von Helmholtz, Hermann. 1870. *Die Lehre von den Tonempfindungen als physiologische Grundlage für die Theorie der Musik*. Braunschweig, 3rd edition.

von Helmholtz, Hermann. 1954. *On the Sensation of Tone*. Dover, New York. translated by A. J. Ellis; translation of the 4th German edition, 1877.

von Kempelen, Wolfgang. 1791. *Mechanismus der menschlichen Sprache nebst der Beschreibung seiner sprechenden Maschine*. J. V. Degen, Vienna.

Voutilainen, Atro. 1994. Designing a parsing grammar. Technical Report 22, University of Helsinki.

Wang, Michelle and Julia Hirschberg. 1992. Automatic classification of intonational phrase boundaries. *Computer Speech and Language*, 6:175–196.

Willems, Nico, René Collier, and Johan 't Hart. 1988. A synthesis scheme for British English intonation. *Journal of the Acoustical Society of America*, 84:1250–1261.

Williams, Briony. 1987. Word stress assignment in a text-to-speech system for British English. *Computer Speech and Language*, 2:235–272.

Williams, Briony. 1994. Welsh letter-to-sound rules: Rewrite rules and two-level rules compared. *Computer Speech and Language*, 8:261–277.

Winer, B., D. Brown, and K. Michels. 1991. *Statistical Principles in Experimental Design*. McGraw-Hill, New York, NY.

Wright, Charles, Mary Jo Altom, and Joseph Olive. 1986. Diagnostic evaluation of a synthesizer's acoustic inventory. *Journal of the Acoustic Society of America, Suppl. 1*, 79:s25.

Xu, Yi. 1993. *Contextual Tonal Variation in Mandarin Chinese*. Ph.D. thesis, The University of Connecticut.

Yarowsky, David. 1994a. A comparison of corpus-based techniques for restoring accents in Spanish and French texts. In *Proceedings of the Second Annual Workshop on Very Large Corpora*, pages 19–32, Kyoto. ACL.

Yarowsky, David. 1994b. Decision lists for lexical ambiguity resolution. In *32nd Annual Meeting of the Association for Computational Linguistics*, pages 88–95, Las Cruces, NM. ACL.

Yarowsky, David. 1996. *Three Machine Learning Algorithms for Lexical Ambiguity Resolution.* Ph.D. thesis, University of Pennsylvania.

Yarowsky, David. 1997. Homograph disambiguation in text-to-speech synthesis. In Jan van Santen, Richard Sproat, Joseph Olive, and Julia Hirschberg, editors, *Progress in Speech Synthesis.* Springer, New York, NY, pages 157–172.

Zhang, Jialu. 1986. Acoustic parameters and phonological rules of a text-to-speech system for Chinese. In *Proceedings of the 1986 International Conference on Acoustics, Speech and Signal Processing*, volume 3, pages 2023–2026. IEEE.

Name Index

Subject Index